Terrestrial Environmental Sciences

Series editors

Olaf Kolditz
Hua Shao
Wenqing Wang
Uwe-Jens Görke
Sebastian Bauer

More information about this series at http://www.springer.com/series/13468

TianXiang Yue · Erik Nixdorf
Chengzi Zhou · Bing Xu
Na Zhao · Zhewen Fan · Xiaolan Huang
Cui Chen · Olaf Kolditz
Editors

Chinese Water Systems

Volume 3: Poyang Lake Basin

Springer

Editors
TianXiang Yue
Institute of Geographical Sciences
and Natural Resources
Beijing, China

Erik Nixdorf
Department of Environmental Informatics
Helmholtz Centre for Environmental
Research - UFZ
Leipzig, Germany

Chengzi Zhou
Department of Environmental Informatics
Helmholtz Centre for Environmental
Research - UFZ
Leipzig, Germany

Bing Xu
School of Environment
Tsinghua University
Beijing, China

Na Zhao
Institute of Geographical Sciences and
Natural Resources
Beijing, China

Zhewen Fan
Jiangxi Remote Sensing Information
System Center
Nanchang, China

Xiaolan Huang
School of Geography and Environment
Jiangxi Normal University
Nanchang, China

Cui Chen
Department of Environmental Informatics
Helmholtz Centre for Environmental
Research - UFZ
Leipzig, Germany

Olaf Kolditz
Department of Environmental Informatics
Helmholtz Centre for Environmental
Research - UFZ
Leipzig, Germany

and

Technical University Dresden
Dresden, Germany

ISSN 2363-6181 ISSN 2363-619X (electronic)
Terrestrial Environmental Sciences
ISBN 978-3-030-07395-4 ISBN 978-3-319-97725-6 (eBook)
https://doi.org/10.1007/978-3-319-97725-6

This Springer imprint is published by the registered company Springer Nature Switzerland AG
The registered company address is: Gewerbestrasse 11, 6330 Cham, Switzerland

Preface

Situated in the rugged beauty of rural Jiangxi Province, China's largest freshwater lake, Poyang Lake, is not only famous for being the scene of historic dynastic struggles like the Battle of Lake Poyang in 1363 AD but also for its highly diverse subtropical flora and fauna, including remarkable assemblages of endemic species such as the freshwater finless porpoise. From October to March each year, vast areas of marsh and small water areas appear in the alluvial plain of Poyang Lake are a habitat of international importance for hundreds of thousands of migratory birds from Siberia and Northern China. In addition, Poyang Lake provides a vital source of resources to millions of people in Jiangxi Province living either at its shoreline or in the hinterland of the basin.

Although having a relatively higher water quality compared to other Chinese lakes, the water resources of Poyang Lake are threatened due to land reclamation, sand dredging, overexploitation, and the exposure of the aquatic system to pesticides from agricultural sources. Moreover, existing and proposed dam projects and changing climatic conditions already adversely impact the seasonal water balance of Poyang Lake. Since the 1990s, the Chinese government has made substantial efforts to address the problem of polluting freshwater resources. More recently, in April 2015, the Action Plan for Prevention and Control of Water Pollution was launched which requires that by 2020, the implementation of adequate protection measures will have improved the water quality of 70% of the water areas of the seven key basins including Yangtze River Basin, which contains Poyang Lake, to be "excellent" or "good" according to the Chinese standard.

In his speech at the 19th National Congress of the Communist Party of China last October, President Xi Jinping demanded the adoption of a holistic approach to conserving natural resources and protecting the environment. Considering that a sustainable management of aquatic resources must be built upon a thorough assessment of status and dynamics of the water resource in relation to human impacts, a research-oriented, multi-disciplinary approach is required. Transferring this approach to Poyang Lake, a research symposium called "sustainable water management and ecosystem restoration in the Poyang Lake Basin," held in Nanchang in November 2014, allowed bringing together German and Chinese scientists from different disciplines such as hydrology, ecology, climate research, and information science as well as involving authorities and stakeholders from the Poyang Lake Basin. Consequently, the Poyang Cooperation group was formed in 2015 to build a joint Sino-German network to encourage synergy between the research groups and to facilitate research on environmental aspects of the Poyang Lake Basin.

This book compiles scientific results achieved by members of the aforementioned joint research group. In addition, experts from renowned research facilities such as Jiangxi Normal University, the Leibniz-Institute of Freshwater Ecology and Inland Fisheries, and Tomsk Polytechnic University made valuable contributions to this book volume. Although far from covering all current environmental research on Poyang Lake and its basin, the present book provides a comprehensive overview in English about current environmental research topics at Poyang Lake. This is of particular interest for those readers who are not able to access the great volumes on the hydrology of Poyang Lake written in Chinese by Shengrui Wang (Chinese Research Academy of Environmental Science) or the combined work of Wenbin Zhou and Jinbao Wan (Nanchang University) together with Jiahu Jiang (Nanjing Institute of Geography and Limnology).

Summarizing, we hope that the positive outcomes and experiences of the joint research recorded here can also prove to be useful for both scientists and funders aiming to design and implement water resource protection projects in other areas and practitioners looking for a modern methodological framework to assess, document and present the state of an aquatic system. Furthermore, this book is the third volume in the Springer series *Chinese Water Systems* that presents the application of state-of-the-art approaches in environmental research to Chinese water systems of national and international importance, namely:

- Song, Yonghui, et al. (eds.). *Chinese Water Systems: Volume 1: Liaohe and Songhuajiang River Basins*. Springer, 2018.
- Sachse, Agnes et al. (eds.). *Chinese Water Systems: Volume 2: Managing Water Resources for Urban Catchments*. Springer, 2018.

Few books are without errors, and this book is likely no exception. Should you discover errors that should be corrected, we would be grateful if you let us know to help improve this book.

Leipzig, Germany
June 2018

Erik Nixdorf

Acknowledgements

The book project is part of the Sino-German cooperation group project. "A modelling platform prototype for environmental system dynamics", which is funded by the Sino-German Centre for Science Promotion. The Sino-German Center for the Science Promotion (CDZ) is a joint research venture funded by the German Research Foundation (DFG) and the National Natural Science Foundation of China (NSFC). The funding under grant GZ1167 (T533D810) is greatly acknowledged.

Chinesisch-Deutsches Zentrum für Wissenschaftsförderung
中德科学中心

HELMHOLTZ ZENTRUM FÜR UMWELTFORSCHUNG UFZ

TECHNISCHE UNIVERSITÄT DRESDEN

TECHNISCHE UNIVERSITÄT DARMSTADT

HOCHSCHULE WEIHENSTEPHAN-TRIESDORF UNIVERSITY OF APPLIED SCIENCES

DLR

The second part of the book is introducing the Research Centre for Environmental Information Science (RCEIS) initiated by the Helmholtz Association of German Research Centers and the Chinese Academy of Sciences. The following Helmholtz Centers and Institutes of the Chinese Academy of Sciences contributed to RCEIS. The funding of RCEIS by the Network Fund of the Helmholtz Association (HIRN-0002) is greatly acknowledged.

HELMHOLTZ ZENTRUM FÜR UMWELTFORSCHUNG UFZ

JÜLICH FORSCHUNGSZENTRUM

KIT Karlsruher Institut für Technologie

中国科学院地理科学与资源研究所 Institute of Geographic Sciences and Natural Resources Research, CAS

DLR

Contents

Contributions

We appreciate the contributions to the third CWS volume by:

- Lars Bilke (Helmholtz Centre for Environmental Research, Germany)
- Qinghua Cai (State Key Laboratory of Freshwater Ecology and Biotechnology, Institute of Hydrobiology, CAS, China)
- Cui Chen (Helmholtz Centre for Environmental Research, Germany)
- Xiaoyu Dong (State Key Laboratory of Freshwater Ecology and Biotechnology, Institute of Hydrobiology, CAS. Shenzhen Academy of Environmental Sciences, China)
- Zhengping Du (Institute of Geographic Sciences and Natural Resources Research, CAS, China)
- Zemeng Fan (Institute of Geographic Sciences and Natural Resources Research, CAS, China)
- Zhewen Fan (Jiangxi Remote Sensing Information System Center, China)
- Na Fang (Jiangxi Normal University, China)
- Fengzhi He (State Key Laboratory of Freshwater Ecology and Biotechnology, Institute of Hydrobiology, CAS, China. Leibniz-Institute of Freshwater Ecology and Inland Fisheries, Freie Universität Berlin, Germany)
- Zhenpeng Hu (Jiangxi Mountain-River-Lake Engineering Academic Committee, Nanchang University, China)
- Xiaolan Huang (Jiangxi Normal University, China)
- Juliane Huth (German Aerospace Center, Germany)
- Sonja C. Jähnig (Leibniz-Institute of Freshwater Ecology and Inland Fisheries, Germany)
- Minfei Jian (Jiangxi Normal University, China)
- Olaf Kolditz (Helmholtz Centre for Environmental Research, TU Dresden, Germany)
- Claudia Kuenzer (German Aerospace Center, Germany)
- Harald Kunstmann (Karlsruhe Institute of Technology, Germany)
- Xinghua Le (Jiangxi Mountain-River-Lake Development Office, China)
- Musheng Liu (Jiangxi Mountain-River-Lake Development Office, China)

- Ming Lu (Institute of Geographic Sciences and Natural Resources Research, CAS, China)
- Xu Lu (Jiangxi Normal University, China)
- Carsten Montzka (Forschungszentrum Juelich, Germany)
- Caiying Ni (Jiangxi Normal University, China)
- Erik Nixdorf (Helmholtz Centre for Environmental Research, Germany)
- Karsten Rink (Helmholtz Centre for Environmental Research, Germany)
- Wenjiao Shi (Institute of Geographic Sciences and Natural Resources Research, CAS, China)
- Stepan Shvartsev (Tomsk Polytechnic University, Russia)
- Evgeniya Soldatova (Tomsk Polytechnic University, Russia)
- Pengfei Song (Jiangxi Normal University, China)
- Xiaoling Sun (State Key Laboratory of Freshwater Ecology and Biotechnology, Institute of Hydrobiology, CAS. Southern University of Science and Technology, China)
- Zhanxue Sun (East China University of Technology, China)
- Changqing Yan (Shandong University of Science and Technology, China)
- Wenjing Yang (Jiangxi Normal University, China)
- Qinghui You (Jiangxi Normal University, China)
- Tianxiang Yue (Institute of Geographic Sciences and Natural Resources Research, CAS, China)
- Dan Zhang (Jiangxi Normal University, China)
- Yapeng Zhao (Institute of Geographic Sciences and Natural Resources Research, CAS, China)
- Siying Zhao (Jiangxi Normal University, China)
- Yi Zhou (Jiangxi Normal University, China)

Part I
Introduction

Chapter 1
Background Information about Poyang Lake Basin

Erik Nixdorf and Chengzi Zhou

1.1 Administration

The by maximum annual extension biggest Chinese freshwater lake—Poyang Lake is located in the southeastern part of China (Fig. 1.1) [1]. The total area of Poyang Lake Basin is about 162 000 km^2. 96.7% of the basin is located within the provincial borders of Jiangxi Province and 1.8% belongs to Huangshan prefecture in Anhui Province. Furthermore, a smaller proportion of 0.7% of the catchment area is located in Fujian Province within the prefectures Nanping and Longyan. Additionally, 0.5% of the basin belongs to Chengzhou in Hunan Province and 0.3% to Xuzhou prefecture in Zhejiang Province. In other terms, 94.2% of Jiangxi Province is located within Poyang Lake Basin, which means that data provided for Jiangxi Province is suitable to represent the characteristics of the Poyang Lake Basin.

Jiangxi Province is with an area of 166 919 km^2 the 13th largest province of China. Jiangxi is bordered by Anhui Province to the north, Zhejiang Province to the northeast, Fujian Province to the east, Guangdong Province to the south, Hunan Province to the west and Hubei Province to the northeast. The mountain ridges of the watershed define most of the provincial border. However, parts of the northern border to Hubei and Anhui Province are formed by the water course of Yangtze River.

Jiangxi Province is divided into 11 prefectures, which are further delineated into 100 counties. Except of a large part of Pingxiang District in the west of Jiangxi Province, the southern parts of Ganzhou Prefecture and the northern parts of Jiujiang Prefecture, which drain directly to Yangtze, all prefectures are located within the Poyang Lake Basin. Nanchang is the capital of Jiangxi, which is, in dependence on the lake's water level, between 50 and 60 km away from the shoreline of Poyang Lake. Most large cities in Jiangxi Province (indicated as red dots in Fig. 1.1) are located near the major rivers.

E. Nixdorf (✉) · C. Zhou
Helmholtz Centre for Environmental Research, Leipzig, Germany
e-mail: erik.nixdorf@ufz.de

T. Yue et al. (eds.), *Chinese Water Systems*, Terrestrial Environmental Sciences,
https://doi.org/10.1007/978-3-319-97725-6_1

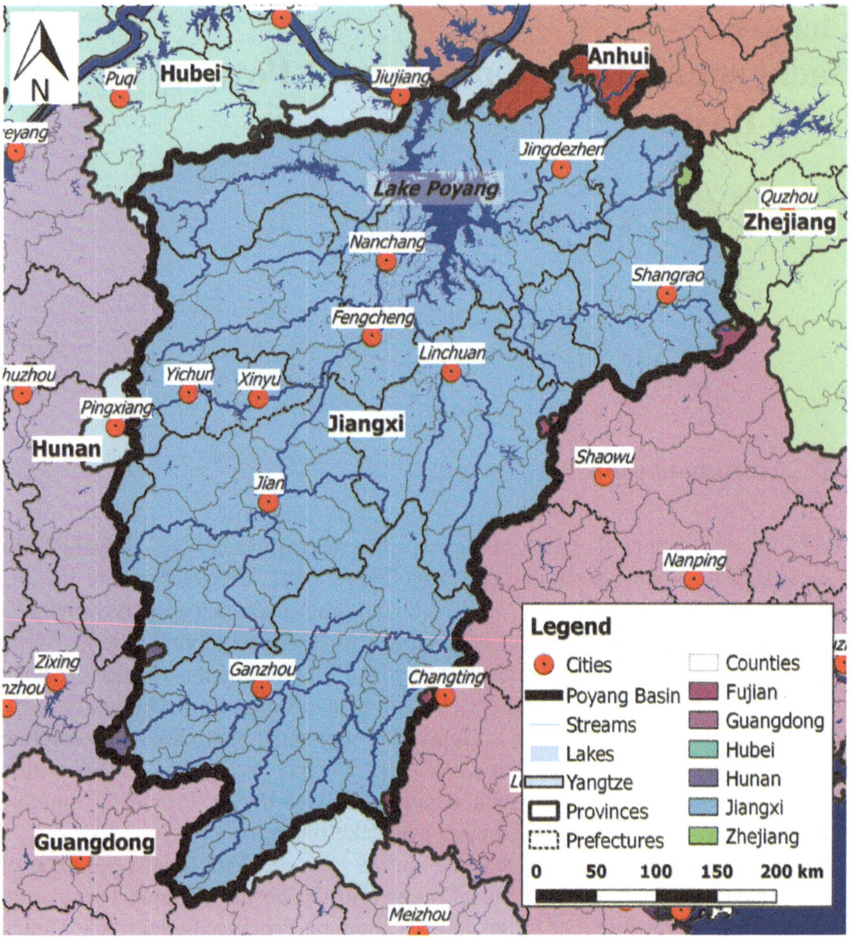

Fig. 1.1 Administrative division of the Poyang Lake Basin

1.2 Physical Topography

Jiangxi has beautiful landscapes with green mountains and clear waters. The main mountain ranges are distributed along three sides of the province border of Jiangxi (Fig. 1.2): The Mufu Mountains, Jiuling Mountains and Luoxiao Mountains on the west; Huaiyu Mountains and Wuyi Mountains on the east; and Jiulian Mountains and Dayu Mountains in the south. Huanggang Mountain at the border to Fujian province is the highest peak with a height of 2158 m above sea level. However, in total, mountainous areas higher than 1000 m are associated to less than 2% of the basin area. Aside of the mountain ranges, the main topographical patterns of Jiangxi Province vary from hilly landscapes with intercepting valleys in the south to flat

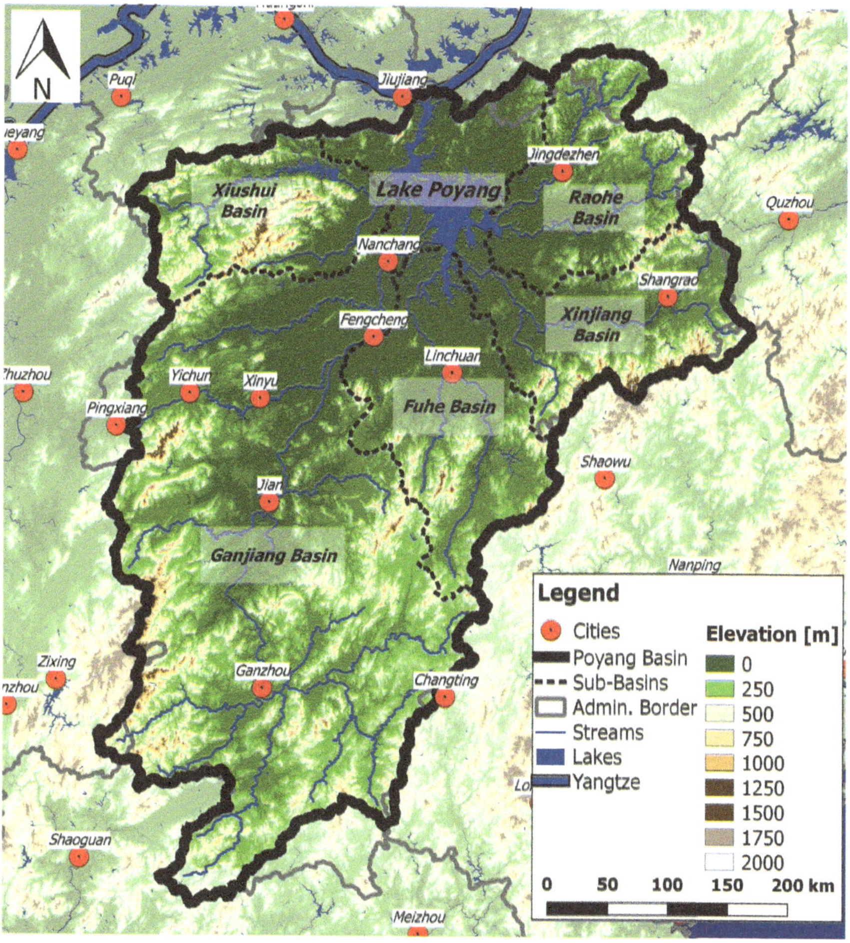

Fig. 1.2 Physical topography of Jiangxi Province. The elevation data is derived from the STS–99 Shuttle Radar Topography Mission [3], which provides data in a spatial grid resolution of about 90 m

alluvial plains in the lower reaches of the primary watercourses in the northern part of the province, where also Poyang Lake is located [2]. The mean altitude of Poyang Basin is 245 m above sea level. More than 70% of areas have an altitude of less than 400 m, which shows the lack of larger plateaus in the catchment.

Forest is the dominant land cover type within Poyang Basin, comprising 49.6% of the total area in 2009. The total standing forest stock is 445 mio. m^3 [4]. The large forest areas are mainly concentrated on elevated areas in the mountain ranges of the catchment (Fig. 1.3).

Another main landcover type is farmland constituting approximately about 29.7% of the area in the Poyang Lake Basin. Mainly the wide valleys formed by the main rivers and the Poyang Lake plain in the central-northern part of the catchment are

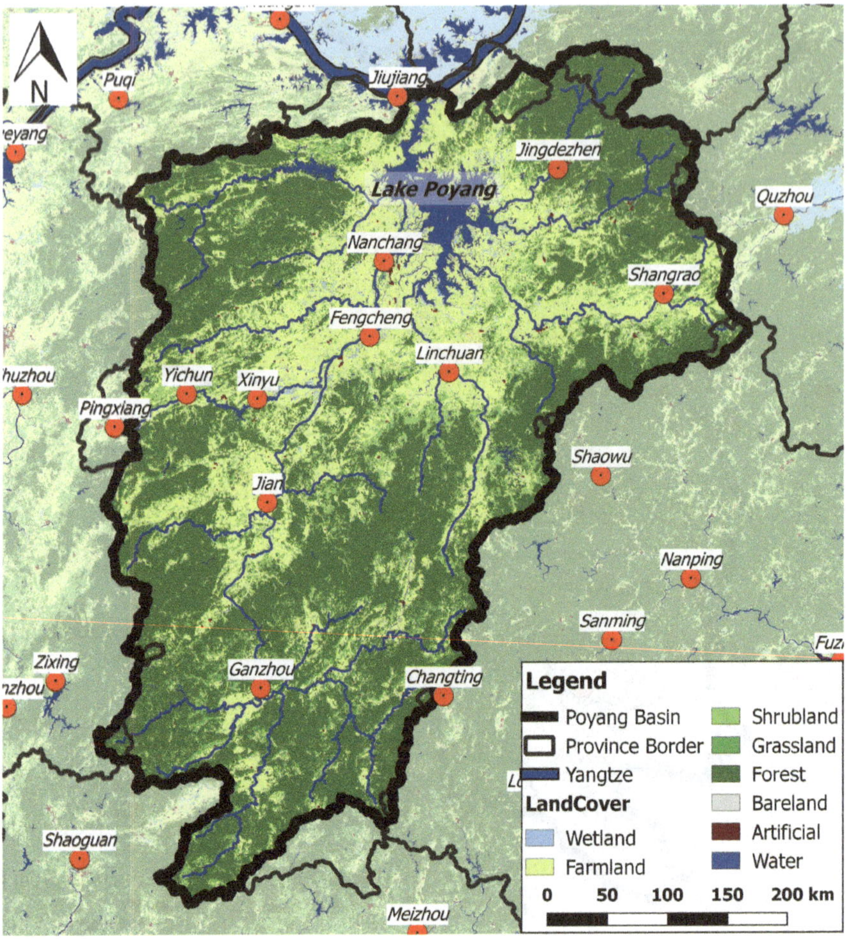

Fig. 1.3 Land cover in Jiangxi Province in 2009 derived from the GlobCover 2009 dataset [5]

extensively cultivated by agricultural activities making Jiangxi Province to one of China's main grain production bases south of the Yangtze River. Shrubland and grassland cover 11.5 and 5.6% of the land area mostly in areas with medium altitude. Although 2.5% of the area is covered by water bodies and 0.6% by wetland according to the input dataset, these shares are subject to seasonal changes due to the water level dynamics of Poyang Lake. During the past twentieth century, many levees were constructed to protect existing farmland from being flooded and to convert wetland areas into farmland used for grain production. However, wetland reclamation was banned in 1986 and laws have been subsequently introduced to promote wetland restoration. Only small areas of the catchment are delineated as bareland (0.4%) which comprises of sand dunes around Poyang Lake but may include other areas e.g. land being prepared for construction activities. Although artificial structures cover only about 0.1% of land in Poyang Lake Basin, cities and towns continue to grow which will in turn require larger shares of available land in the future.

1.3 Climate

Poyang Lake Basin has a humid subtropical climate with short but relatively cool winters and hot and humid summers. Average annual temperature in the capital Nanchang is 17.7 °C. Temperatures can be close to freezing point during winter. Snowfall is very rare in the plain area of Poyang Lake but, more likely, a snow layer can be formed in the mountainous areas of the basin. In spring, temperatures rise fast and reach high average summer temperatures of more than 30 °C in low altitude areas like Nanchang.

Annual precipitation shows a wet and a dry season with a short transition period in between. In Nanchang, annual average precipitation is about 1500 mm per month. Precipitation increases quickly from January to June. June is with 300 mm precipitation the most precipitous month of the year. Average monthly rainfall decreases sharply in July to values of about 150 mm per month. However, Jiangxi province can be impacted by tropical cyclones bringing days of heavy thunderstorms and rainfall in late summer. The dry season begins in September and lasts through December. December is the month with the lowest rainfall of the year with an average rainfall of less than 50 mm in Nanchang (Fig. 1.4).

The distribution of annual average precipitation in Poyang Lake Basin depends mostly on the altitude. Rainfall in mountains areas is much higher than in the hilly and plain area of the basin (Fig. 1.5). Rainfall rates are highest in the northeast part of Jiangxi and decrease further west due to the Wuyi Mountains acting as a protective barrier against the inflow of cold air from the northwest and retain warm moist air originating from the sea. In 2.6% of the basin, annual precipitation can reach more than 2000 mm with maximum values of 2225 mm in the Wuyi Mountains. In contrast, about 87% of the basin receives annual rainfall between 1450 and 1850 mm per year. In some of the river valleys in the southwest part of the basin and in the northern region precipitation descend to 1415 mm.

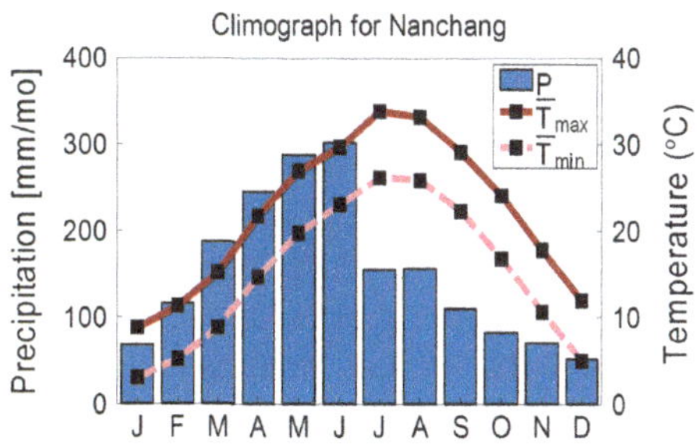

Fig. 1.4 Climograph of monthly temperature and precipitation in Nanchang

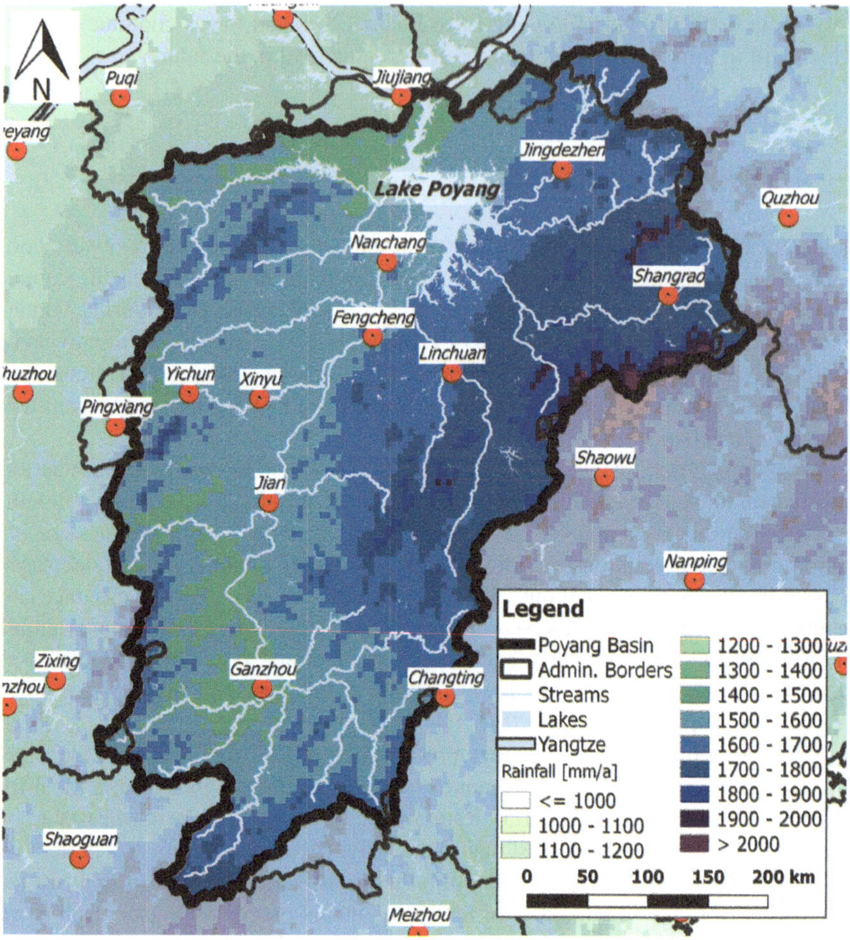

Fig. 1.5 Spatial distribution of average annual precipitation in Poyang Lake Basin

1.4 Cultural Geography

In 2015, Jiangxi Province is home to more than 45 mio. people (Table 1.1). The share between urban and rural population is 52% and 48%, respectively, which means, that more people live in cities than in the countryside. In 1987, when Jiangxi had a population of more than 31 mio. people, only 16.75% of population lived in urban whereas 83.25% lived in rural areas. 99% of the population is Han Chinese due to receiving successive waves of migration from Northern China through the ages.

The population distribution in the Poyang Lake Basin is very uneven (Fig. 1.6).[1] Although the overall population density is about 270 P/km^2, the maximum

[1] http://sedac.ciesin.columbia.edu/data/collection/gpw-v4/citations.

Table 1.1 Population and water supply in the prefecture-level cities of Jiangxi [4]

Prefecture city	Population (1000 P)	Pop. density (P/km^2)	GDP (RMB/P)	Water supply (100 bn. m^3)
Nanchang	5272	737	75879	3.06
Jingdezhen	1635	312	47216	0.83
Pingxiang	1895	496	48133	0.75
Jiujiang	4816	253	39505	2.35
Xinyu	1163	369	81354	0.78
Yingtan	1150	324	55568	0.74
Ganzhou	8527	217	23148	3.24
Ji'an	4890	194	27168	3.00
Yichun	5502	296	29457	4.27
Fuzhou	3984	212	27735	2.35
Shangrao	6701	295	24633	3.19
Average	4139	336	43617	2.23

population density is more than 60 000 P/km^2 in downtown Nanchang. Many factors such as climate, landform, land cover, resources, economy and transportation influence the population distribution. In the mountainous regions of Jiangxi, the population density is typically less than 100 P/km^2 due to less farmland and inconvenient transportation opportunities. Outside of the mountain areas, population density often exceeds 500 P/km^2.

The capital Nanchang is with more than 5 mio. Inhabitants the largest city in Jiangxi and the Poyang Lake Basin. It is the center of economy, transport and culture of Jiangxi Province. Since decades, Nanchang Metropolitan Area is experiencing a strong population growth, which lead to an expansion of the city into former wetland areas of Poyang Lake.

1.5 Economy and Water Management

Jiangxi was one of the nation's most affluent regions before trade patterns changed by opening of treaty ports to the Western Powers in the mid-19*th* century.[2] In 2015, Jiangxi's GDP was about 1 672 bn. RMB. The total value of foreign exports and imports of goods was about 33 bn. USD and 9 bn. USD, respectively.

The GDP of agriculture goods was 177.3 bn. RMB in 2015. Although the GDP from agriculture lags behind industry and services, Jiangxi is one of the three largest rice-producing provinces in China. Rice growing areas account for more than 60% of the overall cultivated area in Jiangxi. Paddy fields cover 85 to 90% of the area

[2]https://www.britannica.com/place/Jiangxi#toc71324.

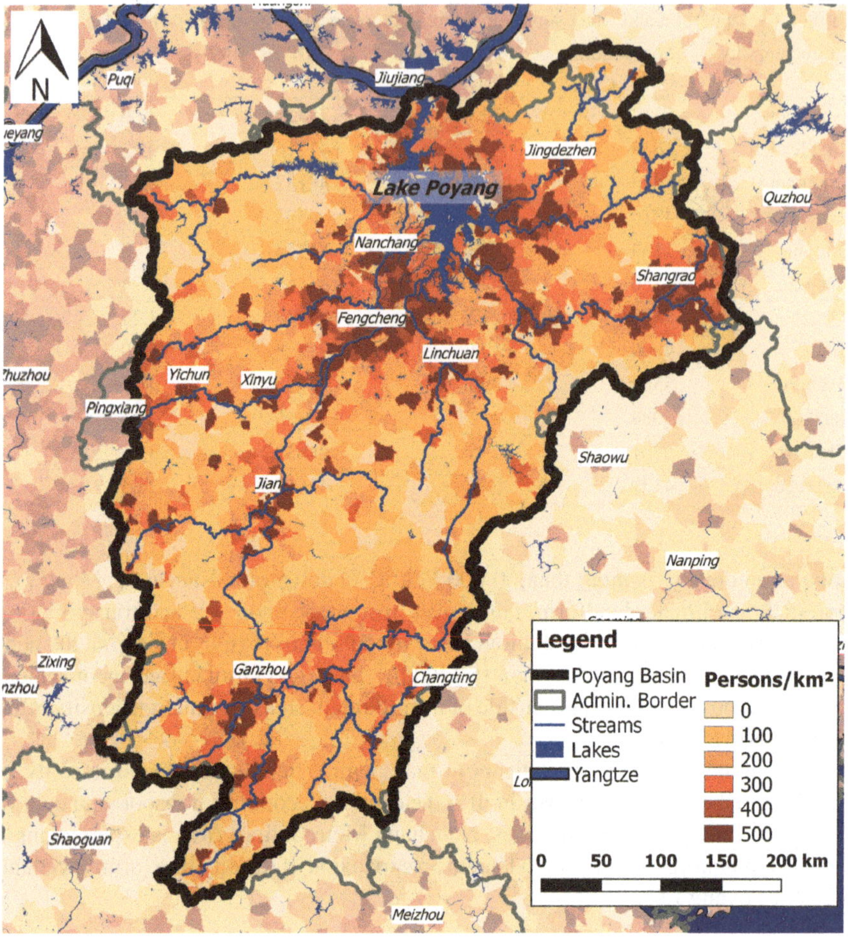

Fig. 1.6 Population density in Jiangxi Province in 2015

planted with grains.[3] Other food crops produced in Jiangxi include cotton, sugar cane, fruits and oil-bearing crops. In 2015, 21.48 mio. tons grain, 1.1 mio. tons cotton and 50 356 tons tea were produced. The total cultivated freshwater area of the whole province is 43.32 thousand hectares with more than 150 different species of fish being raised. The GDP of the industry in Jiangxi was 841 bn. RMB in 2015. Jiangxi's core industries include chemical industry, steel industry, cement industry and pharmaceutical industry. By the end of 2015, the GDP of the service sector was about 654 bn. RMB. Banking, insurance and real estate are the key service industries in Jiangxi [4]. In addition, some areas such as the northern part of Wuyi Mountains are famous scenic tourist areas [1].

[3] http://europe.chinadaily.com.cn/epaper/2015-07/17/content_21307209.htm.

Jiangxi Province has rich deposits of minerals, making it is one of the provinces with highest matching degree of mineral resources in China. Copper, Tungsten, Uranium, Tantalum, Rare Earths, Gold and Silver are called "the seven gold flowers of Jiangxi". Among others, Jiangxi is the largest producer of copper in China with most production coming from the Dexing mining district in the east of Jiangxi Province.

The province's total water supply was about 24.42 bn. m^3, accounting for 12.3% of the total annual water resources (Table 1.1). The vast majority of 23.6 bn. m^3 is supplied by surface waters followed by groundwater supply of in total 824 mio. m^3. Compared with 2014, the province's total water supply decreased by 1.349 bn. m^3 in 2015. In 2015, the total water use in Jiangxi was equal to the supply with about 24 bn. m^3. About 14 bn. m^3 of water were used for agriculture, which was about 59% of the totally used water volume. Annual agricultural water demand highly depends on the precipitation. For instance, the demand decreased by 1.3 bn. m^3 in comparison to 2014, because of unusual high rainfall during the growing season. The second largest water user was the industry with 6 bn. m^3, which was 25% of the total water use. Residential use demanded about 2.8 bn. m^3 of supplied water [6].

1.6 River Network

There are more than 2 400 rivers of various size in Jiangxi province, which have a combined total length of about 18 400 km. Most of them enter Poyang Lake, which in turn empties into Yangtze River. The five major rivers are Gan River, Fu River, Xin River, Xiu River, and Rao River [4]. Gan River with a length of 751 km is the longest river in Jiangxi and the second largest tributary of Yangtze River in terms of water volume. Coming from the west part of Wuyi Mountains Gan River flows through the entire length of the province from south to north, passing cities from Ganzhou to Hukou before pouring into the Yangtze River. Thirteen major tributaries flow into Gan River, which is navigable on more than 500 km of its watercourse.[4]

The Chinese Standard for Surface Water Quality divides the surface waters of China into 6 classes according to their chemical and physical water properties with class I being the best and class >5 being the worst [8]. The water quality of 80 rivers in the province with a total river length of 6 241 km is evaluated annually by governmental organisations of Jiangxi province [6] based on data from 307 monitoring sections. In 2015, 2.9% of all investigated river sections could achieve the highest surface water quality whereas rivers with poor quality had a share of 7.1% (Fig. 1.8). The majority of rivers showed a water quality associated with class II of the standard. The polluted river sections are mainly distributed in the Gan River (e.g. Qingshan section), Fu River (e.g. Luoxi section and Yunshan section), Rao River (e.g. Le'an and Hanjia crossing section) and Xiu River (North Liao River and Jing'an section) with

[4]http://ziliaoku.jxwmw.cn/system/2008/11/12/010079359.shtml.

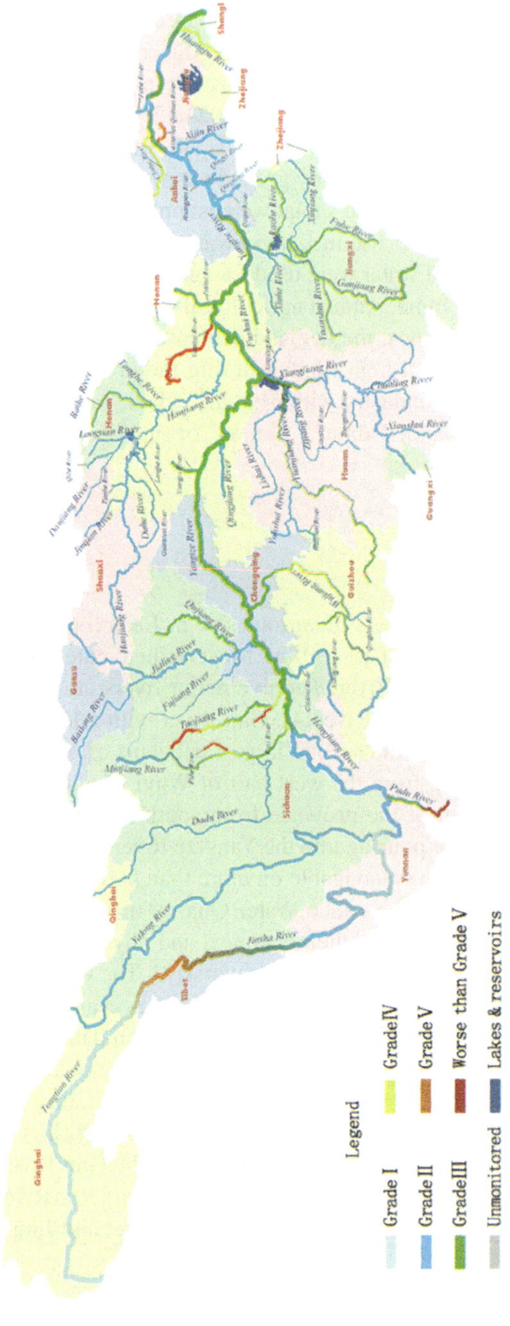

Fig. 1.7 Surface water quality in Yangtze River Basin [7]

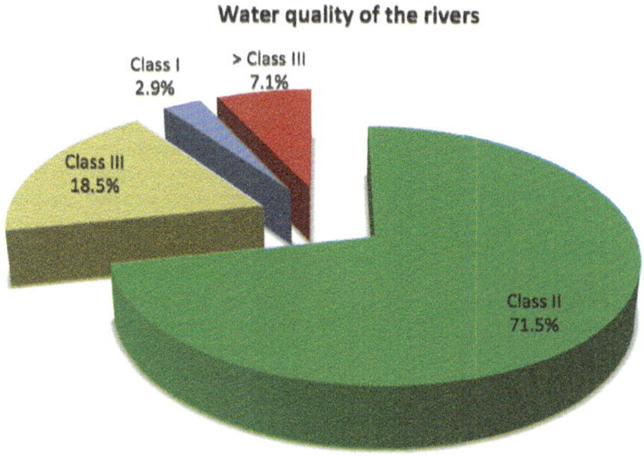

Fig. 1.8 Surface water quality of the rivers in Poyang Basin [6]

main pollutants being ammonia and phosphorus [6]. However, focusing on the entire Yangtze River Basin, the rivers of the Poyang Lake Basin have an above average water quality (Fig. 1.7).

1.7 Reservoirs and Floodplain Lakes

Due to humid subtropical climate, the precipitation pattern in Poyang Lake differs between rainy season and dry season. Therefore, many reservoirs were constructed in the last century to supply drinking water, to prevent water damages, to generate electrical energy and to manage the rivers effectively. Jiangxi Province has 29 large reservoirs and 251 medium-sized reservoir with a total storage volume of 13.165 bn. m^3. Among them, the large reservoirs have a total storage capacity of 10.121 bn. m^3 [6]. The main reservoirs are Hongmen, Shangyoujiang, Zhelin, Jiangmen and Tuolin Reservoir. Zhelin Reservoir, which is located at Xiu River in the northeast part of Poyang Lake Basin is the largest in Jiangxi Province with a storage capacity of 7.9 bn. m^3. The second largest reservoir is Hongmen reservoir in Rao River basin with a capacity of about 2 bn. m^3. Its lake with a surface of about 40 km^2 is a renowned tourist site and a natural reserve for birds.

The water quality of 53 reservoirs of large and medium-sized reservoirs in the province is evaluated on a regular base, including 14 large-scale reservoirs and 39 medium-sized reservoirs. In 2015, all reservoirs achieved class III water standard or better during the whole year, including flood season and non-flood season. Additionally, the eutrophication level is less than in reservoirs and lakes in other regions of China (Fig. 1.9) [1]. However, all major reservoirs are threatened of being continuously filled by incoming sediment loads due to high soil erosion in many upstream areas [9].

Poyang Lake is located in the middle and lower reaches of Yangtze River and has a shoreline of about 1200 km. The lake receives water mainly from five rivers: Xiu

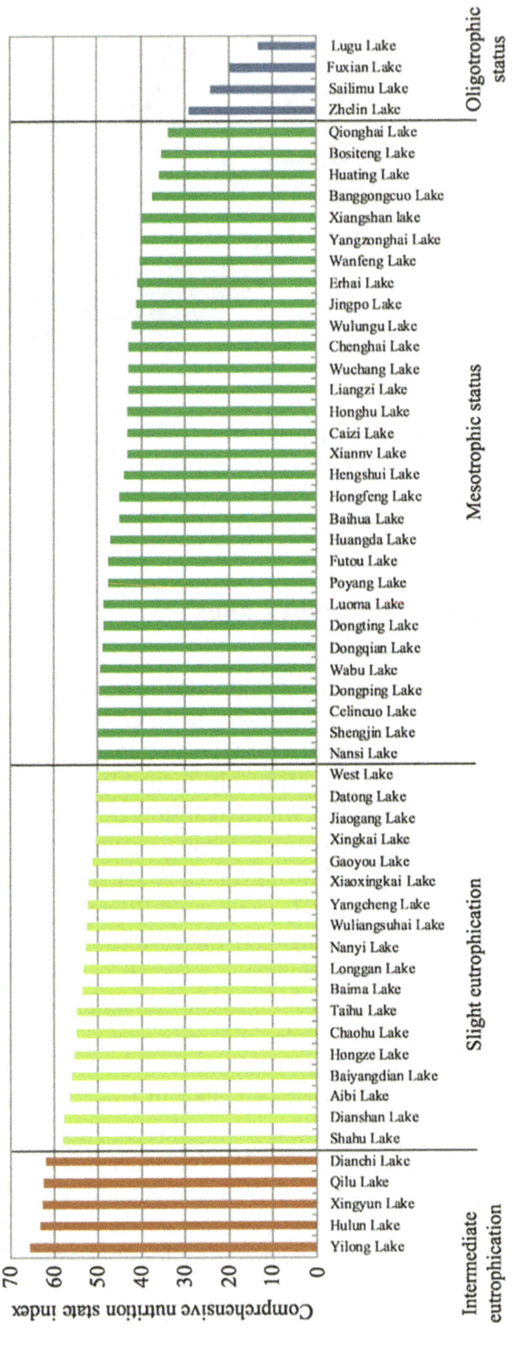

Fig. 1.9 Categorisation of Chinese water bodies by their trophic status in 2016 [1]

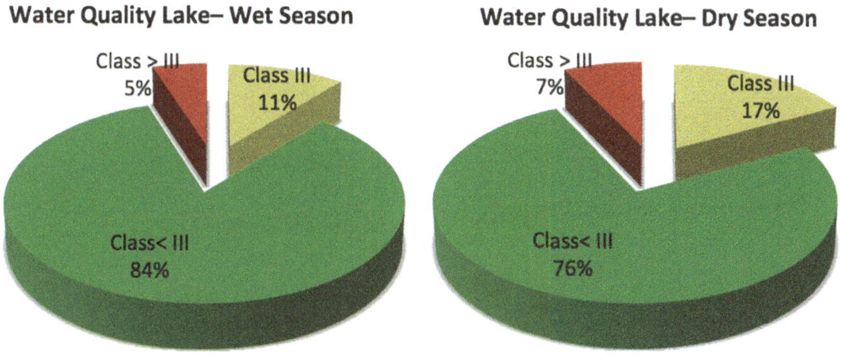

Fig. 1.10 Water quality of Poyang Lake during wet season and dry season

River, Gan River, Fu River, Xin River, and Rao River and discharges into the Yangtze River through a channel in its north part (Fig. 1.2). In response to the annual cycle of precipitation, Poyang Lake can expand to a large water surface of 3 800 km^2 and a volume of 32 bn. m^3 in the wet season, but shrinks to little more than a river during the dry season and thereby exposes extensive floodplains and wetland areas [10].

The modern Poyang lake was formed less than 3000 years ago after the Yangtze River switched to a more southern water course causing the Gan River to back up and form Poyang Lake. Poyang Lake reached its largest size during the Tang Dynasty, when its surface area was about 6 000 km^2. Until the late Ming and early Qing Dynasty it was ultimately gradually evolved into the shape of today's lake [11].

The hydrologic processes of Poyang Lake are complex. Its hydrologic regime is mainly driven by both the five main tributaries and the Yangtze River to which the Poyang Lake drains about 150 bn. m^3 per year [12]. Additionally, the hydrologic system is increasingly disturbed by hydraulic construction measures. The opening of the Three-Gorges-Dam (TGD) in 2003 significantly influenced the hydrological regime in Poyang Lake by reducing length and occurrence of reverse flow conditions [13]. The opening is considered of being one factor of generally decreasing water levels in Poyang Lake during the last decade [14]. If water levels in Poyang Lake continue to decline, the reducing groundwater stores could become a serious concern for the wetland ecosystem, too. This lead to the proposed construction of a new dam at the outflow of Poyang Lake in order to compensate the side effects of the TGD and to rise the lake water level during dry season.

The water quality of Poyang Lake varies with its water volume. Most of Poyang Lake achieves a water quality standard of class III or better (Fig. 1.10). However, water quality deteriorates during dry season with about one fourth of the water body being classified as having class III water quality or worse. In recent years, water quality is continuously deteriorating as the rivers draining to Poyang Lake face several pollution pressures such as an continuous inflow of nutrients and fertilizer residuals from its extensively cultivated shorelines [15] as well as acidity and heavy metals from mining areas [16] and industrial sites [17].

Fig. 1.11 Inside the large wetland areas of Poyang Lake

1.8 Wildlife

Jiangxi Province is characterized by its wide coverage of vegetation and diversity of wildlife. In Jiangxi Province live more than over 600 kinds of vertebrates, including over 170 species of fish, 40 species of amphibians, 70 species of reptiles and 270 species of birds, which all account for more than 20% of the national total of these animal classes.[5] One of the most biodiversity rich areas in Jiangxi Province is the Poyang Lake-wetland system. The large wetlands of Poyang Lake are Asia's largest winter destination for migratory birds [18]. About 95% of the world's Siberian white cranes, 50% of its white-naped cranes and 60% of its swan geese spend winter here every year (Fig. 1.11).[6]

Jiangxi government transformed parts of the wetlands into natural reserves to improve the protection of habitats. One example is the Poyang Lake National Nature Reserve, which is located in the northern part of Poyang Lake covering an area of about 224 km^2. It is one of the first six important wetlands, which joined the Ramsar Convention in China in 1983. It forms a multi-functional complex wetland ecosystem of unique topography and geomorphology, which provides a good habitat for wild animals, especially birds. In this natural reserve alone live 45 species of mammals, 310 species of birds, 48 species of reptiles, 136 species of fish, 227 species of insects, 40 species of shellfishes, 46 species of zooplankton, 50 species of phytoplankton and 476 species of higher plants.[7]

[5] http://www.china.org.cn/english/features/55638.htm.

[6] http://www.china.org.cn/environment/2012-10/17/content_26815948.htm.

[7] http://www.globalnature.org/33539/Nature-Reserve/02_vorlage.asp.

References

1. Ministry of Environmental Protection China. Report on the State of the Environment in China 2016.
2. Li, X., Q. Zhang, and C. Xu. 2014. Assessing the performance of satellitebased precipitation products and its dependence on topography over Poyang Lake basin. *Theoretical and applied climatology* 115 (3-4): 713–729.
3. Jarvis, A., H. I. Reuter, A. Nelson, and E. Guevara. 2008. Hole-filled SRTM for the globe Version 4.
4. Bureau Jiangxi Statistical. 2017. Jiangxi Statistical Yearbook 2016. China Statistics Press.
5. Bontemps, S., P. Defourny, E. Van Bogaert, O. Arino, and V. Kalogirou. 2009. GlobCover 2009: Product description manual, version 1.0. In: *ESA and UCLouvain*.
6. L. Zhu and J. Liu. 2015. Jiangxi water resources bulletin 2015.
7. Ministry of Environmental Protection China. 2015. Report on the State of the Environment in China.
8. State Environmental Protection Administration General Administration of Quality Supervision Inspection and Quarantine of the People's Republic of China. 2002. Environmental quality standards for surface water.
9. Yuan, L., G. Yang, Q. Zhang, and H. Li. 2016. Soil Erosion Assessment of the Poyang Lake Basin, China: Using USLE, GIS and remote sensing. *Journal of Remote Sensing & GIS* 5 (168): 2.
10. Xu, D., M. Xiong, and J. Zhang. 2001. Analysis of hydrological characteristic of Poyang Lake (in Chinese). In *Yangtze River*, 21–23.
11. Q. Tan. 1982. *The Historical Atlas of China*. Beijing, China: Cartographic Publishing House.
12. Zhao, G., G. Hoermann, N. Fohrer, Z. Zhang, and J. Zhai. 2010. Streamflow trends and climate variability impacts in Poyang Lake Basin, China. *Water Resources Management* 24 (4): 689–706.
13. Li, B., et al. 2016. Spatiotemporal variability in the water quality of Poyang Lake and its associated responses to hydrological conditions. *Water* 8 (7): 296.
14. Li, Y., Q. Zhang, A.D. Werner, J. Yao, and X. Ye. 2017. The influence of riverto-lake backflow on the hydrodynamics of a large floodplain lake system (Poyang Lake, China). *Hydrological Processes* 31 (1): 117–132.
15. Duan, W., et al. 2016. Water quality assessment and pollution source identification of the eastern Poyang Lake Basin using multivariate statistical methods. In *Sustainability* 8 (2): 133.
16. He, M., Z. Wang, and H. Tang. 1998. The chemical, toxicological and ecological studies in assessing the heavy metal pollution in Le An River, China. *Water Research* 32 (2): 510–518. https://doi.org/10.1016/S0043-1354(97)00229-7.
17. Xu, B., and G. Wang. 2016. Surface water and groundwater contaminations and the resultant hydrochemical evolution in the Yongxiu area, west of Poyang Lake, China. *Environmental Earth Sciences* 75 (3): 184. ISSN: 1866-6280. https://doi.org/10.1007/s12665-015-4778-8.
18. Ji, W., et al. 2007. Analysis on the waterbirds community survey of Poyang Lake in winter. *Geographic Information Sciences* 13 (1–2): 51–64.

Chapter 2
The Poyang Lake Cooperation Group

Olaf Kolditz and Tianxiang Yue

2.1 Aims and Scope

The general goal for this initiative is to build a Sino-German research network by leading Chinese scientists with their research groups and German experts in the field of environmental informatics, hydrology, climatology and remote sensing (satellite born earth observation). This initiative will form the basis for intensive exchange of research methods and knowledge, and is intended to the development of bilateral research project proposals e.g. to the Ministry of Science and Technology of the People's Republic of China (MOST), National Natural Science Foundation of China (NSFC), German Research Foundation (DFG), German Federal Ministry for Education and Research (BMBF) and/or European Commission (EC).

Several data and modeling platforms have been proposed for accumulation of knowledge by data management and process modeling in recent years. However, a modelling platform for environmental system dynamics (MPESD) needs to be established by fusing data from both remote sensing and ground-based observation. For establishing such MPESD, it is essential to solve theoretical and technical problems e.g. of multi-scale modeling, limited computational power and real-time visualization, to supplement insufficient ground-observation data, and to develop new methods for earth surface modelling (ESM). ESM would take approximate global information (e.g. remote sensing images) as its driving field and locally accurate information (e.g. ground observation data) as its optimum control constraints.

O. Kolditz (✉)
Helmholtz Centre for Environmental Research, Leipzig, Germany
e-mail: olaf.kolditz@ufz.de

O. Kolditz
Technical University Dresden, Dresden, Germany

T. Yue
Institute of Geographic Sciences and Natural Resources Research (CAS),
Beijing, China

T. Yue et al. (eds.), *Chinese Water Systems*, Terrestrial Environmental Sciences,
https://doi.org/10.1007/978-3-319-97725-6_2

The scientific objective of this Sino-German cooperation proposal is to develop a modelling platform prototype for Big Data applications on the basis of giving solutions of the theoretical and technical problems to explore interactive mechanisms of land-use change and water-resource change under climate change and rapidly increasing human activities. It is important to better understand and based on sound knowledge to precisely simulate trends and scenarios of land use and water resources as well as the driving forces behind the changes. The specific research objectives include, (1) developing multi-scale data models, and analyzing effects of the scale transformation on sensitivity of variables and on predictability of ecosystems, (2) developing fast numerical algorithms, e.g. algorithms and methods for parallel computing with GPU and MPI on clusters and national supercomputing platforms, (3) establishing a platform prototype for environmental system dynamics, and (4) intensive exchanging information, jointly publishing results of joint research activities, exchanging scientific staff and students, and jointly organizing technical meetings, symposia and training courses.

2.2 General Description of Overall Project

The general objective of this initiative is to establish a Sino-German research network for developing a simulation platform of environmental system dynamics. The platform prototype will be first being developed for the Poyang Lake Basin in China. Land-cover, land use and water resources changes as well as their interactive mechanisms are to be studied under consideration of driving forces such as climate change and human activities. To this purpose a model-oriented approach will be developed. The German and Chinese partners will receive scientific practice of the newly developed platform including innovative methods (conceptual models and BigData concepts) and modern technologies (computer platforms) being first applied to the Poyang Lake Basin and transferable to other environmental systems.

The general objective will be approached by conducting the following four model-oriented objectives:

1. Models for simulating water resources changes: to establish process-based and distributed hydrological models, taking sub-basin as its hydrological response units, and to simulate future development of water resources under different climatic scenarios.
2. Models for detecting land-cover changes: to detect land-cover changes with higher spatial and temporal resolutions by fusing Moderate Resolution Imaging Spectroradiometer (MODIS) data with higher temporal resolution, with TM data with higher spatial resolution.
3. Models for modeling ecosystem-change driving forces: to develop a model for interpolating observation data from meteorological stations and a model for downscaling data from General Circulation Models (GCMs) under consideration of spatial non-stationarity of climatic elements, and to set up a dynamic model for

simulating spatial distribution of human population on the basis of quantizing related social-economic factors and natural factors as well as population growth.
4. A modelling platform prototype: to integrate the models for water-resource change, land-cover change and driving forces of the changes, to simulate the interactive mechanisms taking Poyang Lake Basin as an example, and to realize dynamic visualization of the land-cover and water-resource change as well as their interactions.

This initiative is coordinated by Olaf Kolditz (Helmholtz-Centre for Environmental Research and Technische Universität Dresden, Germany) and Prof. Dr. Tian-Xiang Yue (Institute of Geographical Sciences and Natural Resources Research, Chinese Academy of Sciences, China) who are the acting directors of the Sino-German Research Centre for Environmental Information Science (RCEIS, see Chap. 19.1). Four working teams are proposed in terms of the four specific objectives (Fig. 2.1).

1. Working team 1 is to pay attention to the models for water-resource change and its theoretical issues, led by Christoph Schüth (Technische Universität Darmstadt, Germany) and Jinjie YU (Institute of Geographical Sciences and Natural Resources Research, Chinese Academy of Sciences, China).
2. Working team 2 is to focus on the models for land-cover change detection and its theoretical issues, guided by Bing XU (Qing-Hua University, China) and Carsten LORZ, (Hochschule Weihenstephan, Germany).
3. Working team 3 is to be involved with the models for ecosystem-change driving forces and its theoretical issues, headed by Thomas Ulrich BERENDONK (Dresden University of Technology, Germany) and Zhe-Wen FAN (Office of Mountain-River-Lake Development Committee of Jiangxi Province, China, the registered unit with National Natural Science Foundation of China).
4. Working team 4 is to address theoretical problems of the modelling platform prototype and its completion, led by Olaf KOLDITZ (Helmholtz-Centre for Environmental Research/TU Dresden, Germany) and Tianxiang YUE (Institute of Geographical Sciences and Natural Resources Research, Chinese Academy of Sciences, China).

The Sino-German cooperation group is to prepare and to publish a research book about the Poyang Lake Basin, which will be compiled by all the working team leaders. The research book on the modelling platform prototype of environmental element dynamics will include theories of the models based on and the processes of developing the models. How the theoretical problems, such as multi-scale data fusion, computational efficiency, large memory requirements and interactive visualization, are tackled for the modelling platform prototype will be part of the research book. The conclusions of interactive mechanism between land-cover change and water-resource change as well as the change trends and scenarios from an application of the modelling platform prototype in Poyang Lake Basin will be submitted to local government of Jiang-Xi province for a reference of its decision making. The Poyang Lake basin research book will be the scientific product of the envisaged project and will be published in the book series on "Terrestrial Environmental Sciences" (http://www.springer.com/series/13468).

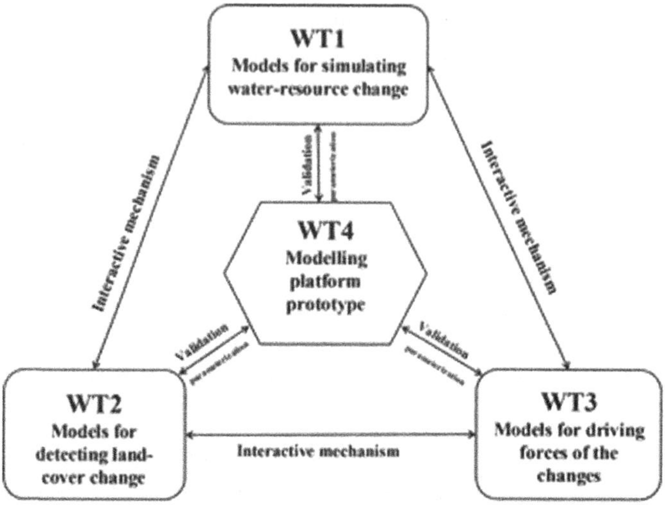

Fig. 2.1 Structure of the cooperation group "A modelling platform prototype for environmental system dynamics—Poyang Lake"

2.3 Future Plans

As a result of the Sino-German Symposium "Sustainable water management and ecosystem restorations in Poyang Lake Basin" (GZ1100) in Nanchang in November 2014, the Chinese and German partners agreed on forming a Sino-German Cooperation Group on developing a modeling platform for describing the evolution of aquatic environments. This project proposal was funded jointly by DFG an NSFC in 2015 (GZ1167). The cooperation group is conducting an intensive exchange program cumulated in two workshops in Germany (2015 in Leipzig) and China (2016 in Beijing-Tianjin) so far. During the latter workshop the Chinese and German partners decided to participate in the open topic call on joint Sino-German projects by German Science Foundation (DFG) together with the Natural National Science Foundation of China (NSFC) in a concerted way by preparing a bundle of interconnected proposals. These project proposals are interlinked on the one hand side but also individual free-standing in research concept and methodology in order to guarantee an individual evaluation of each research proposal.

Understanding the functioning and evolution of aquatic ecosystems is of essential importance for rehabilitation of water quality and restoration of eco-system functions as well as maintaining both in future. The main goal of the Sino-German Cooperation Group "Modeling Platform for Environmental Systems Dynamics" is to further improve this comprehensive system understanding and developing a methodology as well as scientific tools based on the Poyang Lake case study. To this purpose, the cooperation group is combining hydrological and ecological aspects of the system lake under the consideration of related surface processes which are of particular interest in the dynamic aquatic ecosystem Poyang Lake.

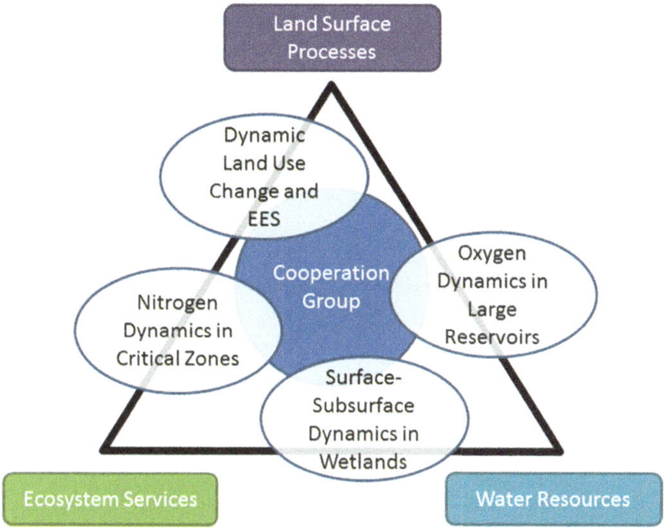

Fig. 2.2 Future concept of the cooperation group

The bundle of project ideas continues the general research concept of the Cooperation Group by combining aspects of land surface processes; water resources management and ecosystem services in an integrated way (Fig. 2.2). On the other side, the DFG#NSFC proposals focus on specific research questions concerning systems dynamics in the more general context (see project ideas below). Moreover the investigation areas cover different climate and land use zones, Poyang Lake and Taihu (humid subtropics), Fengman Reservoir (semi-arid continental).

- Land Use Dynamics: EcoMAPS-Developing a methodology for supporting Ecosystem Services in agricultural focus regions in China and Germany-Monitoring, Assessment and Pareto-optimal Solutions. Dynamic Land Use Change: Developing an Approach on Land Use Change and Ecosystem Services in two study regions in Germany and China—High Resolution Monitoring, Assessment, Visualization and Optimization (Carsten Lorz and Tianxiang Yue),
- Surface-Subsurface Dynamics: Analysing the effects of surface water-subsurface water interaction dynamics on water availability, matter fluxes and ecology in the unique lake wetland system of Poyang Lake, China (Erik Nixdorf and Yunliang Li),
- Oxygen Dynamics: Predicting oxygen dynamics in large reservoirs in four dimensions (Karsten Rinke and Wenqi Peng),
- Nitrogen Dynamics: Nitrogen transport and removal in critical zones and river networks of agricultural subtropical watersheds in China (Michael Rode, Christoph Schüth and Qi Zhang).

Chapter 3
Bibliometric Study of Scientific Literature on Poyang Lake

Chengzi Zhou

3.1 Introduction and Method

Numerous studies have intensively analysed Poyang Lake and its basin during the last decades. In order to track the footprint of research on Poyang Lake a bibliometric analysis was conducted based on entries in the China National Knowledge Infrastructure (CNKI) database and the Scopus database recorded between 1983 and 2017. CNKI has one of the world's largest full-text information volume library "CNKI digital library" and is the biggest databank of journals, articles, books, statistical yearbooks and conference reports in China [1] including different fields and topics such as environmental science, engineering, computer science, medicine, social sciences, energy and chemistry. Most of the articles in the database are written in Chinese with English titles and abstracts. Scopus is the largest abstract and citation database of peer-reviewed English literature including scientific journals, books and conference proceedings.[2] The Keyword "Poyang Lake" was searched in the CKNI database in Chinese and in the Scopus database in English language. Search results at both databases included scientific articles as well as books and dissertations which means that further manual filtering was required to focus on research articles only. In addition, articles published at the websites of research institutions such as Chinese Academy of Sciences were not listed in the database of CKNI and Scopus.

[1] http://cnki.net/gycnki/gycnki.htm.
[2] https://www.elsevier.com/solutions/scopus.

C. Zhou (✉)
Helmholtz Centre for Environmental Research, Leipzig, Germany
e-mail: Chengzi.zh30@gmail.com

T. Yue et al. (eds.), *Chinese Water Systems*, Terrestrial Environmental Sciences,
https://doi.org/10.1007/978-3-319-97725-6_3

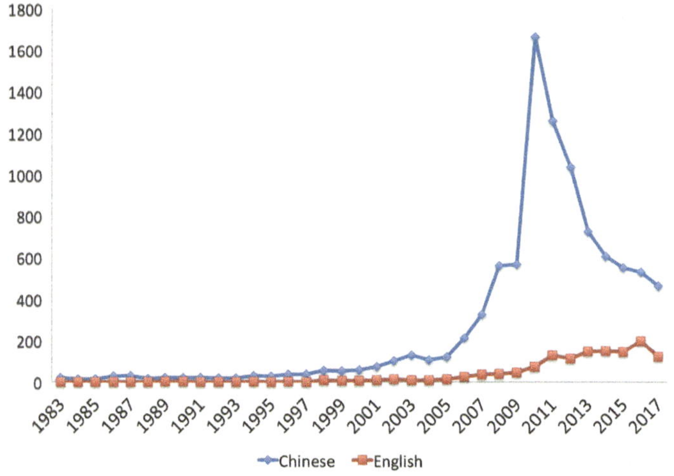

Fig. 3.1 Chinese and English publications about Poyang Lake from 1983 to 2017

3.1.1 Publications

In total, about 9600 Chinese publications and 1300 English publications published between 1983 and 2017 were accomplished by the search query. In the earliest year of records 1983, only 22 (Chinese) publications were accomplished covering the topics land use, fish and economy in Poyang Lake Basin. In contrast, the number of records increased to 1740 publications in 2010 and 730 publications in 2016. It is worth to mention that the number of annual publications in English was lower than 5 until 2005. During the 2000s, numbers of recorded publications increased at both database, particularly considering the publications in Chinese. Examplarily, 514 Chinese publications were found in the CNKI database in 2009, while until 2010, the number has threefolded. However, surprisingly, the number of Chinese annual publications decreased between 2010 and 2017. In 2017, 531 Chinese publications were accomplished from the CNKI database. In contrast, the number of annual publications in English is still increasing with more than 60 recorded English publications in 2017 (Fig. 3.1).

From 1991 to 2017, 1078 Chinese dissertations were published including 924 master dissertations and 104 doctoral dissertations. Before 2000, almost no dissertations could be found in the CNKI database. There were only 1 master dissertation and 2 doctoral dissertations in 2002, while these numbers increase to 114 master dissertations and 12 doctoral dissertations in 2016. However, the maximum of Chinese dissertations was reached in 2012 for the master dissertations (161 in total) and 2013 for doctoral dissertations (14 in total). In recent years, about 110 master dissertations and about 10 doctiral dissertations about processes in Poyang Lake were published per year. Altough the search query was conducted in December 2017, dissertations accomplished in 2017 were very low. One reason for that could be the time span

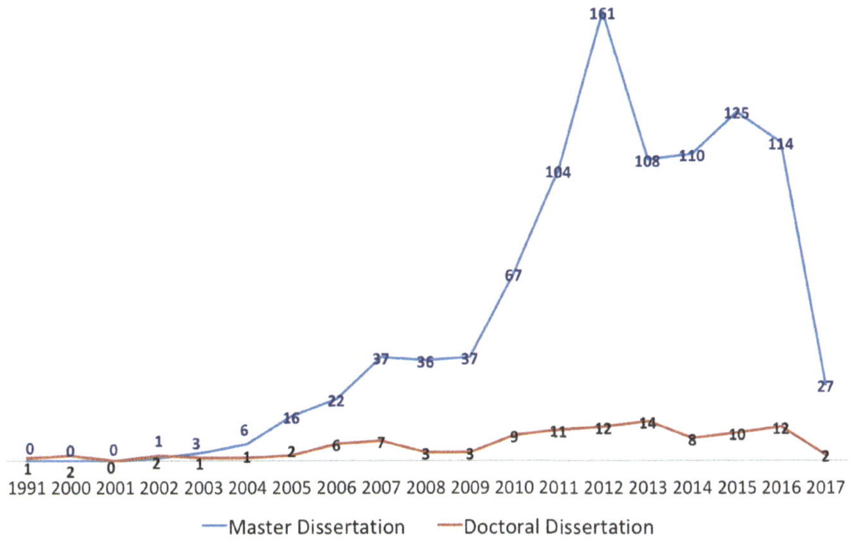

Fig. 3.2 Chinese master and doctoral dissertations from 1991 to 2017

between publishing and being recorded in the NCKI database which may requires several months. It is worth to mention that the first recorded doctoral dissertation at NCKI was written by Gao Lian from Nanjing Normal University in 1991 (Fig. 3.2).

3.1.2 Countries and Institutes

Considering that Poyang Lake is located in China, it is not surprising that most of the published research articles were provided by Chinese research institutes. For English research articles about Poyang Lake, Chinese institutes published 74% of all articles recorded at Scopus (Fig. 3.3). This highlights that many Chinese scientists nowadays are interested to publish their results in renowned international journals. 9.3% of the recorded publications were published by authors (or co-authors) from the United States followed by Australia with 2.1%, France with 1.8% and Germany with 1.7% of total publications. In the period from 1983 to 2017, Nanchang University has with in total 876 Chinese and 124 English articles most publications about Poyang Lake (Fig. 3.4), such as the Chinese publication "Characteristics of distribution, transfer and subtraction of nitrogen and phosphorus contaminants in Poyang lake" [1] and the English publication "Distribution and source of organochlorine pesticides (OCPs) in the sediments of Poyang Lake" [2]. Jiangxi Normal University has the second highest number of records with 745 Chinese and 189 English publications, such as the Chinese publication "Study on Long-term Prediction Model for Drought and Flood in Poyang Lake Basin Based on CEEMD and BP Neural Network" [3] and the

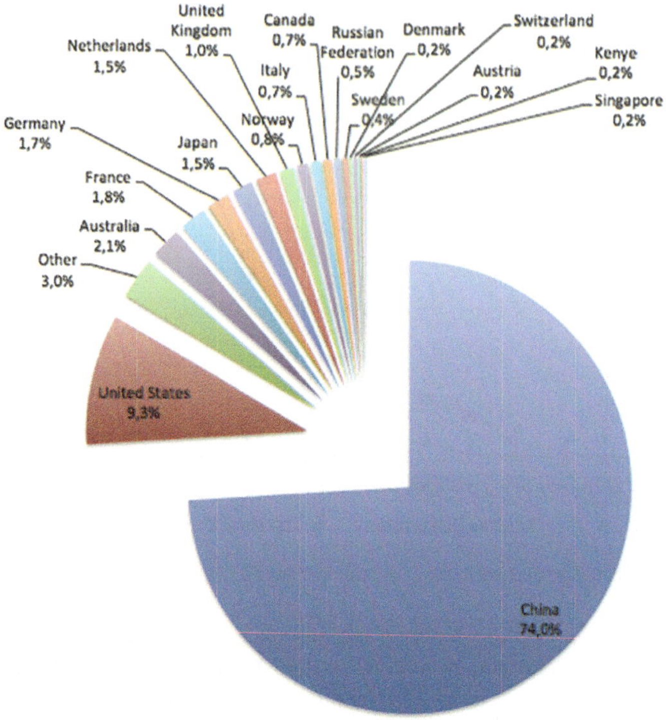

Fig. 3.3 English publications about Poyang Lake by country from 1983 to 2017

English publication "Level, source identification, and risk analysis of heavy metal in surface sediments from river-lake ecosystems in the Poyang Lake, China" [4].

3.1.3 Authors

Regarding publications in Chinese about Poyang Lake (Fig. 3.5), the most accomplished author with 64 articles is Lin Dandan from Jiangsu Institute of Parasitic Diseases contributing highly cited research articles such as "Animal host of schistosoma japonicum and transmission of schistosomiasis in Poyang Lake region" [5]. Xiaoling Chen from the Department of Mapping and Remote Sensing, Wuhan University published most articles in English language about Poyang Lake (Fig. 3.6). Among his 50 contributions are articles such as "RS and GIS based study on landscape pattern change in the Poyang Lake wetland area, China" [6] and "Modeling the impacts of land use/cover change on sediment load in wetlands of the Poyang Lake basin" [7]. Zhang Qi from the Nanjing Institute of Geography and Limnology, Chinese Academy of Sciences contributed 34 publications in English

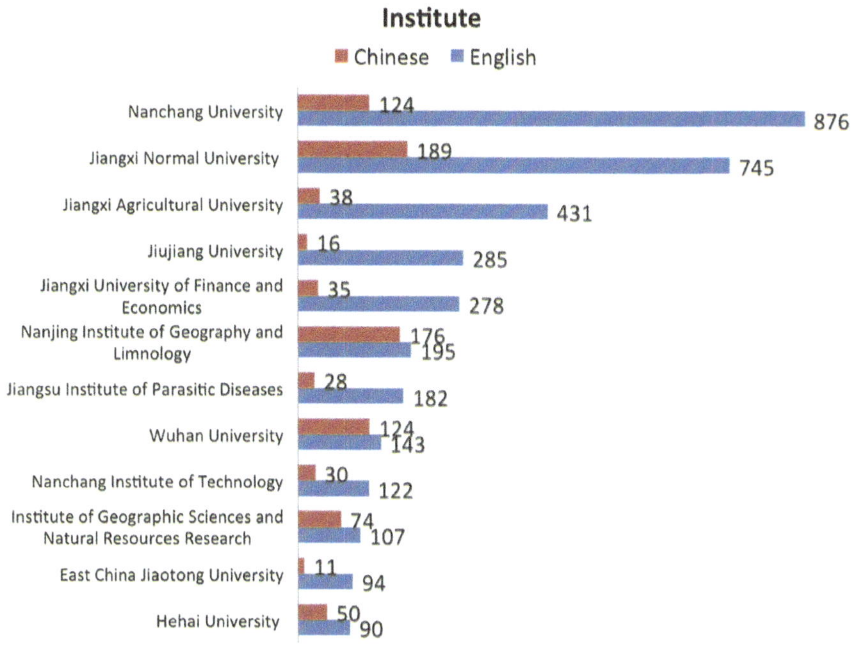

Fig. 3.4 English publications about Poyang Lake by affiliated research institute from 1983 to 2017

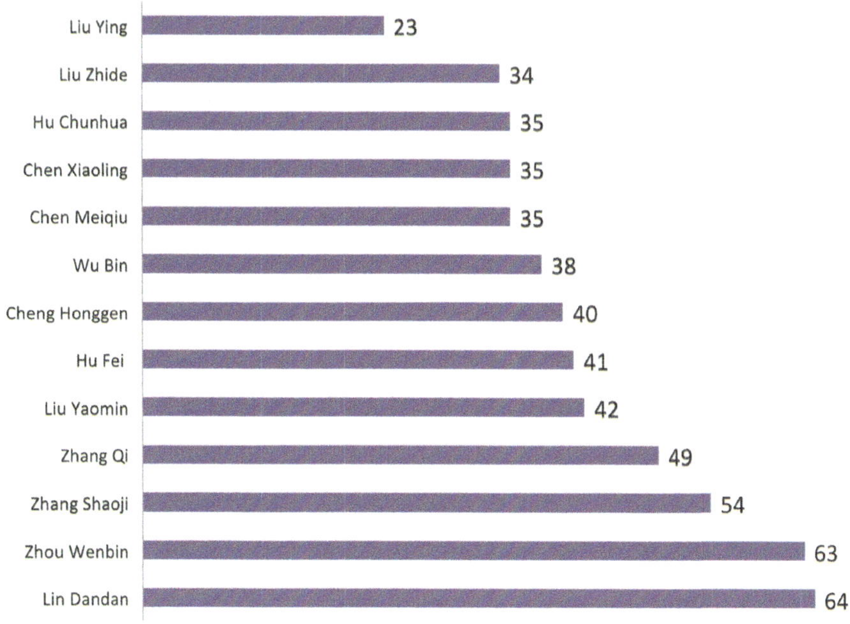

Fig. 3.5 Chinese language publications about Poyang Lake by authors from 1983 to 2017

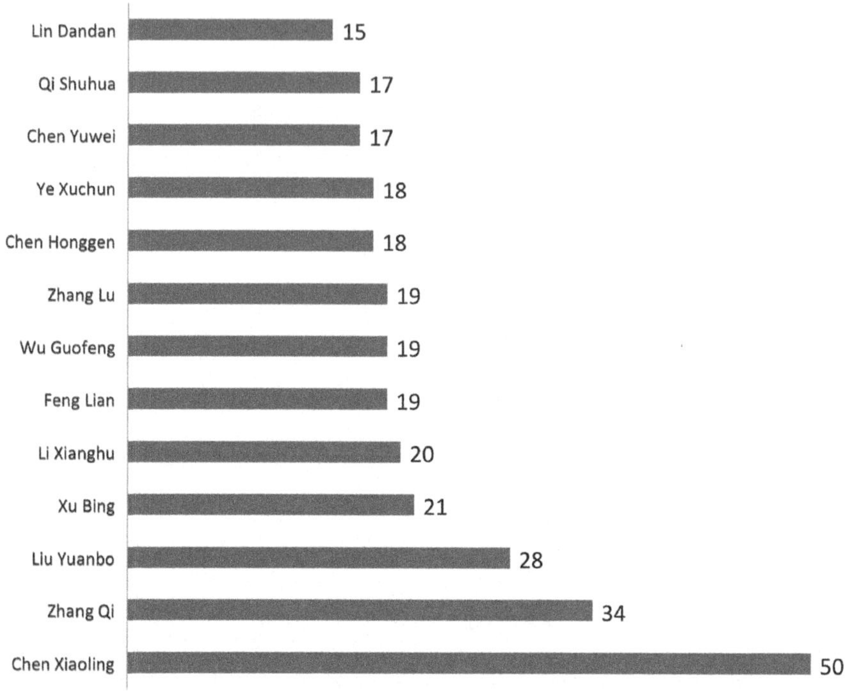

Fig. 3.6 English language publications about Poyang Lake by authors from 1983 to 2017

about Poyang Lake, such as "An investigation of enhanced recessions in Poyang Lake: Comparison of Yangtze River and local catchment impacts" [8]. Additionally, scientists from other countries contributed first author publications about their research on Poyang Lake, too, such as "Landscape analysis of wetland plant functional types: The effects of image segmentation scale, vegetation classes and classification methods" [9], "Flood frequency in China's poyang lake region: trends and teleconnections" [10].

3.2 Most Cited Scientific Publications Until 12/2017

1. No. of Citations: 254
   Title: A strategy to control transmission of Schistosoma japonicum in China
   Authors: Longde Wang, Honggen Chen, Jiagang Guo, Xiaojun Zeng, Xianlin Hong, Jijie Xiong, Xiaohua Wu, Xianhong Wang, Liying Wang, Gang Xia, Yang Hao, Daniel P. Chin and Xiaonong Zhou (2009)
   Short abstract: "Schistosoma japonicum causes an infection involving humans, livestock, and snails and is a significant cause of morbidity in China. After

three transmission seasons, the rate of infection in humans decreased to less than 1.0% in the intervention villages, from 11.3 to 0.7% in one village and from 4.0 to 0.9% in the other ($P < 0.001$ for both comparisons). The rate of infection in humans in control villages fluctuated but remained at baseline levels. A comprehensive control strategy based on interventions to reduce the rate of transmission of S. japonicum infection from cattle and humans to snails was highly effective. These interventions have been adopted as the national strategy to control schistosomiasis in China" [11].

2. No. of Citations: 254 times
   Title: Long-term variations in dissolved silicate, nitrogen, and phosphorus flux from the Yangtze River into the East China Sea and impacts on estuarine ecosystem.
   Authors: Maotian Li, Kaiqin Xu, Masataka Watanabe and Zhongyuan Chen (2007)
   Short abstract: "Variations of dissolved silicate (DSi) flux in the Yangtze River have caused great concern among scientists. Analysis of spatial and temporal variations of DSi indicates that the distribution of DSi concentration (DSiC) is closely related to the occurrence of bedrocks in the river cathchment. On average, the upper Yangtze River and Dongting and Poyang Lake of the middle Yangtze basin serve as the major DSi sinks" [12].
3. No. of Citations: 172 times
   Title: Annual and seasonal streamflow responses to climate and land-cover changes in the Poyang Lake basin, China.
   Authors: Hua Guo, Qi Hu and Tong Jiang (2008)
   Short abstract: "Repeated severe floods and damages in the Poyang Lake basin in China during the 1990s have raised the concern of how the floods have been affected by regional climate variations and by human induced changes in landscape (e.g., draining wetlands around the lake) and land-use in the basin. To address this concern and related issues it is important to know how the climate, land-use and land-cover changes in the region affect the annual and seasonal variations of basin hydrology and streamflow. Results of this study improve our understanding of hydrological consequences of land-use and climate changes, and provide needed knowledge for effectively developing and managing land-use for sustainability and productivity in the Poyang Lake basin [13].
4. No. of Citations: 169 times
   Title: Flood frequency in China's Poyang Lake region: Trends and teleconnections.
   Authors: David Shankman, Barry D Keim and Jie Song (2006)
   Short abstract: "This trend is related primarily to levee construction at the periphery of the lake and along the middle of the Yangtze River, which protects a large rural population. These levees reduce the area formerly available for floodwater storage resulting in higher lake stages during the summer flood season and catastrophic levee failures. The most severe floods in the Poyang Lake since 1950, and ranked in descending order of severity, occurred in 1998, 1995, 1954, 1983, 1992, 1973, and 1977. All of these floods occurred during or immediately

following El Niño events, which are directly linked to rainfall in central China" [10].

5. No. of Citations: 141 times
   Title: Succession of bacterial community structure along the Changjiang River determined by denaturing gradient gel electrophoresis and clone library analysis.
   Authors: Hiroyuki Sekiguchi, Masataka Watanabe, Tadaatsu Nakahara, Baohua Xu and Hiroo Uchiyama (2002)
   Short abstract: "Bacterial community structure along the Yangtze River was studied by using denaturing gradient gel electrophoresis and clone library analysis of PCR—amplified 16S ribosomal DNA with universal bacterial primer sets. DGGE profiles and principal—component analysis demonstrated that the bacterial community gradually changed from upstream to downstream in both 1998 and 1999. Bacterial diversity, as determined by the Shannon index, gradually decreased from upstream to downstream. Clone library analysis of 16S rDNA revealed that the dominant bacterial groups changed from β—proteobacteria and the Cytophaga—Flexibacter- Bacteroides group upstream to high-G+C-content gram-positive bacteria downstream and also that the bacterial community structure differed among the stations in the river and the lakes" [14].
6. No. of Citations: 129 times
   Title: Assessment of inundation changes of Poyang Lake using MODIS observations between 2000 and 2010.
   Authors: Lian Feng, Chuanmin Hu, Xiaoling Chen, Xiaobin Cai, Liqiao Tian and Wenxia Gan (2012)
   Short abstract: "Using Moderate Resolution Imaging Spectroradiometer (MODIS) medium-resolution (250-m) data collected between 2000 and 2010 and an objective water/land delineation method, we documented and studied the short- and long-term characteristics of lake inundation. Significant seasonality and inter-annual variability were found in the monthly and annual mean inundation areas.The changes of the inundation area were primarily driven by local precipitation during non-summer months, while during summer months of July to September when the outflow into the Yangtze River was impeded the effect of precipitation became less significant. These results provide long-term baseline data to monitor future changes in Poyang Lake's inundation area in a timely fashion, for example quantifying the extreme drought conditions during spring 2011" [15].
7. No. of Citations: 123 times
   Title: Effects of the Three Gorges Dam on Yangtze River flow and river interaction with Poyang Lake, China: 2003–2008.
   Authors: Hua Guo, Qi Hu, Qi Zhang and Song Feng (2012)
   Short abstract: "Over the operation period from 2003–2008, data have been collected for preliminary evaluations of actual effects of the Three Gorges Dam (TGD) on the Yangtze River flow and river interactions with downstream lakes and tributaries. These effects are examined in this study, after the climate influence was minimized by comparing hydrological changes between years of similar climate conditions before and after the operation of the TGD. Major results

show that the TGD operation has affected the Yangtze River discharge and water level. The significance of these effects varies seasonally and with different locations along the river" [16].

8. No. of Citations: 120 times
   Title: Interactions of the Yangtze river flow and hydrologic processes of the Poyang Lake, China.
   Authors: Qi Hu, Song Feng, Hua Guo, Guiya Chen and Tong Jiang (2007)
   Short abstract: "Recently available hydrological data from Hukou station at the junction of the Poyang Lake with the Yangtze River along with other data from stations in the Poyang Lake basin have allowed further examination and understanding of the basin effect (basin discharge generated by rainfall) and the Yangtze River blocking effect on variations of the Poyang Lake level and floods at annual to decadal scales. Major results show that the basin effect has played a primary role influencing the level of Poyang Lake and development of severe floods, while the Yangtze River played a complementary role of blocking outflows from the lake. Results of this study provide a utility for improving predictions of the Poyang Lake level and floods, which affect a population of about 10 million" [17].
9. No. of Citations: 106 times
   Title: Modelling spatial-temporal change of Poyang Lake using multitemporal Landsat imagery.
   Authors: Fengming Hui, Bing Xu, Huabing Huang, Qian Yu and Peng Gong (2008)
   Short abstract: "In this study, we assess the feasibility of the use of multitemporal Landsat images for mapping the spatial-temporal change of Poyang Lake water body and the temporal process of water inundation of marshlands. Eight cloud-free Landsat Thematic Mapper images taken during a period of one year were used in this study. The results showed that although the images can be used to capture the snapshots of water coverage in this area, they are insufficient to provide accurate estimation of the spatial-temporal process of water inundation over the marshlands through linear interpolation" [18].
10. No. of Citations: 92 times
   Title: Mathematical modelling of schistosomiasis japonica: Comparison of control strategies in the People's Republic of China.
   Authors: Gail M.Williams, Adrian C.Sleigh, Yuesheng Li, Zheng Feng, George M.Davis, Hongen Chen, Allen G.P.Ross, Robert Bergquist and Donald P.McManusa (2002)
   Short abstract: "We present the first mathematical model on the transmission dynamics of Schistosoma japonicum. The work extends Barbour's classic model of schistosome transmission. It allows for the mammalian host heterogeneity characteristic of the S.japonicum life cycle, and solves the problem of underspecification of Barbour's model by the use of Chinese data we are collecting on human—bovine transmission in the Poyang Lake area of Jiangxi Province in China" [19].

Most cited Chinese literature around Poyang Lake: 1983–2017[3]

1. No. of Citations: 319 times
   Title: Evaluation on functions of Poyang Lake ecosystem.
   Authors: Lijuan Cui (2004)
   Short abstract: "The benefits of Poyang Lake include direct products, ecological functions and wetland characteristics. Three types of methods were used to evaluate the wetland ecosystem service value: direct market evaluation, replacement market and stating partiality method. This paper stressed on the main serve functions of Poyang Lake and provided monetary evaluations on the functions of water storage, flood regulation, soil protection, CO_2 fixation, O_2 release, soil and water purification and habitats. The results revealed that the total monetary value of the main functions of Poyang Lake was 31627×10^{10} Yuan" [20].
2. No. of Citations: 249 times
   Title: Study on Imaging Spectrometer Remote Sensing Information for Wetland Vegetation
   Authors: Qingxi Tong, Lanfen Zheng, Jinnian Wang, Xiangjun Wang, Weidong Dong, Yunman Hu and Shunxing Dang (1997)
   Short abstract: "The paper presents some results of vegetation spectral identification, classification and biomass mapping by hyperspectral imaging spectrometry in Poyang Lake wetland. The study focuses on how to retrieve canopy biophysical characteristics and how to identify vegetation types from hyperspectral image effectively. The main approaches are as follows: (1) Retrieval of apparent surface reflectance from imaging spectrometer imagery, based on Multiheigh technique; (2) Derivative spectral analysis and biomass estimation; (3) Wetland vegetation identification and classification based on spectral waveform matching"[21].
3. No. of Citations: 247 times
   Title: Assessment on Heavy Metal Pollution in the Sediment of Poyang Lake.
   Authors: Xiaofeng Gong, Chunli Chen, Wenbin Zhou, Minfei Jian and Zhenhu Zhang (2006)
   Short abstract: "In order to know the heavy metals content and the degree of the potential ecological risk in the sediment of Poyang Lake in high-water period and lowwater period,based on the detailed survey and analysis of the current state of pollution in Poyang Lake,using the Index of Geoaccumulation and the Potential Ecological Risk Index evaluate the heavy metals pollution of Poyang Lake.The results indicate that Poyang Lake has been polluted by heavy metals in various degrees.According to the Index of Geoaccumulation,the order of the analyzed heavy metals,arranged from highest to lowest pollution degree,is as follows: Cu > Pb > Zn > Cd;The ecological risk to Poyang Lake is: Cd > Pb > Zn > Cu" [22].
4. No. of Citations: 171 times
   Title: Researches on the emergy analysis of Poyanghu wetland.
   Authors: Lijuan Cui and Qinsheng Zhao (2004)
   Short abstract: "The Emergy and material flux of Poyanghu wetland ecosystem

[3]http://epub.cnki.net/kns/default.htm.

were analyzed in this paper by Emergy analysis theory of ecological economy system which was created by H.T.Odum. The article concludes that four steps should be included in the emergy analysis process of wetland ecological benefits: (1) emergy analysis concept ional system should be built up, which can reflect the emergy analysis method; (2) emergy analysis table being drafted out and the emergy value being calculated; (3) emergy indices being estimates; (4)wetland ecological benefits should be expatiated according to analytical tables of emergy indices system and emergy schemes" [23].

5. No. of Citations: 169 times
   Title: An Estimation of Wetland Vegetation Biomass in the Poyang Lake Using Landsat ETM Data. Authors: Rendong Li and Jiyuan Liu (2001)
   Short abstract: "This paper conducted a digital and rapid investigation of the lake's wetland plant biomass using Landsat ETM data acquired on April 16,2000. Using the false color composite derived from the ETM data as one of the main references, the authors planned a reasonable field sampling route for the biomass. Based on geometric correction of both the sampling data and the ETM data to an area equal projection of Albers, the linear relationships among the field biomass and some transformed data from the ETM data. The results show that the sampling data has the best positive correlation to the band 4 data with a coefficient of 0 86. It is shown that the remote sensing technology has many advantages over the traditionally used method engaged in the biomass investigation in a lake" [24].
6. No. of Citations: 146 times
   Title: Modeling study on copper partitioning in sediments, a case study of Poyang lake.
   Authors: Jingsheng Chen, Lin Dong, Baoshan Deng, Liangbi Wan, Min Wang and Zhengliang Xiong (1987)
   The partitioning of copper in sediments from the Poyang Lake was studied. The distribution of copper among different geochemical phases in sediments was controlled by the abundance of the correspondent geochemical phases in sediments. The model proposed by Oakley et al. and Davies-Colley et al. was applied to predict the partitioning of copper among different geochemical phases in aquatic sediments. The conditional equilibrium constants were determined using an artificial system under various pH and temperature conditions. The model was employed to describe the metal partitioning in sediment samples from the Poyang Lake and the results were consistent with that measured in laboratory" [25].
7. No. of Citations: 141 times
   Title: Study on eco-compensation of returning land to lake–take Poyang lake area as a case study.
   Authors: Yu Zhong, Sheng Zhang and Xianqiang Mao (2002)
   Short abstract: "Returning land to lake is one of the important parts of restoring ecological function of the Yangtze River. Whether the project could be successful depends on the compensation to the farmers, who have lost their original benefits and contributed to the ecosystem recovery. This paper takes Poyang Lake area

as a case study and explores the basic issues of the eco-compensation, that is who should pay, who should be paid, how much is the payment and payment vehicle" [26].

8. No. of Citations: 137 times
   Title: Structure of poyang lake wetland plants ecosystem and influence of lake water level for the structure.
   Authors: Zhenpeng Hu, Gang Ge, Chenglin Liu, Fusheng Chen and Shu Li (2010)
   Short abstract: "With an analysis of the hydrologic processes and landform of Poyang Lake, wetland types were classified by using 3S technology, the habitats of wetland plants, structure of major plants community and its distribution rules were investigated in this paper. Poyang Lake has complex system of wetland plants and is extremely rich in biodiversity. The wetland plants are colonel reproduction of plants using roots or stems as the main reproductive organs, so the distribution pattern of plant communities presents the main features of the cluster distribution. In micro-scale, the plant communities distribute based on the moisture gradient by the influence of lake's landform, the elevation, the water lever and exposed time of the beaches. In landscape scale, the distribution of plant communities present community mosaic structure by the influence of soil moisture, groundwater depth and soil structure. In microcosmic-scale, the distribution of plant communities presents community complex structure in some regions by the influence of the micro- to—pography and soil nutrient. Wetland plant's extreme sensitivity to moisture gradient leads the structure to become mutability and fragility. In recent years, the long-lasting low water lever of Poyang Lake causes damage to ecosystem" [27].
9. No. of Citations: 116 times
   Title: Evolution and Analysis of Eutrophication of Typical Lakes in Middle and Lower Reaches of Yangtze River
   Authors: Xiaoying Cheng and Shijie Li (2006)
   Short abstract: "Based on the analysis of the water environment data of the typical lakes in the middle and lower reaches of the Yangtze River, eutrophication refers to the process of lake ecosystem degradation and water quality deterioration due to the enrichment of nutrient elements, the evolution of lakes and ecosystems, The changes are all well coupled with the evolution of lake nutrition level, and then the conceptual model of lake eutrophication is proposed, and the lake states are divided into 10 types, and the eutrophication process of typical lakes in the middle and lower reaches of the Yangtze River is analyzed. Research can also provide corresponding biological targets and chemical targets for the ecological restoration of lakes in the middle and lower reaches of the Yangtze River" [28].
10. No. of Citations: 99 times
    Title: Observed trends and jumps of climate change over Lake Poyang Basin, China: 1961–2003.
    Authors: Hua Guo, Tong Jiang, Guojie Wang, Buda Su and Yanjun Wang (2006)
    Short abstract: "Based on observed data of 14 meteorological stations and six main hydrological stations in lake Poyang basin, the trends and jumps of tem-

perature, precipitation, pan evaporation(PE), reference evapotranapiration(ETr) and discharge are analyzed from 1961 to 2003. The results indicate that the temperature jumped in 1990, then it dominated an upward trend ever since 1990. Significant positive trend was noticed after 1986 in winter. As to precipitation, it changed abruptly in 1990. In 1992, the heavier rainstorm resulted in more precipitation in summer. In 1990s, warm and humid climate tendency was strengthened. The climate change of lake Poyang basin is much significant in the Yangtze River basin" [29].

11. No. of Citations: 99 times
    Title: Ecological environment and sustainable development of Poyang Lake
    Authors: Qiguo Zhao, Guoqin Wang and Haiyan Qian (2007)
    Short abstract: "According to the analysis of the ecological environment of the Poyang Lake Region, such as vegetation, wetland biological diversity, land use, water and soil erosion, soil alkalization, etc. The problems such as serious degradation of the wetland vegetatio, aggravation of water and soil erosion, prevalence of schistosomiasis, decrease in biological diversity, are defined. So some countermeasures and measures are proposed to rational development and protection of ecological environment system of Poyang Lake" [26].

References

1. Tang, G., Y. Lin, Z. Hu, and S. Wang. 2017. Characteristics of distribution, transfer and subtraction of nitrogen and phosphorus contaminants in Poyang Lake. *Resources and Environment in the Yangtze Basin* 26 (9): 1436.
2. Liang, Y., H. Xiao, and X. Liu. 2017. Distribution and source of organochlorine pesticides (OCPs) in the sediments of Poyang. *Environmental Earth Sciences* 76 (12): 622.
3. Wan, Z., M. Jiang, Y. Jiang, W. Hong, and M. Zhang. 2017. Study on longterm prediction model for drought and flood in Poyang Lake Basin based on CEEMD and BP neural network (in Chinese). *Acta Agriculturae Jiangxi* 29 (10): 108–113.
4. Zhang, H., Y. Jiang, M. Ding, and Z. Xie. 2017. Level, source identification, and risk analysis of heavy metal in surface sediments from river-lake ecosystems in the Poyang Lake, China. *Environmental Science and Pollution Research* 24 (27): 21902–21916.
5. Li, D., and S. Zhang. 2002. Geographical environment and spread of schistosomiasis in Poyang Lake area (in Chinese). *Chinese Journal of Epidemiology* 23: 90–93.
6. X. Chen et al. 2006. RS- and GIS-based study on landscape pattern change in the Poyang. Lake wetland area, China. *Geoinformatics 2006: Remotely Sensed Data and Information*, 6419: 64192E.
7. X. Chen et al. 2006. Modeling the impacts of land use/cover change on sediment load in wetlands of the Poyang Lake basin. In *Proceedings of the International Conference Hydrology and Management of Forested Wetlands*, New Bern, 92.
8. Zhang, Q., et al. 2014. An investigation of enhanced recessions in Poyang Lake: comparison of Yangtze River and local catchment impacts. *Journal of Hydrology* 517: 425–434.
9. Dronova, I., et al. 2012. Landscape analysis of wetland plant functional types: The effects of image segmentation scale, vegetation classes and classification methods. *Remote Sensing of Environment* 127: 357–369.
10. Shankman, D., B.D. Keim, and J. Song. 2006. Flood frequency in China's Poyang Lake region: Trends and teleconnections. *International Journal of Climatology* 26 (9): 1255–1266.

11. Wang, L., et al. 2009. A strategy to control transmission of Schistosoma japonicum in China. *New England Journal of Medicine* 360 (2): 121–128.
12. Li, M., K. Xu, M. Watanabe, and Z. Chen. 2007. Long-term variations in dissolved silicate, nitrogen, and phosphorus flux from the Yangtze River into the East China Sea and impacts on estuarine ecosystem. *Estuarine, Coastal and Shelf Science* 71 (1–2): 3–12.
13. Guo, H., Q. Hu, and T. Jiang. 2008. Annual and seasonal streamflow responses to climate and land-cover changes in the Poyang Lake basin, China. *Journal of Hydrology* 355 (1–4): 106–122.
14. Sekiguchi, H., M. Watanabe, T. Nakahara, B. Xu, and H. Uchiyama. 2006. Succession of bacterial community structure along the Changjiang River determined by denaturing gradient gel electrophoresis and clone library analysis. *Applied and Environmental Microbiology* 26 (9): 1255–1266.
15. Feng, L., et al. 2012. Assessment of inundation changes of Poyang Lake using MODIS observations between 2000 and 2010. *Remote Sensing of Environment* 121: 80–92.
16. Guo, H., Q. Hu, Q. Zhang, and S. Feng. 2012. Effects of the three gorges dam on Yangtze river flow and river interaction with Poyang Lake, China: 2003–2008. *Journal of Hydrology* 416: 19–27.
17. Hu, Q., S. Feng, H. Guo, G. Chen, and T. Jiang. 2007. Interactions of the Yangtze river flow and hydrologic processes of the Poyang Lake, China. *Journal of Hydrology* 347 (1–2): 90–100.
18. Hui, F., B. Xu, H. Huang, Q. Yu, and P. Gong. 2008. Modelling spatialtemporal change of Poyang Lake usingmultitemporal Landsat imagery. *International Journal of Remote Sensing* 29 (20): 5767–5784.
19. Williams, G.M., et al. 2002. Mathematical modelling of schistosomiasis japonica: comparison of control strategies in the People's Republic of China. *Acta Tropica* 82 (2): 253–262.
20. Cui, L. 2004. Evaluation on functions of Poyang Lake ecosystem. *Chinese Journal of Ecology* 4: 47–51.
21. Tong, Q., et al. 1997. Study on imaging spectrometer remote sensing information for wetland vegetation. *Journal of Remote Sensing* 1: 82–85.
22. Gong, X., C. Chen, W. Zhou, M. Jian, and Z. Zhang. 2006. Assessment on heavy metal pollution in the sediment of Poyang Lake. *Journal of Environmental Sciences* 4: 732–736.
23. L. Cui and S. Zhao. 2004. Researches on the emergy analysis of Poyang Lake wetland. PhD thesis.
24. Li, R., and J. Liu. 2001. An estimation of wetland vegetation biomass in the Poyang Lake using Landsat ETM Data. *Acta Geographica Sinca* 5: 532–540.
25. Chen, J., et al. 1987. Modeling study on copper partitioning in sediments, a case study of Poyang lake. *Acta Scientiae Circumstantiae* 7 (2): 140–149.
26. Q. Zhao, G. Huang, and H. Qian. 2007. Ecological environment and sustainable development of Poyang Lake. *Acta Pedologica Sinca* 2.
27. Hu, Z., G. Ge, C. Liu, F. Chen, and S. Li. 2010. Structure of poyang lake wetland plants ecosystem and influence of lake water level for the structure. *Resources and Environment in the Yangtze Basin* 19 (6): 597–605.
28. Cheng, X., and S. Li. 2006. Evolution and analysis of eutrophication of typical lakes in middle and lower reaches of Yangtze River. *Chinese Science Bulletin* 51 (7): 848–855.
29. Guo, H., T. Jiang, G. Wang, B. Su, and Y. Wang. 2006. Observed trends and jumps of climate change over Lake Poyang Basin, China:1961–2003. *Journal of Lake Sciences* 18 (5): 443–451.

Chapter 4
Strengthening Integrated Management and Maintaining the Health of Poyang Lake

Zhewen Fan and Zhenpeng Hu

4.1 Trends in the Hydrological Regime of the Wetland Ecosystem

Since the beginning of the 21st century, the hydrology of the Yangtze River has undergone significant changes due to the influence of climate change and human activities. Both the flow of the Yangtze River and the water level of Poyang Lake have been reduced within the last decades. As shown in Fig. 4.1, the average flow at Hankou Station from 1949 to 2002 was 22,809 m^3/s, but 21,231 m^3/s from 2003 to 2014. This means that the average annual discharge decreased by 1,578 m^3/s in the last decade caused by dryer conditions in the Yangtze Basin. The largest decrease of 7,687 m^3/s from the long-run average occurred in October presumably caused by the impoundment of flood waters in the large reservoirs of Yangtze River upstream of Poyang Lake.

Similarly, the precipitation within the Poyang Lake Basin and the average annual water level at Xingzi Staion were calculated for two periods, from 1956 to 2002 and from 2003 to 2015. The precipitation in the Poyang Lake Basin did not show a significant trend between both periods. Compared to that before 2002, the annual precipitation decreased by only 3.34 %. In contrast, the average water level of Poyang Lake at Xingzi Station was 0.98 m lower from 2003 to 2015 than during the period from 1956–2002 (Fig. 4.2). The average water level decreased by 0.95 m during wet season and 1 m during dry season. According to the water level-area relationship of Poyang Lake Basin, the water surface area of the lake is reduced by 470 km^2

Z. Fan (✉)
Jiangxi Remote Sensing Information System Center, Nanchang, China
e-mail: fanmrl@sina.com

Z. Hu
Jiangxi Mountain-River-Lake Engineering Academic Committee, Jiangxi, China

Z. Hu
Nanchang University, Nanchang, China

T. Yue et al. (eds.), *Chinese Water Systems*, Terrestrial Environmental Sciences,
https://doi.org/10.1007/978-3-319-97725-6_4

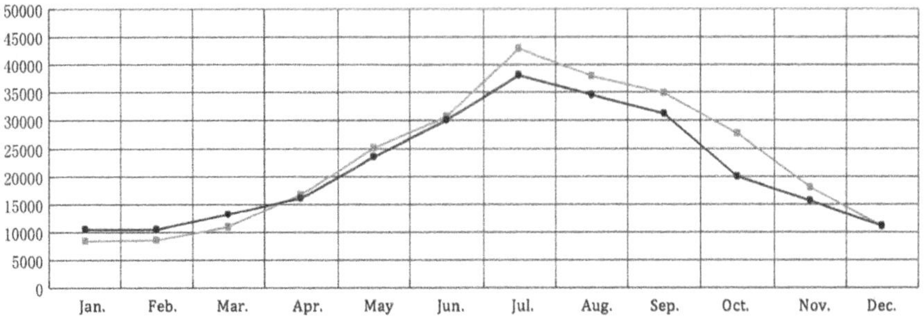

Fig. 4.1 Monthly average discharge (in m^3/s) of Yangtze River at Hankou Station for the periods from 1949–2002 (grey) and 2003–2014 (black)

for wet season and 120 km^2 for dry season after 2003. The long-term low water levels of Poyang Lake continuously reduce the living space of aquatic organisms and deteriorate the wetlands due to lower soil moisture. The significant decrease of discharges in Yangtze River in October during the last decade led to an earlier appearance of low water levels in Poyang Lake and an expansion of the dry season. Consequently, this development brought a series of problems to the development and utilization of water resources and ecosystem health in the wetlands of Poyang Lake.

4.2 Wetland Ecosystem Degradation in Poyang Lake

Changes in the hydrological regime and subversive human activities in rivers and lakes either directly affect or indirectly influence the health of the wetland ecosystems. The mechanism of wetland ecosystem degradation in Poyang Lake and its drivers is shown in Fig. 4.3.

In 1983, the provincial government of Jiangxi organized a first comprehensive scientific investigation of Poyang Lake and conducted the second scientific investigation 30 years later in 2013. In summary, comparing the results of the two inspections, wetland ecosystems are gradually degrading within the last three decades.

4.2.1 Decline of Poyang Lake Water Quality

In the 1980s, the quality of Poyang Lake was dominated by Class II according to the Chinese Standard for Surface Water, which means that the water was in a good physical-chemical status. However, water quality declined slowly in the 1990s along with the increasing industrialization and urbanization of Jiangxi Province. Consequently low water quality worse than Class III started to appear for Poyang Lake in 2003. In 2008, none of the monitoring samples from Poyang Lake achieved

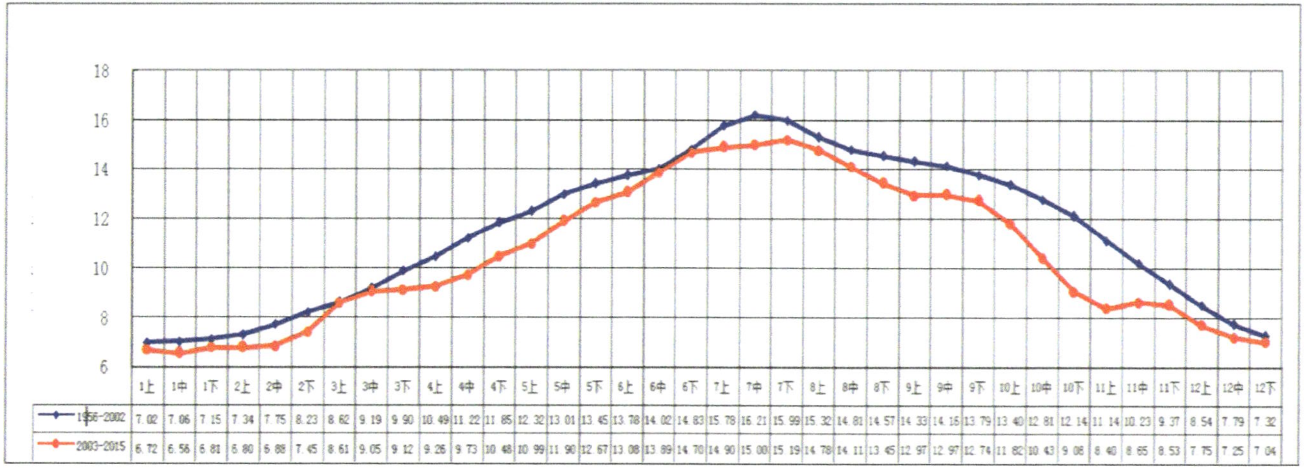

Fig. 4.2 River stage of Poyang Lake for the periods 1956–2002 and 2003–2015. River stages are shown for each third of a month (Beginning of month–middle of month–end of month)

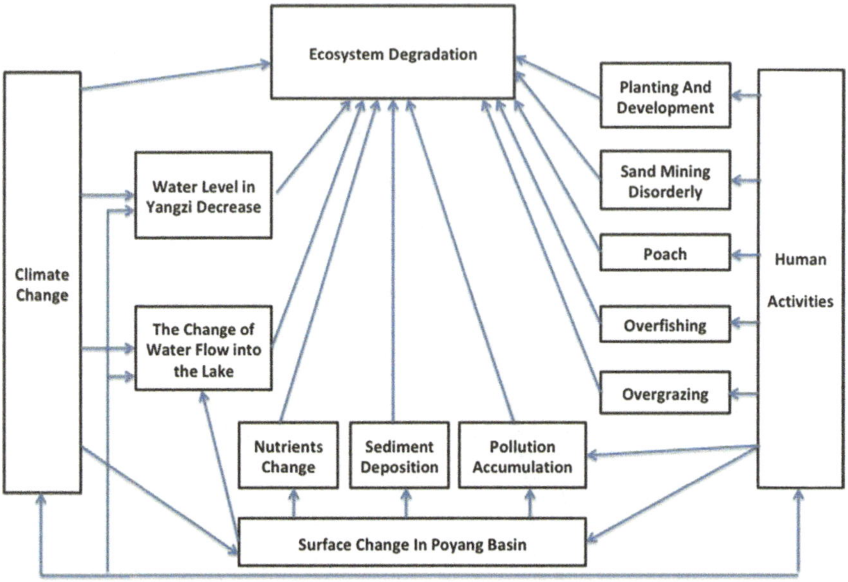

Fig. 4.3 Drivers and mechanisms leading to ecosystem degradation in the wetland ecosystem of Poyang Lake

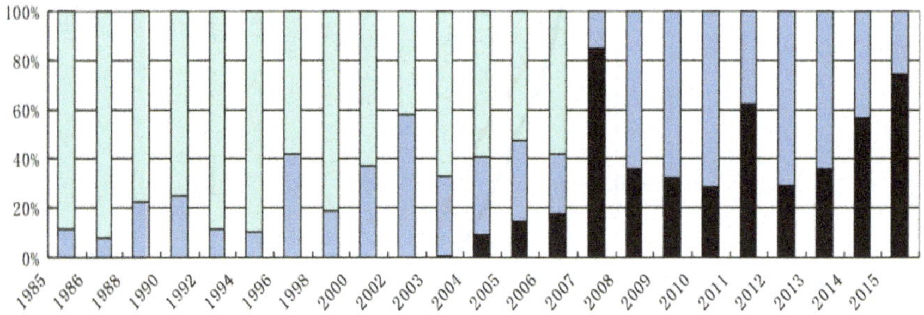

Fig. 4.4 Changes in water quality of Poyang Lake from 1985 to 2015 in accordance to the Chinese Standard for Surface Waters (Black bar: water quality worse than Class III; Dark blue bar: Class III; Light blue bar: Class I–II)

a Class II water quality. In 2015, Class III water only accounted for 25.7% of the total monitored waters (Fig. 4.4). Increasing concentrations of total phosphorus and total nitrogen were the main reason for the deteriorating water quality.

4.2.2 Vegetation Degradation

The vegetation in wetlands absorbs nitrogen, phosphorus, carbon and other nutrients but also heavy metals. The vegetation is the primary producer of useful chemical energy in the ecosystem and plays a fundamental role to provide and maintain the ecosystem services. Over the last three decades several changes in the vegetation patterns have been observed: In 1983, the first scientific investigation estimated a total area of wetland vegetation in Poyang Lake of 2,262 km^2, in which floating plants accounted for about 23.2 % and submerged plants accounted for 49.7 % of the wetland area. In 2013, the survey showed that the total area of wetland vegetation in Poyang Lake was reduced to 1,661 km^2, of which the wet and emergent plant area was about 1463 km^2. Due to the longer period of low lake water levels, more than 150 invasive plant species invaded the wetland and formed a mesogenic meadow dominated by bermudagrass and plants of the genus Verbena. This area covered in total about 198 km^2. Additionally, wetland plants such as Phragmites australis, Nandi, Carex and other communities spread to the lower part of the lake basin, squeezing the living space of the submerged plants. The distribution area of Carex increased from 428 km^2 in the 1980s to 723 km^2. The occurrence of submerged plants such as Hydrilla and Vallisneria generally has been reduced by 1 m of altitude. The distribution area retreated towards the center of the lake, which was only 700 km^2 in 2013. The increasing nutrient concentration of nitrogen and phosphorus not only decreased the water quality of the lake but also led to a rapid expansion of the population of emergent water-resistant plants such as wild rice husks which covered a net area of 116 km^2 in 2013. Similarly, the numbers of environment sensitive species such as Ottelia, Potamogeton maackianus A. Bennett and other environmentally sensitive species gradually reduced with increasing nutrient concentrations. As a prominent example, Potamogeton wrightii Morong was the dominant species of submerged vegetation in 1983 but only showed a sporadic distribution in a small area in 2013. Submerged vegetation plays a very important role in maintaining lake health. At present, the area of submerged macrophytes and emergent plants in Poyang Lake reduced from 1661 km^2 in the first Poyang Lake scientific investigation to less than 1000 km^2, including 116 km^2 of mound community. The resources of wet plants such as Triarrhena lutarioriparia L. Liu and others are dwindling. In the first scientific investigation records, Triarrhena lutarioriparia L Liu was about 3 m. Now it is less than 1 m. The structure of the submerged vegetation community simplified with time. In the 1980s, submerged vegetation communities were rich in biodiversity and consisted of 5–8 species. After the floods of 1998 and 1999 and the drought conditions beginning in the 21st century, the submerged vegetation community began to degenerate and the community structure was simplified to only 3–5 species and the remaining species predominate only certain parts of the wetland water area. For example, Potamogeton crispus is dominant in the water area around Cuoji at the southern branch of Ganjiang River. In the northern part of the lake the community is replaces by wild Wildrice stem. There are almost no other plant species left within the submerged vegetation community.

4.2.3 Reduction of Benthic Animals

The low water level in the Yangtze River led to a sharp rise and fall in the water level of Poyang Lake up to more than 30 cm per day. The drastic drop in water level caused many macrobenthos to die if they would not migrate to deeper waters. In recent years, snails such as Bolinus brandaris and clams are digged more intensively and on larger scales than before. Fishing equipment is getting more and more advanced. Density and biomass of meiofauna living in the lake and on the lakeshore are declining (Table 4.1). Catching Bolinus brandaris and clams not only harms the zoobenthos but also directly damages the submerged vegetation, making it one driver behind the increasing phosphorous concentrations in the aquatic system.

4.2.4 Reduction of Fish Resources

In 2013, 134 species of fish have been recorded in Poyang Lake. From 2012 to 2013, another investigation of fish species in Poyang Lake resulted in the detection of 89 species of fish. Since 2003, the surface area of Poyang Lake has been reduced by 470 km^2 during the flood season, which reduced the available habitats for fish. If the lake level falls below 12 m above sea level, much of the wetland area is not submerged or is inundated with an insufficient water depth which drastically reduced the number of spawning sites. In addition, foliage with sticky eggs is submerged for a short time and the hatching of fish and eggs are immature and no fish fry can be produced due to the low water levels. If the average water level of Poyang Lake is 14.78m, there are 35 fish feeding sites with a total area of about 390 km^2. The average water level from June to September 2006 was 13.89m, 3.1m lower than that of the same period of 2005, which decreased the suitable area of feeding grounds by 26.5%. As another example, in 2013, the average water level of Poyang Lake from June

Table 4.1 Evolution of macrobenthos density and biomass in Poyang Lake

Period (year)	Area	Species	Density (ind/m^2)	Proportion	Biomass (g/m^2)	Proportion
1996–1999	Tong River	31	549	1	116.6	1
1996–1999	Water Channel	41	659	1.2	183.7	1.6
1996–1999	Shallow-Lakes	47	1509	2.75	3318	28.45
2012	Lake	72	349	0.64	65	0.56
2012–2013	Shallow-Lakes	51	1154	2.10	639	5.51

to September was 12.92m, and the area of feeding grounds was reduced by 13.1% in comparison to the previous year. In addition, the overcrowded fisheries and the overfishing of aquatic resources have exacerbated the decay of fish stocks. Electric fishing has been found all over the Poyang Lake region. In 2012, the age structure of the major economic fish in the Poyang Lake Region was 73–100% of the 1–2 age group and this share increased to 83–100% within one year. Consequently, the fish catches show a decrease in age level, a miniaturization in size, and a decline in quality.

4.2.5 *Decentralization of the Winter Migratory Birds*

Migratory birds are at the top of the wetland ecosystem food chain. Reduction of vegetation, benthic fauna and fish resources makes the overwintering processes of migratory birds more sensitive to changes in the hydrological regime of Poyang Lake. Since the start of the annual regular monitoring in 1998, the total number of overwintering migratory birds in the Poyang Lake region has been rising. However, the numbers of rare migratory birds such as cranes and storks have decreased since 2012. The number of migratory birds in the lakes and marsh areas of the Poyang Lake ecosystem has reduced significantly in the last years, which also led to a more patchy and decentralized distribution of the bird habitats. The birds feed on the crops surrounding beach land. In particular, under anomalous hydrological conditions, snow Crane and other species even feed on the crops in the Fu River Hongmen Reservoir and Wuyuan.

4.2.6 *Superimposed Effect of Changing Hydrological Conditions and Human Activities*

As mentioned above, changing hydrological conditions and human activities have both direct impacts on wetland ecosystems and indirect ones through changing the surface conditions of the river basin. Their interaction also has a superposition effect. The reduction of the lake's water level makes human activities such as exploitation and utilization of water and soil resources, sand harvesting and overfishing more simple and easier. For example, the long-term low water level of Poyang Lake and the increase of nutrients in Poyang Lake water body cause rapid expansion of the wild rice. Conversely, the expansion of the wild rice decelerates the water flow which led to an increasing deposition of fine sediments making the lake more shallow and swampy. In addition, the decay of the wild rice adds nutrients into the water, which provide more nutrition for the expansion of this species. The increasing amount of human activities that have a harmful impact on the ecosystem of Poyang Lake causes a continuous decline of the ecological resources, a "tragedy of the commons" that is apparently not preventable without major efforts.

4.3 Strengthening the Integrated Management of Poyang Lake

As shown above, the wetland system of Poyang Lake is in the process of continuous degradation. Regardless of whether the Poyang Lake water conservancy project will be launched, efforts are needed to stop and reverse the current adhesive development. This is a very urgent and important task that must be given high priority by the society and the stakeholders.

4.3.1 Strict Control of Pollution Loads into the Lake

Nitrogen and phosphorus are the main pollutants in Poyang Lake. Sven Erik Joergensen, a Danish scientist, summarized the eutrophication process and management experience of shallow lakes in Europe and concluded that total phosphorus level is a key factor affecting macrophytes and phytoplankton in the lake. If the total phosphorus concentration is less than 0.06 mg/l, the submerged macrophytes in the lake are at a healthy growth state. In contrast total phosphorus concentrations above 0.12mg/l lead to the dominance of algae growth, which inhibits the growth of submerged plants. To maintain the health of Poyang Lake aquatic vegetation, it is necessary to control the total phosphorus concentration in lake water and to achieve an overall surface water quality which meets at least Class III of the standard. Suggested measures to protect the lake water are aiming on the reduction of the pollution load into the lake by, firstly strictly controlling all the regular, fixed discharge of pollution load, secondly strengthening the control of pollution loads from non-point sources:

1. Strengthening urban domestic sewage treatment: Further improve the sewage collection pipe network, improve the efficiency of domestic sewage treatment plants, so that most of the urban sewage can be collected and discharged after an adequate treatment process.
2. Strengthening the treatment of industrial wastewater within the industrial parks: Put an end to the phenomenon of "lawlessness and lax enforcement" to eliminate unmanaged waste water treatment, improve the management of the sewage treatment facilities in the industrial parks, improve the technology and improve the efficiency.
3. Strengthening the treatment of wastewater in the communities: Certain measures should be conducted for the treatment of domestic sewage in villages and towns with a relatively concentrated population around the banks of Poyang Lake,

including installing small-scale sewage treatment plants and using of constructed wetlands for interception of sewage.

4.3.2 Strengthening the Management of Shallow Floodplain Lakes

The Shallow Lake plays a prominent role in increasing biodiversity, maintaining the integrity of wetland ecosystems and enhancing resilience to natural disasters. The protection of the Shallow Lake is of great significance for the protection of the health of the Poyang Lake-wetland system. Hence, the following technical measures are proposed for the lake:

1. Regulate the water level of the Shallow Lake: In accordance with the foraging requirements of migratory birds, regulate the water level of the lake in accordance to a scientific regulation scheme. The Shallow Lake shall be arranged under unified planning according to the topography of the plateau without negatively affecting the wintering of migratory birds.
2. Regulation of hunting: To protect the aquatic vegetation it is strictly forbidden to manually stocking Chinese Down Hair Crabs and to catch snails in the Shallow Lake .
3. Regulation of fishing: It is prohibited to cultivate Chinese Mitten Crab in the Shallow Lake and to catch Bolinus brandaris and clams. The aquatic vegetation should be protected.

4.3.3 Recuperating and Restoring the Ecosystem

Activities need to be carried out to restore the Poyang Lake wetland ecosystem. It is prohibited to illegally dispose sewage, destruct the lakeshore, mine sand from the riverbed without regulation.

4.3.4 Improving the Management System of Poyang Lake

The current management system of Poyang Lake is based on the Jiangxi Poyang Lake Wetland Protection Ordinance issued by the provincial people's congress. The Poyang Lake Wetland Protection and Coordination Bureau under the leadership of the Provincial People's Government is the coordinating council for the lake management and other relevant administrative departments manage one or several tasks of Poyang Lake according to their respective administrative functions. Each department is working independently which disperses management power and law enforcement.

Other disadvantages are weak teams, lack of equipment, lack of synergies, difficulties in safeguarding regular inspection and supervision activities, and a lack of ability to develop and utilize modern monitoring tools. In addition, intersection, displacement, and deficiency in authority is present among departments and regions. It is suggested to improve the management system of Poyang Lake and include all relevant local governments to participate in the Poyang Lake Management Bureau. Based on the current management system, the Poyang Lake Administration is established as a service and implementation agency of the Poyang Lake Integrated Wetland Protection Agency (Fig. 4.5). Each department dealing with wetland management, law enforcement and supervision functions of Poyang Lake should be taken under the responsibility of the Poyang Lake Administration, while their operations should be guided by the relevant departments at higher levels. The major enforcement teams should work together to inspect and control various activities in the lake area. Environmental problems should be solved by the corresponding business law enforcement officers to avoid that until now in some cases, institutions overlap in responsibility and in other cases certain important tasks are not covered by any relevant institutions.

4.3.5 "The River Chief System" as the Starting Point

"The River Chief System" defines the responsibility of local party and government leaders for effective utilization of water resources, environmental protection and aquatic ecology. The main responsibility of this so called river chief inspector is to do his work in accordance to natural science, to strengthen the integrated river basin management, to manage the water system guided by the "Kowloon regulates waterway". He is in charge of forming an effective mechanism for water management, which is governed by a holistic approach covering the terrestrial and aquatic parts of the ecosystem including cooperation with all relevant stakeholders such as the authorities, the local governments and the citizens. Crucial for the success of the "The River Chief System" is to monitor water quantity, water quality and aquatic ecology in river reaches and key waters of counties and cities on a regularly base. Strengthening supervision and law enforcement are also of high importance for a successful management. The "River Chief System" should be implemented to improve the comprehensive management of the river basin to protect the health of Poyang Lake.

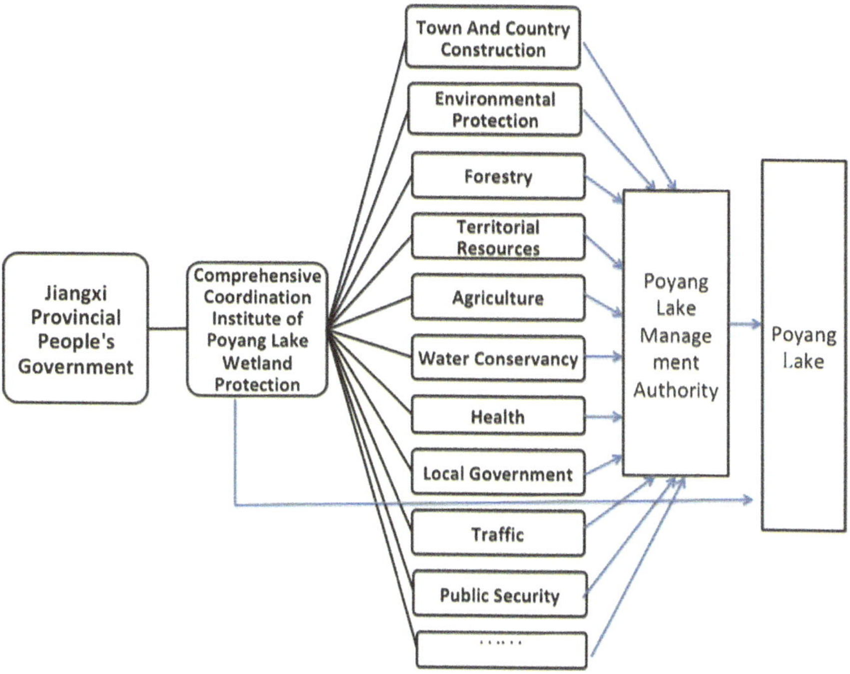

Fig. 4.5 Current system of authorities at the provincial and local level involved in the water resources management of the Poyang Lake wetland system

Part II
Hydro(geo)logy

Chapter 5
Shallow Groundwater of Poyang Lake Area

Evgeniya Soldatova, Stepan Shvartsev and Zhanxue Sun

5.1 Introduction

The water-bearing rocks in the upper part of geological cross-section of the Poyang Lake area are mainly aluminosilicates of different age and composition. The highly water-saturated sediments are associated with the river deltas and channels and the deposits created by them [1]. The deltas and lakeside regions are depressed areas filled by Quaternary alluvial and deluvial sediments with average thickness of about 20–25 m and more. The Quaternary sediments contain gravel, sands, clays and loam. The bedrocks in the surrounding mountains are presented by ancient Proterozoic siltstones, mudstones, slates, tuffaceous sandstones, tuffite, hornfels and polymictic conglomerate of fractured rock intruded by granitoids, which are overlaid by Cretaceous, Paleogene and Quaternary red weathering crust [2, 3]. The relatively low southern and western areas of the Poyang Lake plain consist of undefined strata of Cretaceous-Paleogene sandstones, siltstones, mudstones and their conglomerates, which are locally red-colored and represent consolidated weathering crust [3, 4].

According to the analysis of isotopic composition of H_2O, the shallow groundwater of the Poyang Lake area is of meteoric origin. It was found that the influence of evaporation on the formation of groundwater chemical composition is negligible and observed mainly during the dry season. During this period, water infiltrated through the vadose zone and enriches with heavy oxygen isotopes due to evaporation effects, which lead to a deviation in isotopic composition from the meteoric water line [5]. In addition to this effect, local mixing of the shallow groundwater and surface water subjected to more intensive evaporation may have a further indirect effect on the chemical composition of the groundwater. Xu and Wang [6] could demonstrate with

E. Soldatova (✉) · S. Shvartsev
Tomsk Polytechnic University, Tomsk, Russia
e-mail: 2a61@mail.ru

Z. Sun
East China University of Technology, Nanchang, China

T. Yue et al. (eds.), *Chinese Water Systems*, Terrestrial Environmental Sciences,
https://doi.org/10.1007/978-3-319-97725-6_5

their research on the chemical and isotopic composition of groundwater in Yongxiu area (to the west of Poyang Lake) that a hydraulic connection between the surface water and groundwater exists and that the groundwater in some sites is recharged from the surface water.

5.2 Results

5.2.1 Chemical Composition and Types of the Shallow Groundwater in the Poyang Lake Area

The Total Dissolved Solids (TDS) values of the shallow groundwater in the Poyang Lake area (Fig. 5.1) vary from 25 to 800 mg/L with a mean TDS value of 183 mg/L (Table 5.1). Thus, it can be classified as fresh water [3]. The pH value of the shallow groundwater varies greatly from 4.5 to 7.7, i.e. from acidic to neutral and rarely to slightly alkaline. There is small logarithmic dependence between TDS and pH values (Fig. 5.2). The measured Eh values vary widely (−91–382 mV). However, oxidizing conditions with Eh $>$ 100 mV were observed at the most sampling points. The shallow groundwater is characterized by reducing conditions, which occur mainly in the lower reaches of Gan River and Xiu River.

According to its chemical characteristics, the shallow groundwater belongs to either the HCO_3–Ca–Na or the HCO_3–Na–Ca type. However, human activity and some natural factors lead to an increase of NO_3^-, Cl^- and SO_4^{2-} concentrations and, to a minor extent NH_4^+, K^+, Fe^{3+}, NO_2^-, PO_4^{3-} and F^- concentrations. Most of these elements and compounds are typical components of different mineral and organic fertilizers, sewage water, livestock and household wastes. The highest concentrations of these elements and compounds are found mainly in the shallow groundwater depth up to 10–12 m below surface (Fig. 5.3). This proves indirectly that these components are primarily transported to the groundwater due to the influence of anthropogenic factors.

The most striking data for the anthropogenic impact was found for the nitrogen compounds. The average concentration of NH_4^+ is 0.1 mg/L and its maximum value is 6.4 mg/L (Table 5.1). The maximum concentration of NO_3^- is 206 mg/L with an average of 17.9 mg/L. It is worth to note that the content of NO_3^- in 24 sampling points, which represent 18% of the total number of sampling points, is found to be higher than the World Health Organization standard for drinking water quality [7]. The concentration of NO_2^- reaches 4.3 mg/L while the guideline value according to WHO recommendations [7] is 3 mg/L. Preliminary study of N-compounds sources using dual isotopic approach (δ^{15} N-NO_3^- and δ^{18} O-NO_3^-) demonstrates that the main source of pollution in the Poyang Lake area are manure, sewage and human wastes [8]. The available data do not yet allow separating these two sources from each other. It is also worth to note that nitrate has a high chemical stability and prevails over ammonium despite the fact that reduced forms of N-compounds are abundant

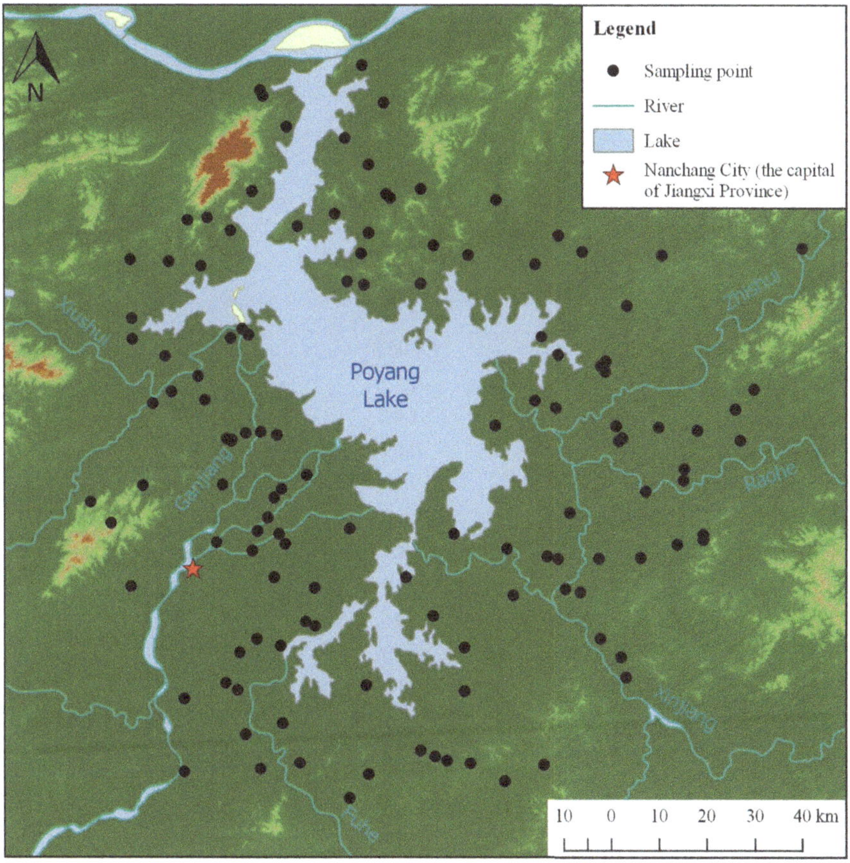

Fig. 5.1 Location of the sampling points in the Poyang Lake area

in pollution sources (manure and sewages). This effect is likely to be caused by the influence of nitrification on the balance of nitrogen species in the water, which penetrates through vadose zone.

Field observations allowed dividing the shallow groundwater into two types: (1) groundwater associated with red earth and (2) groundwater associated with paddy soils [3]. The first type of groundwater is found in natural landscapes, which are slightly modified by agricultural activity and retained their natural features (Fig. 5.4a). The second type of groundwater occurs in areas influenced by intensive agricultural activities for several thousand years with the result that their landscapes are drastically changed (Fig. 5.4b).

Table 5.1 Chemical composition of the shallow groundwater in the Poyang Lake area, mg/L, Eh – mV

Component	Value		
	Min	Max	Average
pH	4.5	7.7	6.23
Eh	–91	382	194
HCO_3^-	2.44	353	50.5
SO_4^{2-}	0.14	148	10.4
Cl^-	0.99	102	14.2
NO_2^-	0.01	4.29	0.022
NO_3^-	0.1	206.3	17.9
PO_4^{3-}	<0.01	1.79	0.054
F^-	<0.05	1.41	0.07
Br^-	0.01	0.69	0.05
Ca^{2+}	1.7	98.2	19.0
Mg^{2+}	0.23	55.2	5.52
Na^+	1.04	58.3	12.1
K^+	0.23	76.0	2.65
NH_4^+	<0.05	6.4	0.1
SiO_2	2.82	77.0	15.7
TDS	25.1	800.3	183
CO_2	3.5	141	18.2
DOC	0.34	7.58	1.15

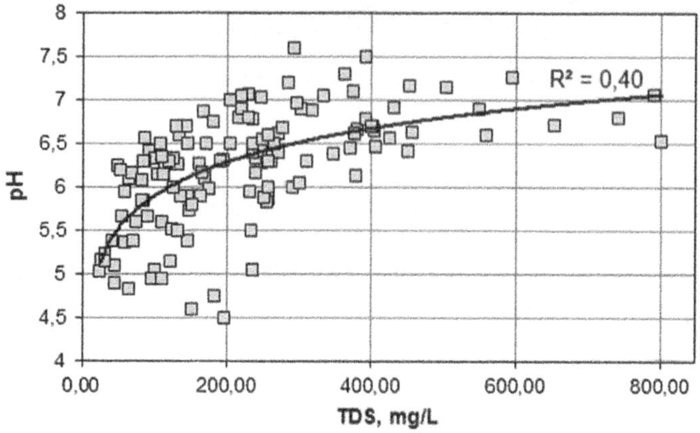

Fig. 5.2 TDS values versus pH values of the shallow groundwater of the Poyang Lake area

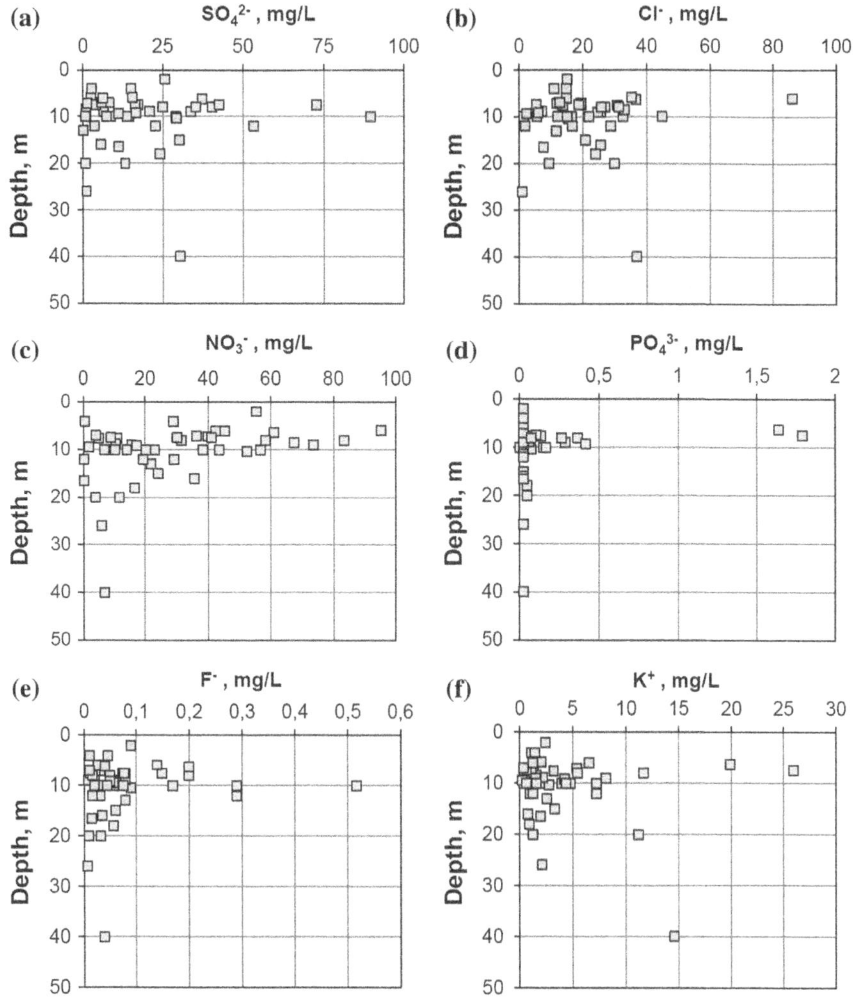

Fig. 5.3 Content of SO_4^{2-} (**a**), Cl^- (**b**), NO_3^- (**c**), PO_4^{3-} (**d**), F^- (**e**) K^+ (**f**) versus the depth of the shallow groundwater occurrence

5.2.2 Chemical Composition of Groundwater Associated with Red Earth

The shallow groundwater associated with red earth normally occurs near the earth surface (depth 2–10 m) in fractured zones of bedrocks and less often directly in the weathering crust. The landscapes are characterized by the presence of the lithomorfic kaolinite weathering crust with abundance of Fe hydroxides and a rare presence of Al hydroxides (red earth). Mineralogical analysis of secondary minerals demon-

Fig. 5.4 Specific landscapes of the Poyang Lake catchment: **a** red earth (natural landscapes); **b** agricultural landscapes with paddy soils

Table 5.2 The mineral saturation index for the shallow groundwater associated with different landscapes of the Poyang Lake area. Legende: Goe: Goethite, Lep: Lepidocrocite, Man: Manganite, Fem: Fe-montmorillonite, Gro: Grossular, Dio: Diopside, Oli: Olivine, Amp: Amphibole, Hyd: Hydroxyapatite

Ground water type	Goe	Lep	Man	Fem	Gro	Dio	Oli	Amp	Hyd
Associ. with red earth	6.49	1.35	−2.12	−6.07	−21.62	−21.77	−32.08	−46.95	−9.24
Associ. with paddy soils	9.76	3.71	−1.74	25.45	−9.76	−14.69	−25.92	−23.31	−1.80

strated that in the samples of the red earth, the predominant minerals are kaolinite with goethite admixture. Research of Hong et al. [9, 10] have also shown that the main minerals in the red earth are Fe oxides and clays, including kaolinite and less often smectite. Positive saturation indexes were calculated for kaolinite and Fe hydroxides and less often for Al hydroxides meaning that these minerals can precipitate from the groundwater associated with red earth (Table 5.2 and Fig. 5.5). At the same time, the shallow groundwater of red earth is unsaturated with calcite, magnesite, montmorillonite, illite and primary minerals (diopside, olivine, amphibole, etc.). The kaolinite weathering crust is formed by acid and intermediate magmatic rocks, crystalline schists, gneisses, and sandstones. The weathering crust is colored reddish, reddish-brown, and brown due to Fe^{3+} hydroxides. This special combination of water, weathering crust, soils with low humus content and vegetation creates landscapes with seasonally humid deciduous and coniferous forests [3].

The salinity of the shallow groundwater of red earth is rather low. The TDS value ranges from 25 to 134 mg/L with an average of 74.7 mg/L (Table 5.3). The pH value varies from 4.9 to 6.6, which means that the shallow groundwater is slightly acidic and rarely acidic [3].

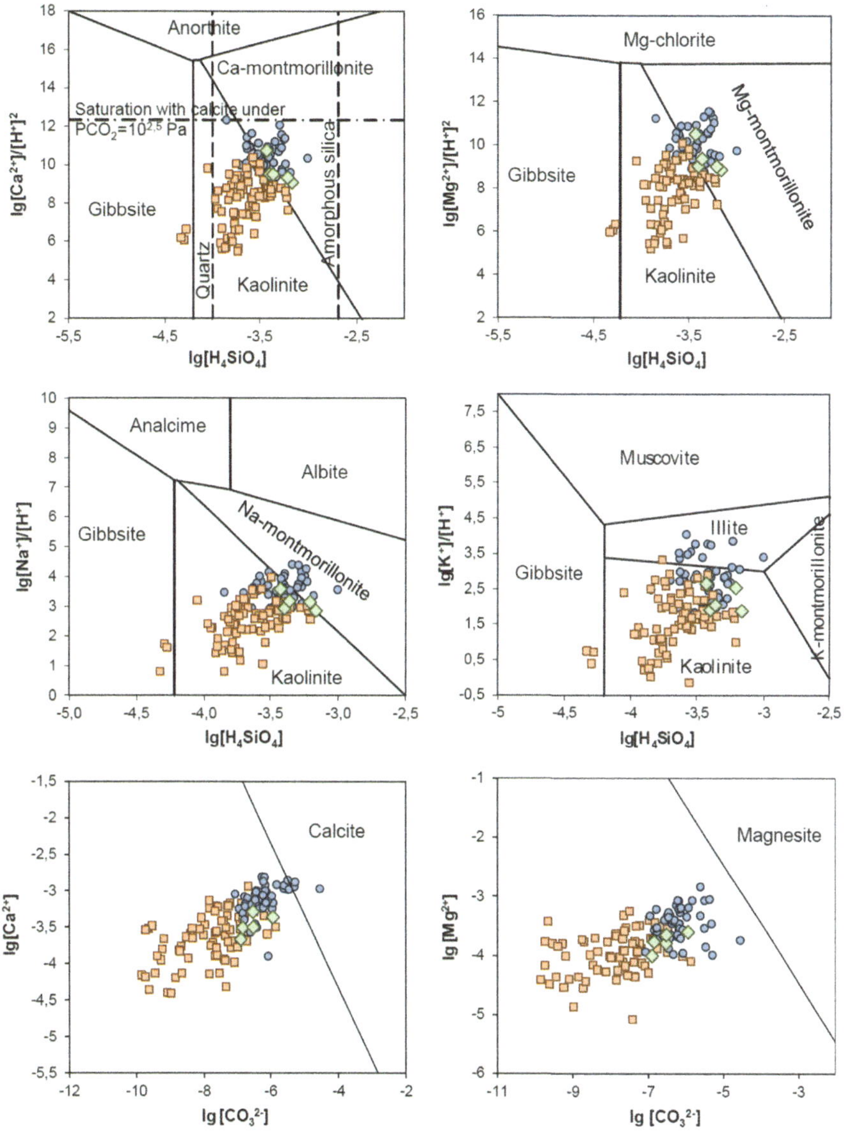

Fig. 5.5 Equilibrium of the different types of shallow groundwater under study with minerals at T = 25 °C and p = 1 atm. Orange and blue circles indicate shallow groundwater associated with red earth and paddy soil, respectively. Green rhombi visualize shallow groundwater with Eh < 0 associated with paddy soils

The geochemical conditions are oxidizing. Such conditions are responsible for low concentrations of Fe and DOC, which is usually mineralized with the formation of CO_2 and H_2O under oxidizing conditions. The process of DOC mineralization

Table 5.3 Average chemical composition of the shallow groundwater associated with different landscape types, mg/L, Eh – mV

Component	Groundwater associated with		Component	Groundwater associated with	
	Red earth	Paddy soils		Red earth	Paddy soils
pH	5.8	6.8	Mg^{2+}	3.21	9.32
Eh	218	164	Na^+	6.29	16.6
HCO_3^-	22.5	146	K^+	1.29	2.09
SO_4^{2-}	4.65	24.3	NH_4^+	0.06	0.08
Cl^-	4.81	20.1	SiO_2	13	21.1
NO_2^-	0.02	0.03	TDS	74.7	318
NO_3^-	6.89	7.46	DOC	1.24	1.51
PO_4^{3-}	0.03	0.09	CO_2	20.5	15.5
F^-	0.04	0.12	Al	0.04	0.02
Ca^{2+}	10.9	43.3	Fe_{tot}	0.07	0.08

provides slightly acidic environment in the groundwater. Thus, the red earth landscape is associated with slightly acidic ultrafresh groundwater with TDS values less than 200 mg/L. The main anion in this groundwater is HCO_3^-, and the main cations are Ca^{2+} and Na^+. However, in some sampling points, high concentrations of Mg^{2+}, Na^+, K^+, NO_3^-, SO_4^{2-}, and Cl^- could be measured [3].

The characteristic features of this groundwater are caused by its shallow occurrence, widespread occurrence of low-soluble ancient magmatic and metamorphic rocks, water exchange, and evergreen broad-leaved forests of South-East China, which prevent intense water evaporation. In particular, the character of mineral equilibrium (Table 5.2 and Fig. 5.5) in the water-rock system is controlled by water exchange which is active in the red earth and defined by the rate of subsurface water flows (locally, up to 20 L/(s km^2) [3]). Active water exchange controls the short-term interaction between water and bedrocks and provides low groundwater salinity. Kaolinite and hydroxides precipitate from the groundwater and subsequently bind Si, Al and Fe. Other elements continue to migrate in the groundwater, thus, the cationic composition of the groundwater is defined by the type of water-bearing rocks. The source of HCO_3^- is atmospheric precipitation and hydrolysis of aluminosilicates with subsequent neutralization of OH^- via the reaction:

$$OH^- + CO_2 = HCO_3^-$$

The deficit of organic matter in this groundwater type draws attention. The DOC concentration is a mere 0.3–1.3 mg/L, while biological productivity of subtropical landscapes is estimated at 2000–2500 t/(y km^2) [11]. One of the reasons is the deforestation of the study area. Natural forest landscapes are replaced by agricultural fields. Moreover, large amount of organic matter decompose during dry seasons leading to a decrease of DOC content and pH within the groundwater. In addition, partial pressure of CO_2 in the soil increases up to $10^{-1.5}$–$10^{-2.5}$ atm, which is a hundred times more compared to the atmospheric values.

5.2.3 Chemical Composition of Groundwater Associated with Agricultural Landscapes

As long as the main occupation of the local population in the study region is agriculture, it affects the chemical composition of shallow groundwater, interaction in water-rock system and character of landscapes.

The landscapes in the region are much affected by widely used irrigation, flooding, and slope terracing. By now, the lowlands and alluvial plains in the region as well the river terraces are cultivated almost completely. Agricultural fields also cover gently sloping hills and low mountains. The landscapes in the region represent specific agricultural fields subjected by the long-term use of fertilizers, the transformation of water exchange and the disturbance of water regimes.

X-ray phase analysis showed that the mineral composition of the red earth samples and the samples of paddy soils are different. In samples of paddy soils, the overall content of kaolinite and goethite decreases, while the proportion of montmorillonite and illite in the total content of secondary minerals increases. Thermodynamic calculations [3] have demonstrated that in terms of equilibrium conditions in the water-rock system, the groundwater associated with agricultural landscapes differs from the groundwater associated with red earth (Table 5.2, Fig. 5.5). While the groundwater associated with red earth is in equilibrium with kaolinite and hydroxides, the groundwater associated with agricultural landscapes is saturated with montmorillonites, less often with illite, kaolinite and calcite.

Thus, natural red earth is replaced by dark-colored anthropogenic soils and montmorillonite weathering crust within the agricultural landscapes (Fig. 5.4b). However, these soils also may have patchy reddish-grey or yellowish-brown color due to admixture of natural kaolinite weathering crust with Fe hydroxides.

The formation of montmorillonite and calcite requires longer time of water-rock interaction than the formation of kaolinite and hydroxides according to thermodynamic characteristics. Consequently, the water-rock system within agricultural landscapes is at the later stages of mineral formation. The groundwater of agricultural landscapes is usually saturated with montmorillonite, whereas the groundwater of natural landscapes of red earth is in equilibrium mainly with kaolinite. However, not only the time of water-rock interaction can affect the formation of secondary minerals. Thermodynamic simulations demonstrate that the increase in pH may result from fertilization, which obviously is more intensive within agricultural landscapes. This pH growth leads to the formation of clay minerals at lower concentrations of main aqueous clay-forming components than under natural conditions [8].

The mean TDS value of the shallow groundwater associated with agricultural landscapes are more than 4 times higher than the one of the groundwater associated with red earth due to longer time of water-rock interaction (Table 5.2). The pH value fluctuates from 6.3 to 7.7 with a mean of 6.8. Hence, the groundwater of agricultural landscapes is dominated by near neutral environment and is more saline and alkaline than the groundwater associated with red earth. It is characterized mainly by the HCO_3–Ca chemical type, though the cation and anion content in the groundwater of paddy soils is several times higher than in the groundwater of red earth. All these features (high TDS and pH values, high concentration of SiO_2) cause the groundwater associated with agricultural landscapes to be in equilibrium with montmorillonite and closer to saturation with calcite than the groundwater of red earth. In contrast to the red earth, some areas in the lower courses of the Gan River and Xiu River associated with agricultural landscapes showed Eh values in the shallow groundwater below 0 mV. The values of Eh vary from −68 to −91 mV. Such gley environment (slightly reducing environment without hydrogen disulfide) is likely to be formed by the infiltration of excess amount of organic compounds into the groundwater in combination with lower rate of water exchange. Organic substances are oxidized by the dissolved oxygen, which reduces the oxidation-reduction potential and increases the CO_2 concentration. Under these conditions, the shallow groundwater has a low content of NO_3^- (0.20–0.24 mg/L), but relatively high contents of Fe^{2+} (17, 0–64, 4 mg/L), SiO_2 (22.6–41.2 mg/L), F^- (0.33–1.0 mg/L), NH_4^+ (2.0–6.4 mg/L), DOC (1.0–5.5 mg/L), CO_2 (17.6–88.0 mg/L) compared with water being in an oxidizing environment.

The formation of gley environment has a significant effect on the equilibrium between groundwater and minerals. Thermodynamic calculation showed that the shallow groundwater with low Eh values is saturated with siderite, daphnite, nontronite due to high content of ferrous iron.

5.2.4 Trace Elements in the Shallow Groundwater of the Poyang Lake Area

Content of trace elements in the shallow groundwater of the Poyang Lake area is comparable with values given by Zeng [12], except of the content of Li and Be which were, on average, one order of magnitude less than the reference values (Table 5.4). However, average concentrations of Mn, Fe, Ni, Sr, and Mo calculated for the shallow groundwater occurred in quaternary deposits of high thickness in the lower courses of Gan River and Xiu River are several times higher than the ones adduced by Zeng [13].

High concentrations of Fe and Mn are distributed all over the groundwater in Poyang Lake area. The highest contents are observed in area where gley environment occurs, i.e. in the lower courses of Gan River and Xiu River. The formation of gley environment leads to an accumulation of Fe and Mn in dissolve phase in the groundwater. Although WHO does not establish the guideline values for Fe and Mn, US Environment Protection Agency in National Secondary Drinking Water Regulations [14] refers to features of these elements in high concentrations to cause the deterioration of organoleptic properties of the water (metallic taste, color staining). The US EPA recommends level of 0.3 mg/L for Fe and 0.05 mg/L for Mn as a limit below which negative effects are not noticed by sense organs.

High concentrations of Fe in the shallow groundwater in the lower reaches of Gan and Xiu Rivers correlate well with a high concentrations of As, which exceeds the

Table 5.4 Trace elements content in the shallow groundwater of the Poyang Lake area, mg/L

Chemical element	Concentration, μg/L			Average by Zeng [12]	Recommendation of WHO [7]
	Min	Max	Average		
Li	0.1	29.4	1.1	12.9	–
Be	<0.003	0.790	0.033	0.13	–
B	1.25	39.35	5.40	7.3	2400
Al	0.3	983.0	15.0	–	–
Sc	0.11	1.11	0.37	–	–
Ti	0.068	48.0	0.87	1.26	–
V	<0.1	25.8	0.4	0.2	–
Cr	0.07	7.0	0.88	1.12	50[a]
Mn	0.2	4967	30.1	80.0	–
Fe	10.0	64400	70.0	110.0	–
Co	0.004	9.240	0.299	0.23	–
Ni	<0.05	9.37	00.78	0.46	70
Cu	<0.5	14.6	0.3	0.5	2000
Zn	<0.001	0.142	0.006	0.003	–
Ga	0.002	0.980	0.023	–	–
Ge	<0.001	0.180	0.008	–	–
As	0.02	79.12	0.63	0.68	10[b]
Se	<0.5	27.2	0.3	–	40[a]
Rb	0.61	37.32	4.76	–	–
Sr	5.3	973.8	89.7	29.6	–
Y	0.01	12.75	0.19	–	–
Zr	<0.005	0.420	0.004	–	–
Mo	<0.02	7.69	0.18	0.07	–
Ag	<0.005	0.049	0.006	–	–
Cd	<0.005	0.973	0.033	0.041	3
Sn	<0.005	1.230	0.009	–	–
Sb	0.001	3.140	0.073	–	20
Cs	<0.005	2.256	0.050	–	–
Ba	19.0	284.0	70.0	–	1300
La	<0.005	20.404	0.129	–	–
Ce	<0.005	36.205	0.190	–	–
Pr	<0.0005	4.2677	0.0310	–	–
Nd	<0.0005	16.0810	0.1280	–	–
Sm	<0.005	3.599	0.025	–	–
Eu	<0.005	0.796	0.015	–	–
Gd	<0.0005	3.3880	0.029	–	–
Tb	<0.005	0.447	0.004	–	–
Dy	<0.005	2.430	0.022	–	–

(continued)

Table 5.4 (continued)

Chemical element	Concentration, μg/L			Average by Zeng [12]	Recommendation of WHO (2017)
Ho	<0.005	0.458	0.005	–	–
Er	<0.005	1.157	0.011	–	–
Tm	<0.0005	0.1441	0.0021	–	–
Yb	<0.0005	0.8976	0.011	–	–
Lu	<0.005	0.135	0.002	–	–
Hf	<0.0005	0.0200	0.0006	–	–
Ta	<0.005	0.010	0.002	–	–
W	<0.005	5.055	0.013	–	–
Re	<0.001	0.360	0.005	–	–
Au	<0.005	0.070	0.002	–	–
Hg	<0.050	1.400	0.016	0.010	6
Tl	<0.005	0.258	0.014	–	–
Pb	<0.05	22.01	0.20	0.59	10[b]
Bi	<0.003	0.083	0.002	–	–
Th	<0.005	0.699	0.002	–	–
U	<0.005	0.903	0.042	–	30[a]

Notice not available,
[a]the guideline value is designated as provisional because of scientific uncertainties [7],
[b]the guideline value is designated as provisional on the basis of treatment performance and analytical achievability [7]

guideline value for drinking water (0.01 mg/L according to WHO [7]) at 7 sampling points within the study area. Thus, it is concluded that reducing conditions result in a mobilization of As. Arsenic in the lower course of Gan and Xiu Rivers is presented in its trivalent form, rather than As^{5+} [15, 16] because of reducing or near neutral environment. Oxides of Fe and Mn are dissolved releasing arsenate that is rapidly reduced to arsenite under these conditions [17]. In the sampling points with Eh less than 0 mV, As^{3-}, which has a higher bioavailability [16], dominates the oxidation state of arsenic [15]. Reduction of iron oxy-hydroxides to Fe^{2+} and decomposition of organic matter may also contribute to mobilization and reduction of As (V) to As (III) as it has been reported by Chen and Ahmed et al. [18, 19]. The higher content of organic matter in comparison with the groundwater of the other areas and gley environment may result from specific combination of natural and anthropogenic factors, such as low topography and overuse of organic fertilizers [8].

Occasional minor exceedances of the drinking water quality standards established by WHO [7] and NSDWR [14] are also observed for Pb and Al. Liao et al. [20] have reported that concentrations of Pb exceed the provisional guideline value established by WHO [7] to the north and to east from the Poyang Lake. They assumed that the Pb in the shallow groundwater of the Poyang Lake area mainly originates from the dissolution of Pb-bearing minerals and from industrial and domestic activities.

5.3 Conclusions

Groundwater of red earth refers to the HCO_3–Ca–Na type, has low TDS values (25–130 mg/L) and is mainly slightly acidic (pH varies from 4.5 to 6.6). Due to active water exchange, this groundwater type is in equilibrium with kaolinite and hydroxides, which give the kaolinite weathering crust a reddish shade. Since at this stage of the water-rock system evolution there are no geochemical barriers for the accumulation of the main cations, their content in groundwater is determined by the type of dissolved host rocks. Such groundwater develops mainly on sites with a hilly topography, which contributes to the formation of active water exchange.

Within agricultural landscapes, where the water exchange rate decreases, there is the shallow groundwater with higher TDS values (160–600 mg/L) and pH values (6.3–7.7) in comparison with the groundwater associated with red earth. The equilibrium with minerals in this case shifts toward montmorillonites and illite due to new stage of water-rock interaction established. In gley environment the shallow groundwater is saturated with siderite, daphnite, nontronite and possibly other ferruginous aluminosilicates.

Regardless of the landscape type, rate of water exchange, type of host rocks, relatively high concentrations of pollutants were identified in certain sampling points (both associated with red earth and agricultural landscapes) due to influence of anthropogenic factors. Prevailing pollutants are NO_3^-, Cl^-, SO_4^{2-}. High concentrations of NH_4^+, K^+, NO_2^-, PO_4^{3-}, F^- in comparison with background concentrations are registered to a minor extent. This pollution leads to an increase in TDS values up to 800 mg/L and a change in the groundwater chemical type to Cl–NO_3, NO_3, SO_4–HCO_3, SO_4–Cl–HCO_3 by anionic composition and K–Na–Ca, Ca–K, Ca–Na–K by cationic composition. It is also worth to note that content of N-compounds, and less often As, Pb, and Hg exceed guideline values for drinking water established by WHO. The high concentrations of pollutants in the shallow groundwater more common for the areas in Fu River, Gan River and Xiu River basins but also in the lakeside area.

Acknowledgements The research of N-compounds behavior is funded by Russian Science Foundation (RSF), Project No17-77-10017. Analysis of the shallow groundwater chemical composition and further data processing were carried out at Tomsk Polytechnic University within the framework of Tomsk Polytechnic University Competitiveness Enhancement Program grant.

References

1. Jiangxi Provincial Bureau of Geological Exploration and Exploitation for Mineral Resources. 2013. Hydrogeological map of Poyang Lake area. Published Report. Nanchang, PR China.
2. Jiangxi Bureau of Geology and Mineral Resources. 1984. Regional Geology of Jiangxi Province. Beijing, 922.

3. Shvartsev, S. et al. 2016. Evolution of the groundwater chemical composition in the Poyang Lake catchment, China. *Environmental Earth Sciences* 75: 1239. https://doi.org/10.1007/s12665-016-6065-8.
4. Jiangxi Provincial Bureau of Geological Exploration and Exploitation for Mineral Resources. 1996. Geological map of Jiangxi Province. Published Report. Nanchang, PR China.
5. Polyakov, V., and V. Ferronsky. 2012. *Isotopes of the earth's hydrosphere*. Netherlands: Springer.
6. Xu, B., and G. Wang. 2016. Surface water and groundwater contaminations and the resultant hydrochemical evolution in the Yongxiu area, west of Poyang Lake, China. Environmental Earth Sciences 75 (3), 184. ISSN: 1866-6280. https://doi.org/10.1007/s12665-015-4778-8.
7. World Health Organisation. 2017. *Guideline for drinking water quality*, Geneva.
8. Soldatova, E., et al. 2017. Source and behavior of nitrogen compounds in the shallow groundwater of the Poyang Lake basin, China. *Journal of Contaminant Hydrology* 202: 59–69.
9. Hong, H., Y. Gu, K. Yin, K. Zhang, and Z. Li. 2010. Red soils with white netlike veins and their climate significance in south China. *Geoderma* 160: 197–207. https://doi.org/10.1016/j.geoderma.2010.09.019.
10. H. Hong, G. J. Churchman, Y. Gu, K. Yin, and C. Wang. 2012. Kaolinitesmectite mixed-layer clays in the Jiujiang red soils and their climate significance. *Geoderma* 173–174: 75–83. https://doi.org/10.1016/j.geoderma.2011.12.006.
11. Perelman, A., and N. Kasimov. 2000. *Landscape Geochemistry*, 1999. Moscow: Astrea [in Russian].
12. Zeng, Z. 1997. The background features and formation of chemical elements of groundwater in the region of the middle and lower reaches of the Yangtze River (in Chinese). *Acta Geologica Sinica* 71: 80–89.
13. Shvartsev. Z., I. Mazurova, E. Soldatova, and N. Guseva. 2015. Size fractionation of trace elements in the surface water and groundwater of the Ganjiang River and Xiushui River basins, China. *Environmental Earth Sciences* 27.
14. US EPA. 2003. National secondary drinking water regulations. Code of Federal Regulations.
15. Putilina, V.S., I. Galitskaya, and T.I. Yuganova. 2011. Behaviour of arsenic in soils, rocks and groundwater. Transformation, adsorption/desorption, migration. *Analyst Review, Novosibirsk* [in Russian].
16. Agency for Toxic Substances and Disease Registry. 2007. Toxicological profile for Arsenic. U.S. Department of Health and Human Services, Public Health Service. Atlanta.
17. Gräfe, M., and D.L. Sparks. 2006. Solid phase speciation of arsenic. In: Managing arsenic in the environment. From soils to human health, ed. Naidu, R. et al. Collingwood, Australia: CSIRO, 75–92.
18. Chen, J. 2012. *Decontamination of heavy metals: Processes, mechanisms, and applications*. New York: CRC Press.
19. Ahmed, F., M.H. Bibi, H. Ishiga, T. Fukushima, and T. Maruoka. 2010. Geochemical study of arsenic and other trace elements in groundwater and sediments of the Old Brahmaputra River Plain, Bangladesh. *Environmental Earth Sciences* 60: 1303–1316.
20. Liao, F. et al. 2017. Distributions, sources, and species of heavy metals/trace elements in shallow groundwater around the Poyang Lake, East China. *Exposure and Health*. https://doi.org/10.1007/s12403-017-0256-8.

Chapter 6
Modelling Seasonal Groundwater Flow Dynamics in the Poyang Lake Core Region

Erik Nixdorf

6.1 Introduction

According to the Chinese Environmental Quality Standards for Surface Water (GB3838-2002), Poyang Lake, the biggest fresh water lake in China, has an overall water quality which is better than in most other large Chinese Lakes, e.g. Tai Lake or Chao Lake. However, water quality is continuously deteriorating as the lake faces several pollution pressures such as a continuous inflow of nutrients and fertilizer residuals from its extensively cultivated shorelines [1] as well as acidity and heavy metals from mining areas [2] and industrial sites [3]. Similarly, groundwater resources, which are primarily extracted by shallow private wells around the lake and provide about 8% of the drinking water supply [4], are under pressure by high nitrate [5] and heavy metal concentrations [6].

Conducted oxygen isotope studies showed that 77% of the water inflowing the Poyang Lake via its stream system is baseflow [7] giving the hydrologically well-connected groundwater system an important role in restraining floodwater during the flood season and in ensuring river run-off supply during the dry season [8]. Although these results further indicate that, intermediate water storage as well as transformation and degradation processes in the subsurface are likely to affect the water and solutes budget in Poyang Lake, previous research mostly focused on seasonally changing water quality due to dilution effects and did not consider subsurface impacts in existing conceptual and numerical models [9]. One reason for this development is the current lack of available continous data records from monitoring wells, piezometers and soil sensors [10] forcing previous groundwater investigations to focus on single point measurements in private shallow water wells (e.g. [6, 11, 12]). In this context, a fundamental issue to solve is the role of the widely fluctuating groundwater-surface water interface determining the water quality and quantity in Poyang Lake [13].

E. Nixdorf (✉)
Helmholtz Centre for Environmental Research, Leipzig, Germany
e-mail: erik.nixdorf@ufz.de

T. Yue et al. (eds.), *Chinese Water Systems*, Terrestrial Environmental Sciences,
https://doi.org/10.1007/978-3-319-97725-6_6

Physically-based numerical modeling of the shallow groundwater resources in the Poyang Lake Region provides an opportunity to approximate the impact of changing water levels on the exchange of matter and water fluxes between the lake and the groundwater system of the wetlands. Numerical codes using different concepts to couple the surface-subsurface flow systems [14] such as OpenGeoSys [15], Hydro-GeoSphere [16] or MODFLOW [17] have been successfully applied on various spatial-temporal scales to assess the interaction processes between both compartments (e.g. [18–21]). Although the functionality of most of the fully coupled models is impressive, high computational costs, constraints on data availability and a challenging parametrization limit their applicability on regional scales and make simpler coupling schemes more widely applied [21].

In view of the aforementioned limitations, we chose the numerical code OpenGeoSys to set up a model of the groundwater regime in the Poyang Lake Core Region for variable surface water levels by integrating surface water dynamics into the model as boundary conditions. In this context the main objectives of this study are:

1. To develop a methodology to embed temporally variable water surface areas obtained from Landsat satellite images and information on surface water levels into the groundwater simulation
2. To assess the dynamics of groundwater levels and groundwater flow directions in the Poyang wetlands for high and low water levels in Poyang Lake
3. To quantify the seasonal exchange volume flux between Poyang Lake and the wetlands
4. To identify the parameters driving exchange rates and groundwater regime in the lake-wetland-groundwater system of Poyang Lake.

6.2 Materials and Methods

6.2.1 *Modelling Concept*

Numerical flow modelling is a powerful tool to analyse the influence of different impact factors on the groundwater regime. In this study, the groundwater regime in the Poyang Lake Core Region was modelled using the software OpenGeoSyS, which was embedded in a developed framework of various pre- and post-processing steps to integrate free available GIS input data and to analyse the modelling results (Fig. 6.1).

The investigation of the groundwater regime focused on the Poyang Lake Core Region, which consists of a dozen of counties with a total area of about 19,833 km^2 [22]. The formation of hydrostratigraphic units in the Poyang Lake area is complex with all strata except Devonian, Lower Jurassic, and Upper Tertiary being exposed in some areas of the basin [23]. However, unconsolidated sediments of alluvial origin, which mostly consist of Quaternary sand, form the most abundant aquifer type [4] in the riparian floodplains and the lacustrine plain forming Poyang Lake (Fig. 6.2).

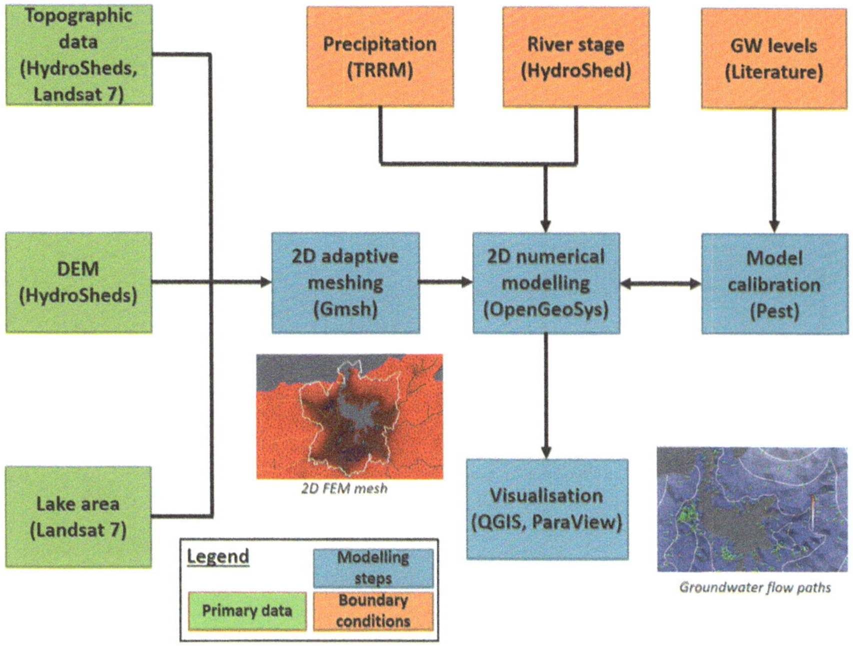

Fig. 6.1 Schematic flowchart of the methodology adopted for this study. The brackets in the green boxes indicate the data source. For an explanation of the abbreviations see the running text

According to their yield and water quality characteristics, the unconsolidated aquifers of the Poyang Lake Core Region are roughly divided into three areas: The Northern District between Xiushui and Duchang, the Southern District from Xishan-Suihong until Fuyang and the Central District around Nanbei area [24]. In general, the depth of groundwater in the Poyang Lake region is less than 10 m, indicating a relative abundance of groundwater resources. Although groundwater stored in other aquifer types such as karstic aquifers or rock fissures is an additional important water resource in some areas of the Poyang Lake Basin, these aquifer types are rare in the Poyang Lake Core Region (Fig. 6.2).

The main groundwater flow directions is proposed to follow the general topography from the bordering mountains and associated fractured rock aquifer towards the low-lying alluvial aquifers associated with river systems and the ambient wetland areas of Poyang Lake [26]. Unfortunately, the lack of a dense continuously equipped monitoring well network hinders a more precise characterization of the spatial-temporal dynamics of the groundwater regime [10].

Although focusing the investigations on the Poyang Lake Core Region is promising due to its simpler hydrogeology, defining appropriate hydrological boundary conditions along the borderline of the region is difficult and may lead to a wrong representation of the flow field. Hence, the groundwater model was set up for the entire Poyang Lake Basin with its borders being defined as no-flow-conditions. However, mesh resolution, model calibration and result analysis focused on the Lake Poyang Core Region only (see Sect. 6.2.3.1).

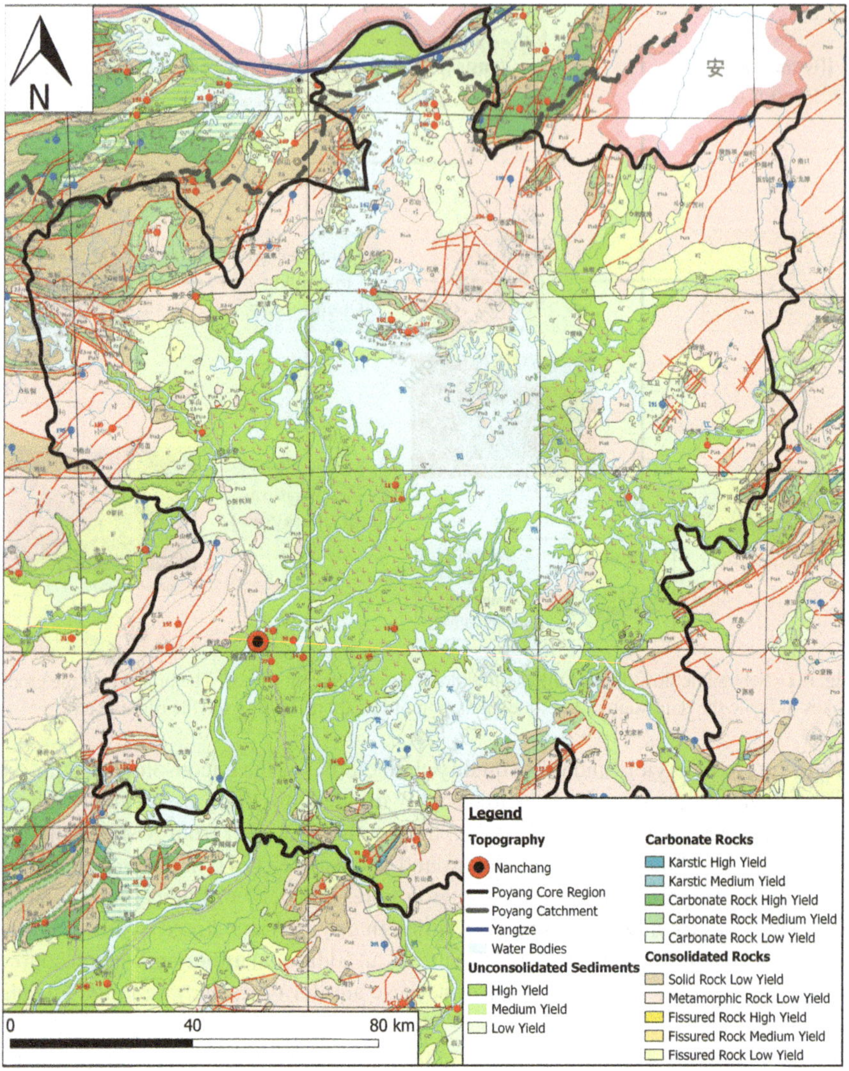

Fig. 6.2 Hydrogeological map of the Poyang Lake Core Region delineating the aquifer system by rock type and yield (modified after [25])

Numerical groundwater flow simulations were conducted for two periods: At low water levels in Poyang Lake in Dec 1999 and at high lake water levels in September 2000. The year 2000 was chosen as the study aims to show the seasonal impact of water level changes prior to the construction of the Three-Gorges-Dam in 2003. Additionally, the chosen time period ensures temporal data consistency with other input datasets such as the used DEM provided by the HydroSHEDs project [27] which based on the SRTM satellite mission in February 2000. The particular dates were selected considering the availability of satellite data to determine the extension of the surface water bodies in the Poyang Lake floodplain. For both dates, cloud-free and processed Landsat 7 datasets from the Global Land Survey (GLS) at a resolution of 30 m were provided by the United States Geological Survey (USGS). This ensured the correct extraction of waterbody extensions from the provided maps. Furthermore, Hui et al. [28] showed in their analysis of Landsat images that in the hydrological year 2000 the surface area of Poyang Lake obtained from Landsat images was lowest in Dec 1999 with less than 1500 km^2. In contrast, the lake surface area on the Landsat image of Sep 2000 was almost at the annual maximum with about 3300 km^2. Hence, the selected dates are suitable to represent periods of either significantly low or high water level within the annual cycle.

6.2.2 *Delineation of Water Courses and Shorelines*

A GIS database was set-up including all geographic input data, which were either available as raster or vector data. The shoreline of Poyang Lake and other smaller lakes in the wetlands were extracted from the two Landsat 7 GLS raster datasets by using the low reflectance value of water in the infrared bands of the Landsat 7 ETM bands (e.g. [29, 30]). The delineated patches of waterbodies were converted into polygons.

The boundaries of Poyang Lake Basin and the course of its major rivers were obtained from the HydroSHEDs DEM dataset and converted to polylines. In addition, stream orders were assigned in accordance to the classical stream order system. Each river system was branched down to three different stream orders. The extracted river courses were compared with data from GoogleEarth and OpenStreetMap and corrected, if necessary (Fig. 6.3). In total 30 main rivers were delineated and were included in the set-up of the groundwater flow model.

6.2.3 *Numerical Groundwater Flow Model*

6.2.3.1 Mesh Generation

The open source meshing tool Gmsh [31] was used to generate the Finite-Element Mesh (FEM) required for the groundwater flow model.A Gmsh geofile was generated for each scenario using MATLAB including the points and lines in the meshing environment, which define the river network, the domain boundaries, the inundation area covered by the floodplain lakes and other important features such as the location of the monitoring wells (Fig. 6.4). The main waterbody of Poyang Lake was defined as a part of the domain boundary to prevent groundwater flow beneath the lake (Fig. 6.4). In total, four meshes with a different number of triangular elements were created by Gmsh covering both investigation periods with either a higher

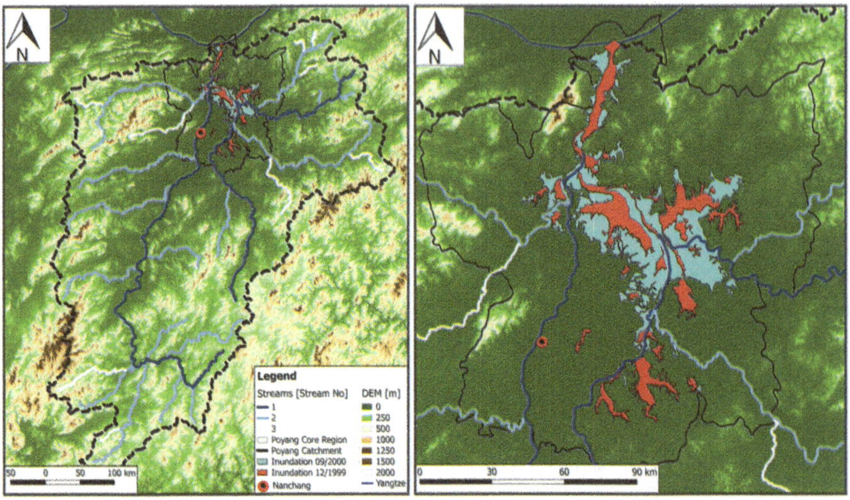

Fig. 6.3 Visualisation of the topographical features used for the groundwater model set-up. The left subplot shows the entire Poyang Lake Basin including the river network (colored by stream order). The right subplot focuses on the Poyang Lake Core Region to show in detail the areas inundated by the Poyang Lake and its floodplain lakes during both investigation periods

Table 6.1 Number of elements in the triangular FEM mesh used for each scenario. The mesh size for elements inside/outside the Poyang Lake Core Region is given in brackets

	Triangular elements per FEM mesh	
Scenario	High water (09/2000)	Low water (12/1999)
Fine (0.5 km /10 km)	152,311	173,340
Coarse (2 km /10 km)	17,156	18,552

("fine") or a lower ("coarse") mesh resolution (Fig. 6.4). A higher resolution means that the mesh element size inside the Poyang Lake Core Region was defined to be 500 m, which was four times higher than for the coarse mesh (Table 6.1). The mesh resolution for elements outside the Poyang Core Region was set to 10,000 m for all scenarios. Locally, mesh resolution may differ from the defined setting, e.g. near to river junctions.

After mesh creation, all elements were assigned by a z coordinate using the provided DEM dataset. The height of the nodes representing waterbodies were adjusted to the actual hydraulic head used as a boundary condition for these nodes (see Sect. 6.2.3.2). In addition, nodes in the interior of a floodplain lake where assigned with a similar elevation to ensure a gradient free water surface (Fig. 6.4). The triangular meshes created by Gmsh were checked for mesh quality and converted to the OGS5 mesh format required for the following numerical simulation. Non-triangular elements were removed in order to improve the post-processing of the simulation (see Sect. 6.2.3.3). All results presented in Sect. 6.3 were derived from the OGS simulation based on the coarse version of the FEM mesh. The fine mesh was used to analyse the sensitivity of simulation results, which is shown in Sect. 6.4.1.

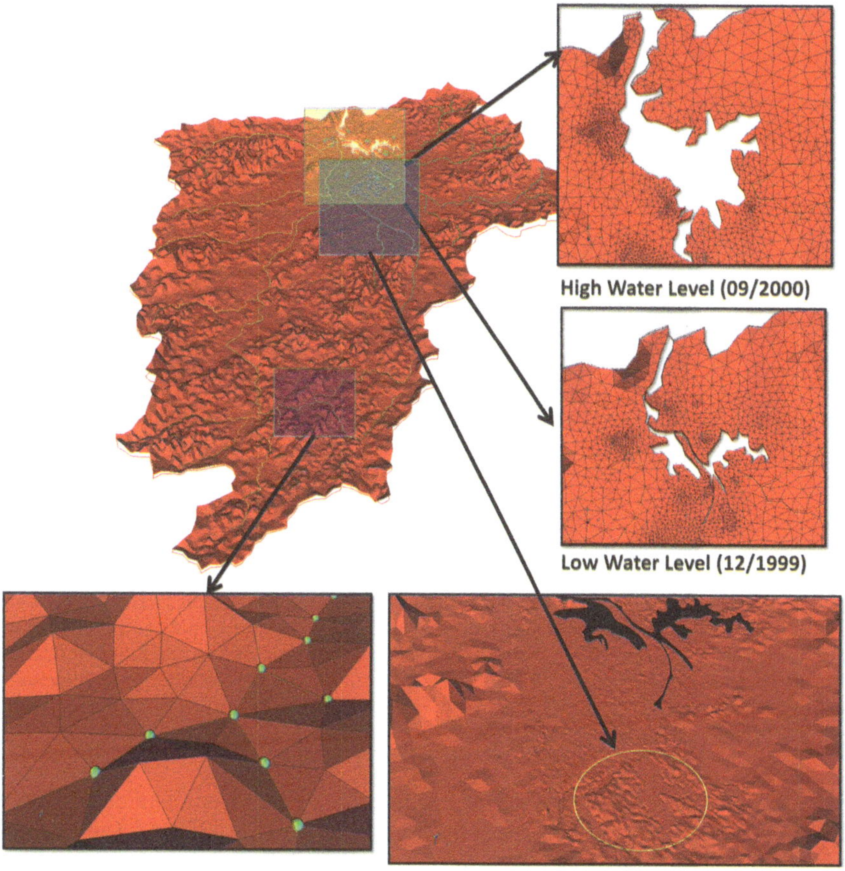

Fig. 6.4 FEM mesh set-up for the groundwater flow model. Upper-left: Mapped FEM mesh of the model domain covering the entire Poyang Lake Basin. Upper-right: Different mesh structures were used for low and high water periods reflecting the dynamic changes of topographical features. Furthermore, mesh element density increases near the floodplain lakes and the rivers within the Poyang Lake Core Region. Lower-left: The generation of a mesh consisting of triangular finite element allows to set boundary conditions geometrically correct on mesh nodes directly, e.g. by using polylines to represent the rivers. Lower-right: The DEM based mapping of the mesh was adjusted by the actual heights of the water bodies, e.g. areas within floodplain lakes were flattened

6.2.3.2 Set-up of the Groundwater Flow Model

After transforming the study area into a 2D triangular mesh, all parts of the domain boundaries, which were not assigned to Poyang Lake, represent the catchment boundaries and were subsequently set as zero-flux boundary conditions in the model set-up. Following the hydrogeological perspective that surface water represents a surface manifestation of groundwater [32], constant head conditions obtained from the DEM dataset were defined for all nodes representing the rivers in the domain. A locally

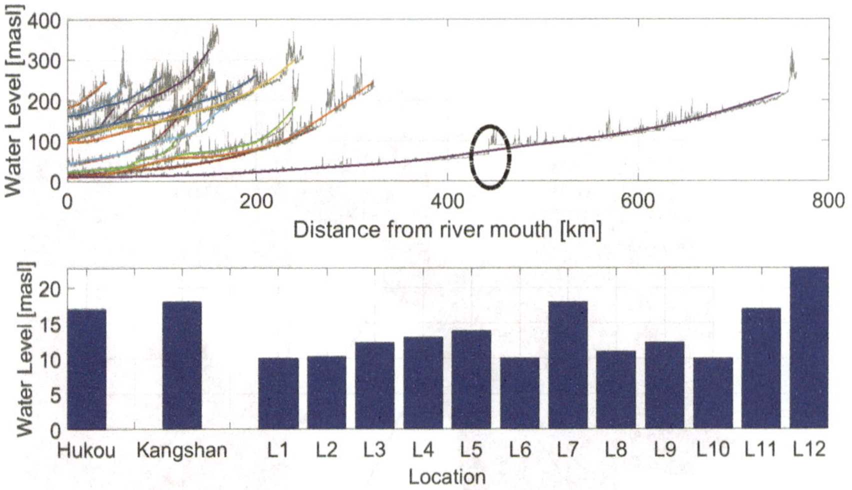

Fig. 6.5 Parametrization of Dirichlet BCs for the numerical simulation. The upper plot shows the height profile of the rivers included in the set-up of the groundwater flow model. The grey lines show the raw height profiles obtained from the DEM and the colored lines the height profiles after applying the smoothing algorithm. The black ellipse illustrates that steep gradients in the river height profile, e.g. caused by a reservoir in the upstream reach of Gan River, could not be represented well by the smoothed curve. The lower plot shows gauged water levels at Hukou and Kangshan station as well as derived water levels at 12 floodplain lakes (L1 to L12) during the investigated high water level period

weighted smoothing algorithm was implemented in MATLAB and applied to smooth the height profiles along each river extracted from the DEM (Fig. 6.5). The smoothing algorithm was adapted by increasing the span of data points included in the local smoothing until the hydraulic gradient reached a uniform direction along the entire river course. This ensured that errors within in the DEM data, e.g. due to the relatively coarse resolution of 90 m, did not lead to an implausible reversal of water flow towards upstream direction. Although the chosen smoothing approach introduces errors in regions with steep slopes, e.g. near reservoirs or in the headwater of rivers, those areas are exclusively located outside the Poyang Lake Core Region (Fig. 6.5). However, it should be noted that the chosen approach neglects existing seasonal variation of stream water levels [33], but provides average water levels being more representative for low water levels considering the date of the SRTM satellite mission.

Seasonal water levels for all existing floodplain lakes except Poyang Lake itself were obtained from the DEM. Here we could take advantage of the fact that the Shuttle Radar Topography Mission was conducted in February 2000, which means that elevations in the Lake Poyang Core Region were retrieved at very low water levels disclosing the surfaces of many temporally flooded areas. Outliers were removed from DEM elevation data for each floodplain lake by using the q_{10} (10% quantile) value of all elevation data points located inside its surface area as the water level (in meters above sea level) for all nodes representing each floodplain lake. Finally, the polylines representing each floodplain lake and being associated with a certain water level (or in other terms hydraulic head) were included as first order BCs in the numerical groundwater flow model.

For each investigation period, water levels in Poyang Lake were obtained from the official record at Hukou station close to the junction of Poyang Lake and Yangtze River as well as at Kangshan station, which is at the upstream end of Poyang Lake (Fig. 6.5). Hydraulic heads for all other nodes representing the shoreline of Poyang Lake were obtained by linear interpolation based on the relative distance of each point between both stations.

Groundwater recharge was computed directly by cell-wise subtraction of evaporation from precipitation. Lacking a sufficient online network of ground-based meteorological stations, monthly average precipitation estimates were obtained from the Tropical Rainfall Measuring Mission (TRRM) satellite, which measured global rainfall intensity with a spatial resolution of 0.25° and a temporal resolution of 3 hours using passive microwave sensors [34]. Global monthly actual evapotranspiration maps were provided in a spatial resolution of 8 km by the satellite remote sensing based evapotranspiration (ET) algorithm of Zhang et al. [35]. The derived recharge data was calculated for the higher grid resolution of the ET dataset by spatially interpolating the precipitation data. For each of the two investigated months (Dec 1999 and Sep 2000), the average of calculated recharge of the respective month and the two previous months was assigned as a spatial non-uniform second-type (flux) boundary condition to all nodes of the groundwater flow model except the nodes belonging to either a river or the surface of a floodplain lake. This temporal averaging procedure aimed to address the delayed response of the aquifer to infiltration due to transport processes in the unsaturated zone as well as the impact of direct runoff.

Regarding the subsurface properties, hydraulic conductivity was defined as homogenous and isotropic for each cell of the model domain due to several reasons. Firstly, although some information about the dominant hydrogeology structures were available, the amount of accessible data from bore logs was insufficient to set-up a structural model of the complex layered hydrostratigraphy in the Poyang Lake Basin. In addition, most of the surficial aquifer of the Poyang Lake Core Region consists of unconsolidated sediments of fluvial or alluvial origin (see Fig. 6.2). In addition, a porosity value of 0.25 was selected for the entire domain which based on the literature value for sand [36].

The simulations of the groundwater flow model were processed one a personal computer with 4 cores using the OpenGeoSys V5.7 modelling software. Model calibration was done using shallow water levels of 54 monitoring wells located in the surficial unconsolidated aquifer of the Poyang Lake Core Region [11] assuming steady-state conditions and an average water level of 5 m below surface based on the observations. Monitoring wells, which either did not refer to the unconsolidated sediments of the floodplain or were partially flooded during wet season, were excluded from the calibration procedure. Hence, 46 wells remain for the calibration procedure. The calibration itself was conducted using the model-independent parameter estimation code PEST and aquifer transmissivity as calibration parameter. Considering that the time of available water level measurements in the monitoring wells is unknown, the calibration was conducted twice for both seasons using the same observed water levels. Resulting transmissivities were compared and the average value was used as parameter for subsequent numerical simulations.

6.2.3.3 Post-processing

The groundwater simulation results are provided by OpenGeoSyS in the VTK format. The software ParaView [37] provides a huge variety of filters for this data format to visualize groundwater heads and groundwater velocities as well as the trajectories of mass-less particles in the velocity field. For a stationary vector field $\mathbf{v}$, the trajectory curve, or streamline, of each particle is given by integrating the velocity field over time along the trajectory:

$$\mathbf{s}(t) = \mathbf{s}_0 + \int_0^t \mathbf{v}(\mathbf{s}(u))du \tag{6.1}$$

where $\mathbf{s}$ is the trajectory, $\mathbf{s}_0$ is the seed source and u the integration time which defines the integration step size. The streamlines were calculated in ParaView using the "StreamTracerWithCustomSource" filter which integrates the above given equation by approximation using integration schemes of Runge-Kutta type (cf. [38]). Points were defined in a spatial distance of 500 m along the shoreline of Poyang Lake and used as seed source of the trajectories. Pore velocities were calculated by dividing the Darcy velocities obtained from the groundwater simulation by the porosity. Subsequently, fluid trajectories were computed in an spatial resolution of 100 m by integrating the pore velocity flowfield in forward and backward direction. The resulting trajectories were processed in MATLAB to detect the location of termination and the corresponding travel time.

For the computation of total flux into Poyang Lake, a function was written in MATLAB to load VTK output files into the MATLAB environment. The total volume flux to Poyang Lake can be mathematically described as a surface integral over the Darcy velocity field:

$$Q = \iint_A \mathbf{v} \cdot d\mathbf{A} \tag{6.2}$$

where $\mathbf{v}$ is the velocity vector and $d\mathbf{A}$ is the vectorial surface element of the surface A. For our modelling domain consisting of triangular elements, the integral can be approximated by a sum over all elements:

$$Q = \sum_{j=1}^{n} \mathbf{v}_j \cdot \mathbf{A_j} \tag{6.3}$$

With j being one of the n elements which has a share A_j of the cross-sectional area defining the Poyang Lake. The velocity vector $\mathbf{v}_j$ describes the groundwater velocity in each element which belongs to the boundary of Poyang Lake. Darcy's Law can be applied to replace the velocity vector in normal direction:

$$Q \approx \sum_{j=1}^{n} -k_j \cdot I_j \cdot A_j = \sum_{j=1}^{n} -k_j \cdot \nabla H \cdot A_j \tag{6.4}$$

where k_j is the hydraulic conductivity and I_j the hydraulic gradient (in normal direction to the lake surface) in each element. The hydraulic gradient can be rewritten as the derivative of hydraulic head ∇H in normal direction to the surface A_j. In a linear triangular FEM element with given hydraulic heads at each of the three nodes, the hydraulic gradient vector is constant within the triangle and can be calculated by interpolation using barycentric coordinates. The hydraulic head H at an arbitrary point $\mathbf{r}$ within a triangle can be described in barycentric coordinates by:

$$H(\mathbf{r}) = \lambda_1 \cdot H_1 + \lambda_2 \cdot H_2 + \lambda_3 \cdot H_3 \tag{6.5}$$

where λ_1, λ_2 and λ_3 are the barycentric coordinates of the point $\mathbf{r}$ within each triangle and H_1, H_2 and H_3 the hydraulic head at each of the vertices. Using the relationship that $\lambda_3 = 1 - \lambda_1 - \lambda_2$, the barycentric coordinates can be calculated from the cartesian coordinates by:

$$\begin{pmatrix} \lambda_1 \\ \lambda_2 \end{pmatrix} = T^{-1} \begin{pmatrix} x - x_3 \\ y - y_3 \end{pmatrix} \tag{6.6}$$

With T^{-1} being the inverse of the transformation matrix

$$T = \begin{pmatrix} x_1 - x_3 & x_2 - x_3 \\ y_1 - y_3 & y_2 - y_3 \end{pmatrix} \tag{6.7}$$

The hydraulic gradient at each point $\mathbf{r}$ of the triangle can be calculated by calculating the gradient of Eq. 6.5:

$$\nabla H(r) = \nabla(\lambda_1 \cdot H_1) + \nabla(\lambda_2 \cdot H_2) + \nabla((1 - \lambda_1 - \lambda_2) \cdot H_2) \tag{6.8}$$

Considering that the gradient of Eq. 6.6 is:

$$\nabla \begin{pmatrix} \lambda_1 \\ \lambda_2 \end{pmatrix} = \nabla \left(T^{-1} \begin{pmatrix} x - x_3 \\ y - y_3 \end{pmatrix} \right) = T^{-1} \nabla \begin{pmatrix} x - x_3 \\ y - y_3 \end{pmatrix} = T^{-1} = \text{const within element} \tag{6.9}$$

Rules for products can be applied on Eq. 6.8 to derive the equation for the hydraulic gradient in each triangle:

$$\mathbf{I} = \begin{pmatrix} I_1 \\ I_2 \end{pmatrix} = \frac{1}{\det(T)} \begin{pmatrix} y_2 - y_3 & y_3 - y_1 & y_1 - y_2 \\ x_3 - x_2 & x_1 - x_3 & x_2 - x_1 \end{pmatrix} \cdot \begin{pmatrix} H_1 \\ H_2 \\ H_3 \end{pmatrix} \tag{6.10}$$

As shown by Eq. 6.10 the vector of the hydraulic gradient in a triangle is independent of the position within the triangle. The scalar component of the hydraulic gradient $\mathbf{I}_j$ perpendicular to the cross-sectoral area A_j of each element is represented by the dot product of the hydraulic gradient vector and the standardized normal of A_j:

$$I_j = \mathbf{I_j} \cdot \frac{\mathbf{A_j}}{|\mathbf{A_j}|} \tag{6.11}$$

Finally, Eq. 6.4, can be further simplified using the definition that the aquifer transmissivity of each element T_j is defined as the product of hydraulic conductivity and aquifer thickness and the fact that the cross-sectional area of each element is defined by its edge length L_j and the effective aquifer thickness d_j:

$$Q \approx \sum_{j=1}^{n} -k_j \cdot I_j \cdot L_j \cdot d_j = \sum_{j=1}^{n} -T_j \cdot I_j \cdot L_j \tag{6.12}$$

This conversion allows to directly benefit from the computation of the model parameter transmissivity during the model calibration procedure (see Sect. 6.2.3.2). Finally, referring to the conceptual model that the aquifer transmissivity is constant over the whole domain, the total flow of water between the wetland subsurface and the Poyang Lake can be approximated by:

$$Q \approx T \cdot \sum_{j=1}^{n} \cdot I_j \cdot L_j \tag{6.13}$$

The above-derived equations were implemented as MATLAB functions to compute the exchange flow for all model set-ups.

6.3 Results

6.3.1 Groundwater Regime

Figure 6.6 shows the resulting spatial distribution of hydraulic head and groundwater velocity fields for high and low water levels. In both cases, groundwater levels, which are equal to the hydraulic head in case of unconfined aquifers, are higher in the hilly areas bounding the western, eastern and southern borders of the Poyang Lake Core Region. Being a subdued replica of topography [39], groundwater levels decrease with high gradients towards the floodplain areas where gradients are more gentle and hydraulic heads rarely exceed 25 m above sea level.

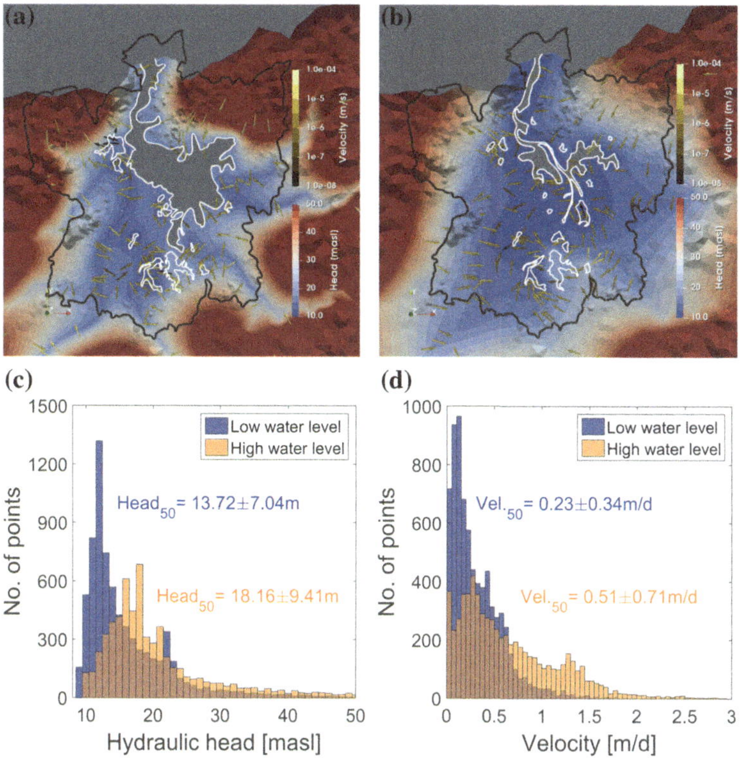

Fig. 6.6 Simulated groundwater regime in the Poyang Lake Core Region. The hydraulic head distribution is visualized as colored surface of the FEM mesh during high (Fig. 6.6a) and low (Fig. 6.6b) water levels in Poyang Lake. Groundwater velocities are visualized as glyph vectors colored by the magnitude of the velocity vector. The white polylines show the shoreline of Poyang Lake and the floodplain lakes. The two lower subplots show histograms of nodal hydraulic head (Fig. 6.6c) and Darcy velocity (Fig. 6.6d) distribution in the Poyang Lake Core Region for each investigation period. Median values of the distribution ± interquartile range are given within the plots

Comparing the different seasonal water levels of Poyang Lake, the estimated median hydraulic head is with 18.16 m during the high water season about 4.5 m higher than during the low water season. At high lake water levels, the frequency distribution of hydraulic heads (Fig. 6.6c) shows a larger proportion of hydraulic heads above 25 m, which represent heads outside the floodplain area rarely impacted by the dynamics of Poyang Lake. Hence, this effect was caused by higher recharge rates during the summer months.

The majority of groundwater velocities were in a range between 0.1 and 10 m/d, which are typical values for alluvial aquifer systems. Areas of higher velocities were associated with higher topographical slopes. Both, the median groundwater velocity of 0.51 m/d and the interquartile range of 0.71 m/d were significantly larger during the high water level season (Fig. 6.6d), although the difference of surface water levels between Kangshan and Hukou station is only 1/6 of the difference at low water

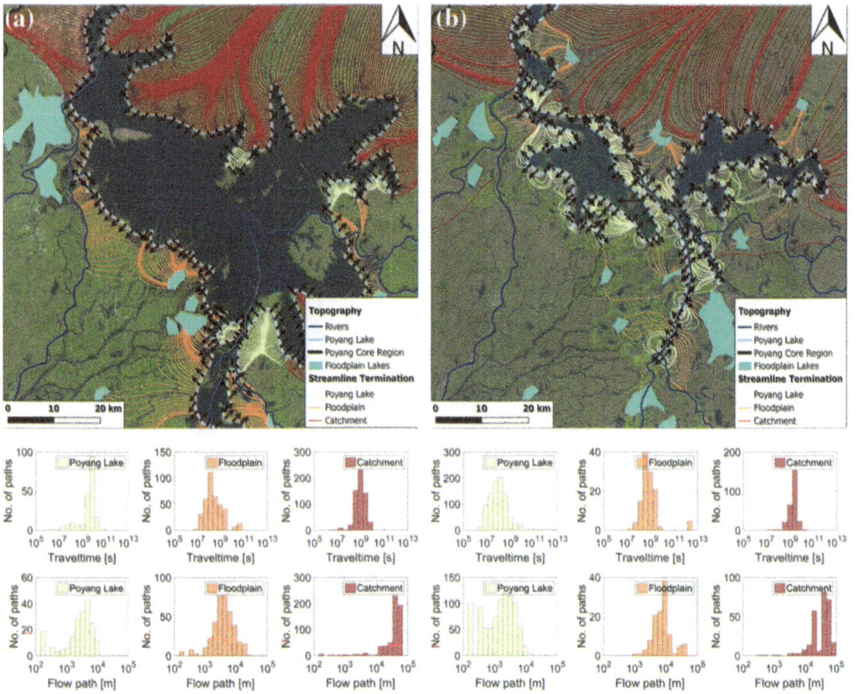

Fig. 6.7 Visualisation of streamline paths and corresponding traveltimes during high (Fig. 6.7a) and low (Fig. 6.7b) water levels in Poyang Lake. The upper plots visualize the streamlines between Poyang Lake and the corresponding location of termination. Different colors indicate whether these location is either Poyang Lake itself, a river or lake within the floodplain or the domain boundary of the Poyang Lake Core Region. The black arrows visualize the direction of flow velocity at the shoreline of Poyang Lake. The background image is taken by the Landsat-7 satellite on 23rd Sep 2000 (Fig. 6.7a) and 10th Dec 1999 (Fig. 6.7b). The lower-left and lower right plots consist of a set of histograms which visualize the traveltime and flow path length distribution for each location of streamline termination for both investigation periods

Table 6.2 Summary statistics of flow path lengths and subsurface traveltimes at low and high water levels in Poyang Lake using the median ± the interquartile range of each dataset

	High water level (09/2000)		Low water level (12/1999)	
Streamline destination	Length [km]	Traveltime [a]	Length [km]	Traveltime [a]
Poyang Lake	3.00 ± 3.01	149.5 ± 114.4	1.70 ± 2.70	2.98 ± 7.54
Floodplain lakes	4.1 ± 3.9	6.07 ± 21.7	8.00 ± 4.99	18.76 ± 38.8
Catchment	41.2 ± 18.2	31.53 ± 45.1	43.1 ± 33.7	69.1 ± 73.5

levels. One explanation is that areas in the wetlands of Poyang Lake with very low hydraulic gradients and velocities are part of the lake during high water level and thus outside of the model domain. In addition, recharge values are larger during the wet period causing steeper hydraulic gradients towards the Poyang Lake resulting in higher groundwater velocities.

6.3.2 Groundwater-Surface Water Interaction

The interaction between Poyang Lake and the wetland area was investigated by computing and analysing the flowpath and corresponding travel times during high (Fig. 6.7a) and low water levels (Fig. 6.7b). The resulting flow paths were subdivided by their respective end-members into three groups:

1. "Catchment": Water is transported along these flow paths from areas outside the Poyang Lake Core Region (which means from the wider catchment area) towards the Poyang Lake.
2. "Floodplain": This type of flow paths visualizes the movement of water between the waterside of Poyang Lake and both, floodplain lakes and rivers in the Poyang Lake Core Region.
3. "Poyang Lake": Water infiltrates from Poyang lake in the subsurface and reenters the lake at another position.

Independently of the flow conditions, the simulation showed that the northern part of the shoreline received its water from hilly areas farther away from the floodplain. However, during low water levels, some of the water reaches the main waterbody from one of the northern floodplain lakes which appeared due to the shrinkage of the lake surface. The location of flow path origins leads to long average flow paths length of 43.1 $\pm$ 33.7 km and 41.2 $\pm$ 18.2 km in dependence of the lake water level (Table 6.2). The relatively small range of seasonal variation was related to the regional groundwater regime in the headwaters which was only marginally affected by the actual lake water level. The about 45% lower average travel times along the flow paths during high water level were caused by the comparatively higher groundwater velocities.

Although the vast majority of subsurface flow paths interacting with the southern, western and eastern lakeshore of Poyang Lake belonged to the types "Floodplain" and "Catchment", properties largely varied in dependence of the water level in Poyang Lake. During high water level, most flow paths in this region described a hydrological connection between Poyang Lake and the rivers and floodplain lakes (orange lines in Fig. 6.7) with an average flow path length of 4.1 $\pm$ 3.9 km. Flow paths of the category "Poyang Lake" were restricted to relatively small areas typically having a winding shoreline and showed an average length of 3.00 $\pm$ 3.01 km.

Under low water level conditions, the interaction between Poyang Lake and the subsurface was, except of the northern shoreline, dominated by hyporheic flow paths of infiltrating lake water with a short average length of 1.70 $\pm$ 2.70 km and an average travel time of less than 3 years. Additionally, a proportion of water was transported from the floodplain lakes towards the lake. Corresponding flow path length and travel times were larger than during high water levels due to the smaller size of Poyang Lake, which lead to longer distances between the main water body and the floodplain lakes.

The seasonally varying patterns of subsurface flow paths between the different hydrological compartments of the investigated system impacted the total exchange rate of water between the Poyang Lake and the shallow groundwater (Fig. 6.8a). For

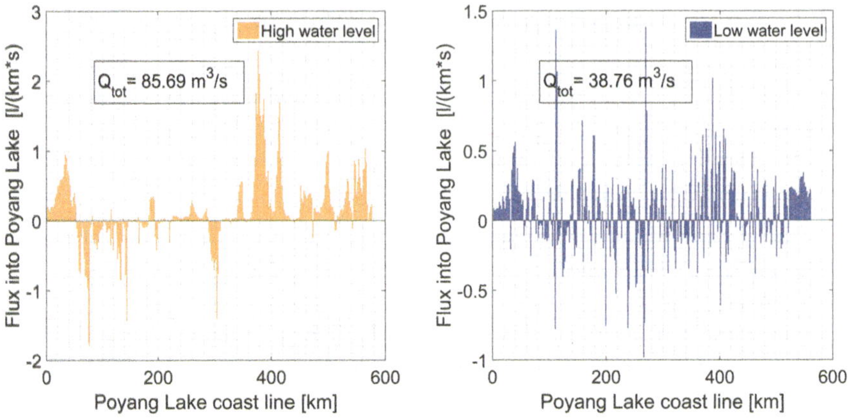

Fig. 6.8 Exchange flux between Poyang Lake and the ambient groundwater system during high (left plot) and low (right plot) water levels. The exchange flux is given in liter per second and km shoreline over the length of the entire coastline of Poyang Lake. The coastline lengths of 0 was defined for the left bank of the Poyang Lake estuary and summation of length was conducted in counter-clockwise direction following the polyline defining the shoreline of the lake. The total exchange flow over the entire shoreline is given within each plot

high water levels in Poyang Lake, the total amount of direct groundwater discharge into the lake summed up to 85.69 m^3/s, which was more than double the discharge of 38.76 m^3/s during low water levels. The key reason for this is the large number of hyporheic flow path during the latter scenario, which decreased the net inflow to Poyang Lake significantly. This is visualized by the large range of shoreline kilometers with alternating fluxes in the right plot of Fig. 6.8b. Although Poyang Lake lost water to the southern wetlands during high water level as visualized in the left plot of Fig. 6.8 as the area of negative fluxes between coastline kilometre 50 and kilometre 300, this loss was more than overcompensated by the inflow of groundwater over large sections along the northern part of the shoreline of Poyang Lake.

6.4 Discussion

6.4.1 Dependence of Model Results on Mesh Resolution

A crucial aspect of any numerical modelling set-up is the definition of an adequate mesh resolution. Ideally, the obtained simulation results do not change if the mesh resolution is further refined. However, this assumption may not be valid if boundary conditions are defined along complex geometries such as in the presented groundwater flow model of the Poyang Lake Core Region. Subsequently, the groundwater flow model was run again using the finer FEM mesh resolution (see Sect. 6.2.3.1).

Table 6.3 Comparison of descriptive statistical measures of the FEM groundwater flow simulation using either the fine or the coarse mesh resolution. The robust statistic measures median ± interquartile range are used to describe each parameter except of total discharge to Poyang Lake

	Fine mesh		Coarse mesh	
Parameter	Low water (12/1999)	High water (09/2000)	Low water (12/1999)	High water (09/2000)
Groundwater velocity (m/d)	0.21 ± 0.34	0.47 ± 0.71	0.23 ± 0.34	0.51 ± 0.71
Hydraulic head (m)	13.32 ± 5.65	17.77 ± 7.89	13.72 ± 7.04	18.16 ± 9.41
Flow path length (km)	0.70 ± 1.90	9.34 ± 38.09	3.11 ± 8.19	13.68 ± 37.27
Traveltime (a)	0.46 ± 3.15	32.51 ± 90.91	8.03 ± 33.95	36.35 ± 174.6
GW discharge (m^3/s)	44.36	61.31	38.76	85.69

The metrics average groundwater velocity, hydraulic head, flow path length, travel time and groundwater discharge were calculated for both, low and high water level of Poyang Lake and compared with the previous results obtained with the coarse FEM mesh (Table 6.3). A distinction of different types of flow paths was not included in the calculation of the statistics.

Regarding groundwater velocity and hydraulic head, average values showed moderate differences between 2.1 and 8.7% for each scenario. In addition, both parameters showed a very similar shape of their frequency distribution for different mesh sizes. In contrast, large differences in flow path length and travel time were observed between both FEM meshes during the low water season. Average flow path lengths were about 0.7 ± 1.90 km, which was only about 22.5% of the length simulated using the coarse mesh. In addition, corresponding travel times even decreased by more than one magnitude. One key reason is that the length of obtained hyporheic flow paths was often less than 2 km (see Fig. 6.7) which is smaller than the actual mesh resolution of the coarse mesh. Hence, refining the mesh helps to improve the representation of local flow systems in the transition area between the open water and the wetland area.

6.4.2 *Modell Sensitivity on Hydraulic Input Parameters*

Studies of Lai et al. [40]and Yao et al. [41] demonstrated that significant bed erosion is taking place in the Poyang Lake which has caused a decrease in water levels of between 1.2 and 2 m at the outlet of Poyang Lake during low-water levels within more than 20 years of observation. Until now, the lowest recorded water level at Hukou station was 5.9 m [42]. Hence, the impact of water level changes at the outflow on the groundwater discharge to Poyang Lake was assessed by varying the downstream hydraulic head boundary between 7 m and 12 m. This means a difference of −2 to +3 m from the actual measured water level at Hukou station of 9 m (above sea level) in December 1999 used in the reference scenario for low water levels in Poyang Lake (see Sect. 6.2.3.2). The upper limit was chosen to analyse the impact of damming Poyang Lake by the proposed construction of a dam in its northern sector.

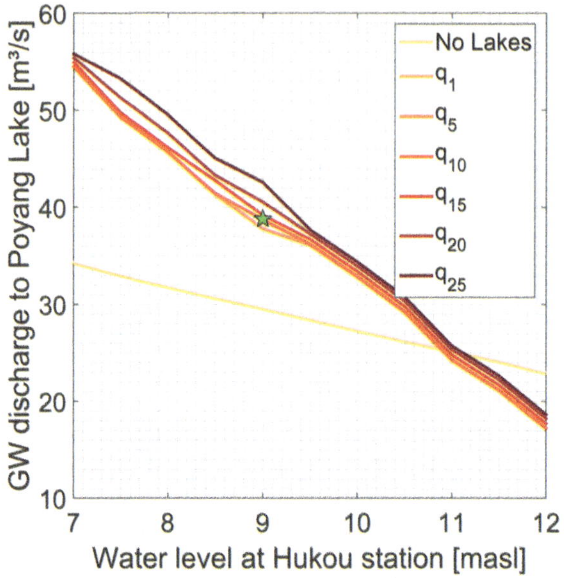

Fig. 6.9 Groundwater discharge to Poyang Lake during low-water season in dependence on the water level at Hukou station. The bundle of curves displays the different quantiles used to calculate the hydraulic head of each floodplain lake. For instance, q_{10}, the reference scenario, represents the 10th percentile of elevation values obtained for the area covered by each floodplain lake (for details see Sect. 6.2.3.2). The label "No Lakes" means that the floodplain lakes were not used as boundary conditions for the groundwater flow simulations. The asterisk indicates the reference set-up which was used for the standard groundwater flow field simulations presented in Sect. 6.3

Considering the large wetland system with its numerous ponds and floodplain lakes, the way in which it is represented as boundary conditions in the model set-up may influence the amount of computed groundwater discharge to Poyang Lake. Subsequently, the groundwater flow field at low water levels was simulated again with OGS using different hydraulic heads for each floodplain lake. These differences in hydraulic heads were implemented in the model set-up by defining the Dirichlet boundary condition describing each floodplain lake using different quantiles of the elevation data distribution covering its surface area. Additionally, a simulation (called "No Lakes" in Fig. 6.9) was conducted excluding all floodplain lakes from being model boundary conditions.

Simulation results (Fig. 6.9) revealed that the groundwater discharge to Poyang Lake significantly depends on the selected water level at Hukou station. Lower water levels result in higher groundwater discharge to Poyang Lake of up to 57 m^3/s whereas groundwater discharge decreased towards a minimum of 17 m^3/s for very high water levels at the estuary of Poyang Lake. Independently of the representation of the floodplain lakes in the set-up, the obtained relationship between modelled groundwater discharge and water level at the outflow turned out to be approximately linear. One reason is that our simple approach of representing the water level dynamics in

Poyang Lake silently assumed a linear decrease from Hukou to Kangshan, which is in contrast to e.g. the results of hydrodynamic model simulations of Poyang Lake [41].

The amount of groundwater discharge into Poyang Lake from the wetland changed significantly by including the floodplain lakes as boundary conditions of the groundwater model. For low outflow water levels (Fig. 6.9), groundwater discharge rates were up to 40% higher presumably due to higher hydraulic gradients towards the Poyang Lake. However, in the absence of floodplain lakes, groundwater discharge to Poyang Lake decreases more gently with increasing outflow water level. According to the model, if outflow water levels of Poyang Lake exceeded 11 m, the existence of floodplain lakes influences the amount of groundwater discharge into Poyang Lake negatively.

Our analysis further revealed that the impact of the chosen DEM data quantile to represent the hydraulic head of each floodplain lake is less important for the variations of groundwater discharge than the question whether to include the floodplain lakes in the simulation at all. For low outflow water levels, higher quantiles, meaning higher water levels in the floodplain lakes, increase the groundwater discharge but this effect becomes less significant for outflow water levels of more than 9 m. One possible explanation is that the DEM values which were obtained during low water levels in Poyang Lake do not show large variations in the area defining the surface of the floodplain lakes. However, the question remains how the amount of floodplain lakes included in the model will affect the groundwater lake interaction, which should be further investigated in future studies.

6.4.3 Impact of Groundwater Discharge on the Water Balance of Poyang Lake

Our 2D numerical investigation of the Poyang Lake Core Region showed that the inflow of groundwater into Poyang Lake depends on the water level of Poyang Lake and the representation of boundary conditions such as floodplain lakes and rivers in the model. Although discharge values tend to be higher for high water levels due to the different flow field characteristics, discharge may reach values of approximately 57.7 m^3/s in Poyang Lake during the dry season if water levels at the estuary are very low (Fig. 6.9). Projected to one year, groundwater inflow to Poyang Lake is less than 2.5% of the total long term average annual surface water inflow to Poyang Lake of 1.15 10^{11} m^3/a [42]. However, during drought years, surface water inflow to Poyang Lake may decrease tremendously down to values of 22 10^8 m^3 per month measured in October 2009 [42], which means that during those extreme conditions direct groundwater inflow could sum up to 7.6% of the surface water inflow. Beside of its relevance for the seasonal groundwater balance, our numerical investigation of the system showed that a considerable number of hyporheic flow paths exist which connect different points of the Poyang Lake shoreline during low water levels in the lake. The resulting changes in water quality that occur if surface water flows along hyporheic flow paths in the wetlands subsurface with a different chemical milieu may have an significant impact on surface water quality and aquatic habitat zonation (e.g. [43, 44]).

One should note that the applied groundwater flow model assumed a two-dimensional groundwater flow field with a homogeneous and isotropic hydrogeological setting. On the other hand, flow paths beneath the lakebed of Poyang Lake have been excluded by the selection of the domain boundaries. Hence, some features such as Tothian flow systems [45] and the upwelling of deep groundwater from the mountainous region in the floodplain, e.g. along one of the 13 main geological faults in the Poyang Lake Core Region [8] could not be disclosed by the applied model but should be investigated in depth in future studies based on the results of this investigation.

References

1. Duan, W., et al. 2016. Water quality assessment and pollution source identification of the eastern Poyang Lake Basin using multivariate statistical methods. *Sustainability* 8 (2), 133.
2. He, M., Z. Wang, and H. Tang. 1998. The chemical, toxicological and ecological studies in assessing the heavy metal pollution in Le An River, China. *Water Research* 32 (2): 510–518. https://doi.org/10.1016/S0043-1354(97)00229-7.
3. Xu, B., and G. Wang. 2016. Surface water and groundwater contaminations and the resultant hydrochemical evolution in the Yongxiu area, west of Poyang Lake, China. *Environmental Earth Sciences* 75 (3), 184. ISSN: 1866-6280. https://doi.org/10.1007/s12665-015-4778-8.
4. Zeng, Z.H., and C.M. Duo. 1997. Regionalization of Poyang Lake area Jiangxi for a rational exploitation and utilization of groundwater resources (in Chinese). *Geological Science and Technology of Jiangxi* 24 (2): 188–191.
5. Soldatova, E., et al. 2017. Sources and behaviour of nitrogen compounds in the shallow groundwater of agricultural areas (Poyang Lake basin, China). *Journal of Contaminant Hydrology* 202: 59–69. https://doi.org/10.1016/j.jconhyd.2017.05.002.
6. Liao, F. et al. 2017. Distributions, sources, and species of heavy metals/trace elements in shallow groundwater around the Poyang Lake, East China. *Exposure and Health*. https://doi.org/10.1007/s12403-017-0256-8.
7. Hu, C., et al. 2013. Seasonal variation of oxygen-18 in precipitation and surface water of the Poyang Lake Basin, China. *Isotopes in Environmental and Health Studies* 49 (2): 188–196. https://doi.org/10.1080/10256016.2013.740480.
8. Zhan, L., et al. 2016. Isotopic signatures of precipitation, surface water, and groundwater interactions, Poyang Lake Basin, China. *Environmental Earth Sciences* 75 (19). https://doi.org/10.1007/s12665-016 6081-8.
9. Yao, X., S. Wang, Z. Ni, and L. Jiao. 2015. The response of water quality variation in Poyang Lake (Jiangxi, People's Republic of China) to hydrological changes using historical data and DOM fluorescence. *Environmental Science and Pollution Research* 22 (4): 3032–3042. https://doi.org/10.1007/s11356-014-3579-z.
10. Ye, X., Q. Zhang, L. Bai, and Q. Hu. 2011. A modeling study of catchment discharge to Poyang Lake under future climate in China. *Quaternary International* 244 (2): 221–229. https://doi.org/10.1016/j.quaint.2010.07.004.
11. Soldatova, E., N. Guseva, G. Wang. 2014. Characteristic features of groundwater pollution in the Poyang Lake catchment. In *fIOPg Conference Series: Earth and Environmental Science* vol. 21, 12023. https://doi.org/10.1088/1755-1315/21/1/012023.

12. Deng, T., et al. 2014. Seasonal variations of arsenic at the sediment water interface of Poyang Lake, China. *Applied Geochemistry* 47: 170–176. https://doi.org/10.1016/j.apgeochem.2014.06.002.
13. Li, Y., J. Yao, and L. Zhang. 2016. Investigation into mixing in the shallow floodplain Poyang Lake (China) using hydrological, thermal and isotopic evidence. *Water Science and Technology* 74 (11): 2582–2598. https://doi.org/10.2166/wst.2016.444.
14. Furman, A. 2008. Modeling Coupled SurfacefntextendashgSubsurface Flow Processes: A Review. *Vadose Zone Journal* 7 (2): 741. https://doi.org/10.2136/vzj2007.0065.
15. Kolditz, O., et al. 2012. OpenGeoSys: an open-source initiative for numerical simulation of thermo-hydro-mechanical/chemical (THM/C) processes in porous media. *Environmental Earth Sciences* 67 (2): 589–599. https://doi.org/10.1007/s12665-012-1546-x.
16. Brunner, P., and C.T. Simmons. 2011. fHydroGeoSphereg: A fully integrated, physically based hydrological model. *Ground Water* 50 (2): 170–176. https://doi.org/10.1111/j.1745-6584.2011.00882.x.
17. McDonald, M.G., and A.W. Harbaugh. 1988. *A modular three-dimensional finitedifference ground-water flow model*, vol. 6. VA: US Geological Survey Reston.
18. Delfs, J.-O., F. Blumensaat, W. Wang, P. Krebs, and O. Kolditz. 2011. Coupling hydrogeological with surface runoff model in a Poltva case study in Western Ukraine. *Environmental Earth Sciences* 65 5, 1439–1457. https://doi.org/10.1007/s12665-011-1285-4.
19. E. Nixdorf and N. Trauth. 2018. Evaluating the reliability of time series analysis to estimate variable riparian travel times by numerical groundwater modelling. In *Hydrological Processes* Dec 2017, 408–420. ISSN: 08856087. https://doi.org/10.1002/hyp.11428.
20. Lange, W.J.D., et al. 2014. An operational, multi-scale, multi-model system for consensus-based, integrated water management and policy analysis. *Environmental Modelling & Software* 59: 98–108. https://doi.org/10.1016/j.envsoft.2014.05.009.
21. Barthel, R., et al. 2012. Integrated modeling of global change impacts on agriculture and groundwater resources. *Water Resources Management* 26 (7): 1929–1951. https://doi.org/10.1007/s11269-012-0001-9.
22. Tang, X., et al. 2016. Changing land use and its impact on the habitat suitability for wintering Anseriformes in Chinafntextquotesinglegs Poyang Lake region. *Science of the Total Environment* 557–558: 296–306. https://doi.org/10.1016/j.scitotenv.2016.03.108.
23. Li, Y., X. Zhang, G. Zhao, J. Yao, and Q. Zhang. 2015. Dynamics of groundwater level and laterally hydraulic connection between lake and groundwater in Poyang Lake area (in Chinese). *Resources and Environmental in the Yangtze Basin* 25 (12): 1894–1902.
24. Zeng, Z. 1999. The distribution feature and development assessment of SiO2 in groundwater in the area of Poyang Lake, Jiangxi Province (in Chinese). *Geology and Prospecting* 35 (2): 37–40.
25. China Geological Survey. 2015. Hydrogeological Map of Jiangxi Province. http://www.ngac.org.cn/Special/2015earthday/sgh.htm. Accessed 13 Mar 2018.
26. Li, Y., Q. Zhang, and X. Li. 2013. Multi-objectives model calibration for distributed hydrological model in the Poyang Lake watershed. *Resour Environ Yangtze Basin* 22 (5): 565–572.
27. Lehner, B., K. Verdin, and A. Jarvis. 2008. New global hydrography derived from spaceborne elevation data. *Eos, Transactions American Geophysical Union* 89 (10), 93. https://doi.org/10.1029/2008eo100001.
28. Hui, F., B. Xu, H. Huang, Q. Yu, and P. Gong. 2008. Modelling spatialtemporal change of Poyang Lake usingmultitemporal Landsat imagery. *International Journal of Remote Sensing* 29 (20): 5767–5784.
29. McFeeters, S.K. 1996. The use of the Normalized Difference Water Index (NDWI) in the delineation of open water features. *International Journal of Remote Sensing* 17 (7): 1425–1432.
30. Mishra, K., and P.R.C. Prasad. 2015. Automatic extraction of water bodies from landsat imagery using perceptron model. *Journal of Computational Environmental Sciences* 2015: 1–9. https://doi.org/10.1155/2015/903465.

31. Geuzaine, C., and J.-F. Remacle. 2009. Gmsh: A 3-D finite element mesh generator with built-in pre- and post-processing facilities. *International Journal for Numerical Methods in Engineering* 79 11, 1309–1331. ISSN: 1097-0207. https://doi.org/10.1002/nme.2579.
32. Winter, T.C., J.W. Harvey, O.L. Franke, and W.M. Alley. Ground water and surface water, A single resource. *U.S. Geological Survey Circular* 1139 (1998), 79.
33. Li, Y., Q. Zhang, J. Yao, A.D. Werner, and X. Li. 2014. Hydrodynamic and hydrological modeling of the Poyang Lake catchment system in China. *Journal of Hydrologic Engineering* 19 (3): 607–616. https://doi.org/10.1061/(asce)he.1943-5584.0000835.
34. Huffman, G.J., et al. 2007. The TRMM multisatellite precipitation analysis (TMPA): Quasi-global, multiyear, combined-sensor precipitation estimates at fine scales. *Journal of Hydrometeorology* 8 (1): 38–55.
35. Zhang, K., J.S. Kimball, R.R. Nemani, and S.W. Running. 2010. A continuous satellite-derived global record of land surface evapotranspiration from 1983 to 2006. *Water Resources Research* 46 (9). https://doi.org/10.1029/2009wr008800.
36. Heath, R.C. 1983. Basic ground-water hydrology. *US Geological Survey Water-Supply Paper* 2220: 84.
37. Henderson, A., J. Ahrens, and C. Law. 2004. The Paraview Guide. Clifton Park, NY. https://www.researchgate.net/profile/Charles_Law/publication/51992227_The_Paraview_Guide/links/551c042e0cf2fe6cbf7621de/The-Paraview-Guide.pdf.
38. Prince, P.J., and J. R. Dormand. 1981. High order embedded Runge-Kutta formulae. *Journal of Computational and Applied Mathematics*, 7 (1), 67–75. ISSN: 0377-0427. https://doi.org/10.1016/0771-050X(81)90010-3. http://www.sciencedirect.com/science/article/pii/0771050X81900103.
39. Haitjema, H.M., and S. Mitchell-Bruker. 2005. Are water tables a subdued replica of the topography? *GroundWater* 43 (6), 781–786. https://doi.org/10.1111/j.1745-6584.2005.00090.x.
40. Lai, X., et al. 2014. Sand mining and increasing Poyang Lake's discharge ability: A reassessment of causes for lake decline in China. *Journal of Hydrology* 519: 1698–1706. https://doi.org/10.1016/j.jhydrol.2014.09.058
41. Yao, J., Q. Zhang, X. Ye, D. Zhang, and P. Bai. 2018. Quantifying the impact of bathymetric changes on the hydrological regimes in a large floodplain lake: Poyang Lake. *Journal of Hydrology* 561: 711–723. https://doi.org/10.1016/j.jhydrol.2018.04.035.
42. Zhang, Z., et al. 2015. Examining the influence of river lake interaction on the drought and water resources in the Poyang Lake basin. *Journal of Hydrology* 522: 510–521. https://doi.org/10.1016/j.jhydrol.2015.01.008.
43. Gomez-Velez, J.D., J. W. Harvey, M. B. Cardenas, and B. Kiel. 2015. Denitrification in the Mississippi River network controlled by flow through river bedforms. *Nature Geoscience* 8: 941. https://doi.org/10.1038/ngeo2567//10.0.4.14/ngeo2567articles/ngeo2567
44. Dwivedi, D., I.C. Steefel, B. Arora, and G. Bisht. 2017. Impact of intrameander hyporheic flow on nitrogen cycling. *Procedia Earth and Planetary Science* 17: 404–407. https://doi.org/10.1016/j.proeps.2016.12.102.
45. Toth, J. 1963. A theoretical analysis of groundwater flow in small drainage basins. *Journal of Geophysical Research* 68 (16): 4795–4812. https://doi.org/10.1029/jz068i016p04795.

Part III
Water Quality and Pollution

Chapter 7
Eutrophication and Water Quality Assessment in the Poyang Lake Wetlands

Na Fang, Qinghui You, Wenjing Yang, Xu Lu, Yi Zhou and Caiying Ni

7.1 Introduction

Wetland is a transition zone between terrestrial and aquatic ecosystems, and plays an important role in climate regulation, flood regulation and biodiversity conservation [1]. Currently, the negative impacts of human activities on wetland are more and more increasing. For example, pollutants from agricultural, industrial and domestic sewage largely degrade wetland water quality and cause eutrophication of the water bodies [2, 3]. Assessing eutrophication state of wetland water bodies facilitates more accurate evaluation of wetland ecological health and further provides the scientific basis for wetland protection and restoration.

Chlorophyll *a* (Chl *a*) concentration in water is an important indicator of phytoplankton biomass [4], and is often used as a key criterion for eutrophication assessment. However, due to the complexity of aquatic ecosystems, multiple criteria in addition to Chl *a* concentration are often used for eutrophication assessment [5]. Previous studies demonstrate that there are close relationships between Chl *a* and other water quality parameters, such as nitrogen and phosphorus concentrations, temperature and pH. Studying the relationships between Chl *a* and other water quality parameters can provide a warning for the occurrence of algae blooms.

As the largest freshwater lake in China, Poyang Lake is a typical flood plain lake with great seasonal water level fluctuation. During low water level season, large beaches appear at Poyang Lake and form a large area of wetland [6, 7]. This vast wetland provides important habitats for many wintering migratory birds. In recent years, human disturbances on Poyang Lake have increased, leading to degraded water quality and more severe eutrophication [8]. So far, eutrophication assessment has con-

N. Fang · Q. You · W. Yang (✉) · X. Lu · Y. Zhou · C. Ni
Jiangxi Normal University, Nanchang, China
e-mail: wenjing0282@163.com

C. Ni
e-mail: ncy1919@126.com

T. Yue et al. (eds.), *Chinese Water Systems*, Terrestrial Environmental Sciences,
https://doi.org/10.1007/978-3-319-97725-6_7

ducted on the main water body of Poyang Lake [9–12]. There are few studies focusing on eutrophication assessment of the wetland. In this study, Chl *a* concentration and Carlson's Trophic State Index (TSI) were used to evaluate the eutrophication status of Poyang Lake wetland, which will provide a scientific basis for wetland protection and management.

7.2 Methods

7.2.1 Data Acquisition and Processing

In this study, 30 typical sites covering the entire lake area of Poyang Lake wetland were sampled from September to October 2016 (Fig. 7.1). Water quality parameters including temperature (T), pH and Chl *a* were measured in the open water of each sampling site by using a portable multi-parameter water quality meter (YSI6600). Water transparency (SD) was measured using Secchi water clarity tube. Three 250 mL water samples were collected at each sampling site, and stored in a portable refrigerator at < 4 °C. The chemical oxygen demand (COD), total phosphorus (TP) and total nitrogen (TN) were measured following the protocols recommended in the Standard Methods for the Analysis of Water and Wastewater (Ministry of Environmental Protection of the People's Republic of China, 2012).

7.2.2 Eutrophication Assessment

Carlson Nutritional Status Index (TSI) was calculated to evaluate the eutrophication status of Poyang Lake wetland based on TP, SD and Chl *a* using the following formula [13]:

$\mathrm{TSI\ (SD)} = 10\ (6\text{-}\ln \mathrm{SD}/\ln 2)$

$\mathrm{TSI\ (Chl}\ a) = 10\ (6\text{-}(2.04\text{–}0.68 \ln \mathrm{Chl}\ a)/\ln 2)$

$\mathrm{TSI\ (TP)} = 10\ (6\text{-}\ln (48/\mathrm{TP})/\ln 2)$

where SD is measured by meter (Secchi depth); the unit for TP and Chl *a* is $\mathrm{mg/m^3}$; TSI (SD), TSI (Chl *a*) and TSI (TP) are the indices for eutrophication calculated based on SD, Chl *a* and TP, respectively. The final TSI value for each sampling site was represented by the mean of TSI (SD), TSI (Chl *a*) and TSI (TP). TSI values varied from 0 to 100. A value < 40 indicates oligotrophy, ≥ 40 and < 50 mesotrophy, and ≥ 50 eutrophy [14].

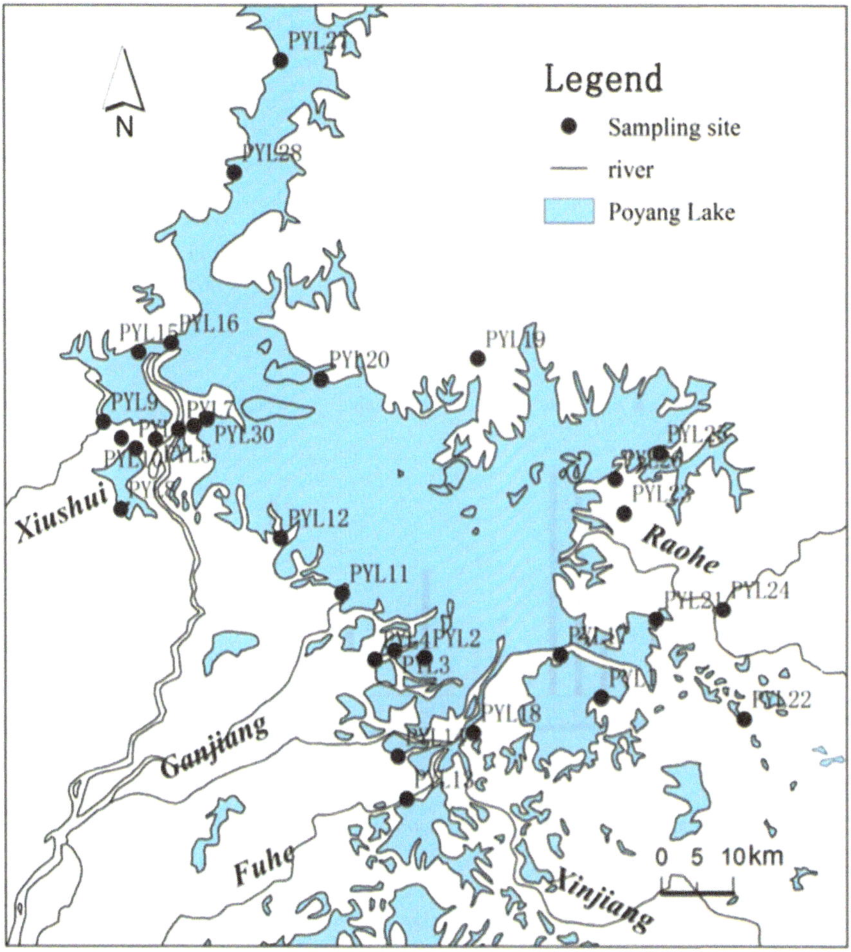

Fig. 7.1 Distribution of thirty sampling sites in Poyang Lake wetland

7.3 Results

7.3.1 Nutrient and Chl a Concentrations

TN concentration in the water bodies of Poyang Lake wetland varied from 0.85 to 5.89 mg/L with an average of 2.17 mg/L. TP concentration ranged between 0.04 and 0.23 mg/L with a mean of 0.15 mg/L (Table 7.1). Published studies indicated that if TN and TP concentrations in lakes were higher than 0.20 and 0.02 mg/L, respectively, algae blooms might occur [15]. Our results showed that TN and TP concentrations in Poyang Lake exceeded the thresholds for algae blooms.

Table 7.1 Minimum, maximum, mean and standard deviation of water quality parameters for Poyang Lake wetland

Parameter	Maximum	Minimum	Mean	Standard deviation
TN(mg/L)	5.89	0.85	2.17	0.91
TP(mg/L)	0.23	0.04	0.15	0.06
SD(m)	0.45	0.02	0.16	0.12
T(℃)	28.85	14.64	19.84	3.09
pH	8.04	7.15	7.68	0.24
Chl *a*(mg/m^3)	28.8	2.20	10.94	7.67
COD(mg/L)	5.46	1.82	3.62	1.01

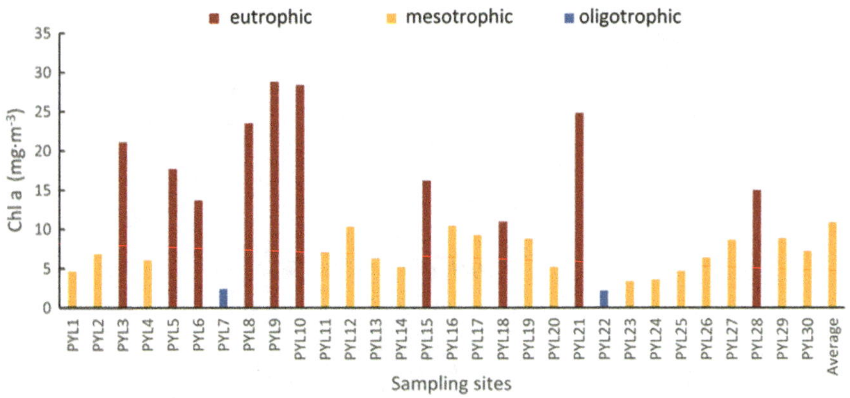

Fig. 7.2 Eutrophication assessment for thirty sampling sites in Poyang Lake wetland based on Chl *a* concentration

Chl *a* concentration in the water bodies of Poyang Lake wetland varied from 2.20 to 28.80 mg/m^3 with an average of 10.94 mg/m^3. According to the criterion for assessing the lake's nutritional status proposed by the International Organization for Economic Cooperation and Development (OECD), Chl *a* concentration > 11 mg/m^3 indicates eutrophy, > 3 mg/m^3 and ≤ 11 mg/m^3 mesotrophy, and < 3 mg/m^3 oligotrophy. Ten sampling sites were evaluated as eutrophic, 18 as mesotrophic and two as oligotrophic based on this criterion (Fig. 7.2).

7.3.2 Eutrophication Assessment Using Carlson's Trophic State Index (TSI)

TSI values for 30 sampling sites in Poyang Lake wetland ranged from 58.53 to 84.24 with an average of 72.04 (Fig. 7.3). TSI values of all sampling sites were higher than 50, indicating that water bodies of Poyang Lake wetland were eutrophic.

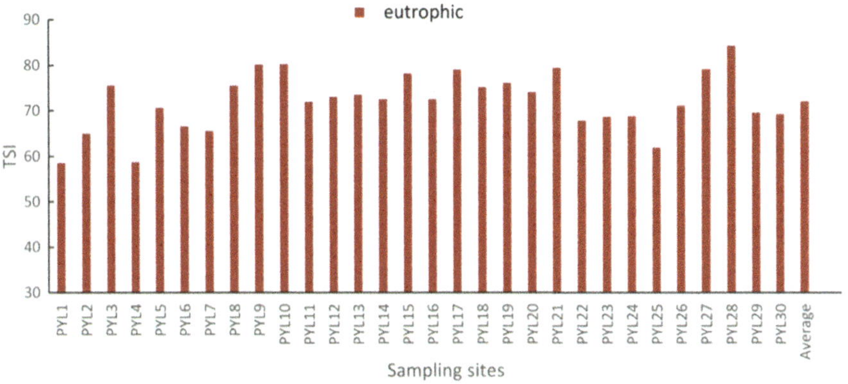

Fig. 7.3 Eutrophication assessment for thirty sampling sites in Poyang Lake wetland based on Carlson's Trophic State Index (TSI)

7.3.3 *Correlations of Chl a Concentration and Other Water Quality Parameters*

Pearson correlation analysis was conducted between Chl *a* concentration and other water quality parameters. All water quality parameters were log-transformed to obtain normal distributions. Significant positive correlation was found between Chl *a*

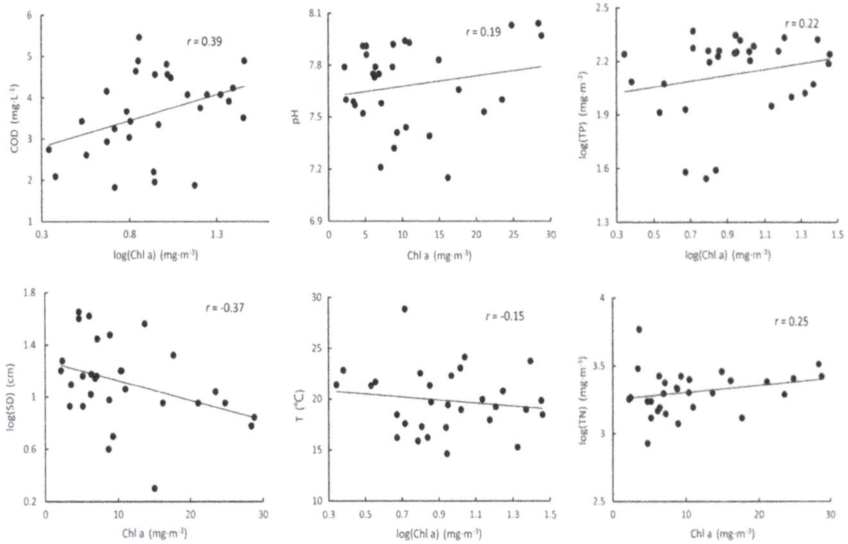

Fig. 7.4 Relationships betwween Chl *a* and other six water quality parameters for Poyang Lake wetland

and COD, with a correlation coefficient of 0.39 ($p < 0.05$). Chl *a* was also found to be significantly correlated with SD ($r = -0.37$, $p < 0.05$). A negative correlation was obtained between Chl *a* and temperature, though the correlation was not significant ($r = 0.15$, $p > 0.05$) (Fig. 7.4).

7.4 Discussion

In this study, both single-criterion and multiple-criteria methods were used to evaluate the trophic state of the water bodies of Poyang Lake wetland. The evaluation results of the two methods are inconsistent for some sampling sites. According to TSI evaluation results, all sampling sites were in the state of eutrophication, whereas the evaluation based on Chl *a* concentration classified ten sites in eutrophic condition, 18 and two in mesotrophic and oligotrophic conditions, respectively. The growth of phytoplankton and Chl *a* concentration are affected by many water environmental factors such as nutrients, temperature, transparency and pH. The TSI method that uses multiple criteria for eutrophication assessment is thus considered superior than the method only taking account of Chl *a* concentration.

This study showed that nitrogen and phosphorus concentrations in Poyang Lake wetland were higher than the thresholds for algae blooms, whereas Chl *a* content was relatively low. The correlations between TN, TP and Chl *a* were not significant, indicating that nutrients might not play a key role for algae growth in Poyang Lake. Two reasons may explain the relatively low Chl *a* concentration in the water bodies of Poyang Lake wetland [16–18]. First, Poyang Lake is a riverine lake with high flow rate, while the growth of phytoplankton requires relatively stable water environment. Second, the turbidity of water in Poyang Lake is high due to the large amount of suspended solids in water caused by the prevalent sand mining activities. Poor lighting condition in the water limits the growth of phytoplankton.

Our results revealed a positive relationship between Chl *a* concentration and COD. Phytoplankton photosynthesis produces a large amount of organic matter, and therefore increases COD in the water [19]. We found that the correlations between Chl *a* concentration and temperature as well as pH were not significant, probably because the variations of temperature and pH in all sampling sites were not great enough to have large impacts on the growth of phytoplankton.

Single-criterion method based on Chl *a* concentration showed that high levels of eutrophication were found at the sampling sites PYL9 and PYL10, followed by PYL21, PYL8 and PYL3, with Chl *a* concentration > 20 mg/m^3. These sampling sites are characterized by low water velocity, well developed aquaculture and increasing nutrient concentrations in water due to the discharge from farmland, villages and sewers. All these factors promote the growth of phytoplankton, and consequently lead to high Chl *a* concentration in the water. In contrast, two sampling sites PYL22 and PYL7 evaluated as oligothrophic were located near river mouths of Ganjiang River, where water flows fast and is highly turbid—an unfavorable environment for the growth of phytoplankton. TSI evaluation results showed that the highest eutrophication appeared at PYL28, PYL10, PYL9 and PYL21, which was consistent with the evaluation results based on Chl *a* concentration. PYL28 is close to the Yangtze River estuary and is surrounded by a large industrial area. In addition, sand mining activity is prevalent in this area and increases water turbidity. The sampling sites with the lowest eutrophication levels were PYL1, PYL4 and PYL25. These sites are located in national and provincial nature reserves, where human activities have been strictly controlled.

This study indicates that water quality of Poyang Lake wetland is strongly influenced by human activities. Agricultural, industrial and domestic pollution as well as sand mining activities have important impacts on eutrophication status of the water bodies of Poyang Lake wetland [3]. Therefore, controlling pollutant discharge, optimizing and upgrading sewage treatment facilities, and limiting sand mining activities are critical for improving the water quality of Poyang Lake. In addition, establishing a systematic water quality monitoring system and increasing residents' awareness of environment protection through publicity and education are also important for the wetland protection and restoration.

7.5 Conclusions

- Nitrogen and phosphorus concentrations in the water bodies of Poyang Lake wetland were higher than the thresholds for algae blooms.
- Water bodies of Poyang Lake wetland were in the state of eutrophication as a whole, especially those suffering strong human disturbances.
- Chl *a* concentration in the water bodies of Poyang Lake wetland was relatively low probably due to high water velocity of the lake. Chl *a* concentration significantly and negatively correlated with water transparency, and positively correlated with COD, whereas the correlations with other water quality parameters such as temperature and pH were not significant.

Acknowledgements This work was supported by National Natural Science Foundation of China (41561097), Major Ecological Security Issues of Jiangxi Province and Monitoring Implementation (JXS-EW-00), Education Department of Jiangxi Provincial Government (150334) and Natural Science Foundation of Jiangxi Province (20142BAB213023).

References

1. Cui B., and Z. Yang. 2001. Research review on wetland ecosystem health (in Chinese). PhD thesis.
2. Han, D., Y. Yang, Y. Yang, and K. Li. 2012. Recent advances in wetland degradation research (in Chinese). *Acta Ecologica Sinica* 32 (4): 1293–1307.
3. Wu, Y., and W. Yang. 2015. Indicators and implementing methods of wetland biodiversity monitoring: Taking Great Lakes coastal wetlands as an example (in Chinese). *Biodiversity Science* 12 (3): 527–535.

4. Lau, S., and S.N. Lane. 2002. Biological and chemical factors influencing shallow lake eutrophication: A long-term study. *Science of the Total Environment* 288 (3): 167–181.
5. Wang, M., X. Liu, and J. Zhang. 2002. Evaluate method and classification standard on lake eutrophication (in Chinese). PhD thesis.
6. Qi, S., Q. Zhang, F. JIiang, and F. Liao. 2014. Study on the effects on the landscape spatial pattern of the wintering birds'habitat from lake level in Poyang Lake Wetland (in Chinese). *Journal of Natural Resources* 29.8.
7. Han, X., X. Chen, and L. Feng. 2015. Four decades of winter wetland changes in Poyang Lake based on Landsat observations between 1973 and 2013. *Remote Sensing of Environment* 156: 426–437.
8. Xia, S., X. Yu, Y. Liu, Y. Jia, and G. Zhang. 2016. Current issues and future trends of Poyang Lake wetland (in Chinese). *Resources and Environment in the Yangtze Basin* 25 (07): 1103–1111.
9. Mao, Y., X. Zhou, and M. Wang. 2014. Study on the eutrophication status in Poyang lake during lower water period (in Chinese). *Journal of Nanchang University (Natural Science)* 38 (6): 596–599.
10. Yu, J., Y. Liu, X. Zhong, and J. Yao. 2009. Evaluation method of eutrophication in Poyang Lake and its leading factors (in Chinese). *Acta Agriculturae Jiangxi* 21 (4): 125–128.
11. Wang, Z. et al. 2017. Eutrophication assessment of water quality in Poyang lake basin based on comprehensive analysis method of combined extension (in Chinese). *Acta Ecological Sinica* 37.12.
12. Wan, J., and W. Yan. 2007. Evaluation methods application in and probing into eutrophication of Poyang Lake area (in Chinese). *Journal of Nanchang University (Natural Science)* 31 (2): 210–214.
13. Carlson, R.E. 1977. A trophic state index for lakes. *Limnology and Oceanography* 22 (2): 361–369.
14. Bott, T.L., D.S. Montgomery, D.B. Arscott, and C.L. Dow. 2006. Primary productivity in receiving reservoirs: Links to influent streams. *Journal of the North American Benthological Society* 25 (4): 1045–1061.
15. Huang, Y. 2001. Research on inland water pollution ecology (in Chinese). PhD thesis, pp. 265–267.
16. Wu, Z., L. Zhang, B. Liu, and Y. Chen. 2014. Spatial distribution of chlorophyll a in Poyang Lake during wet season and its relationship with environmental factors (in Chinese). *Wetland Science* 12 (3): 286–292.
17. Liu, X., B. Liu, Y. Chen, and J. Gao. 2016. Responses of nutrients and chlorophyll a to water level fluctuations in Poyang Lake (in Chinese). *Chinese Journal of Environmental Science* 37 (6): 2142–2148.
18. Chen, X., Y. Zhang, L. Zhang, L. Chen, and J. Lu. 2013. Characteristic of nitrogen and phosphorus in Lake Poyang during high water period (in Chinese). *Lake Science* 25 (5): 643–648.
19. Cao, C., B. Zheng, J. Zhang, M. Huang, and Z. Chen. 2009. Systematicinvestigetion into winter and spring algal blooms in Daning River of Three Gorges Reservoir (in Chinese). *Chinese Journal of Environmental Science* 30 (12): 3471–3480.

Chapter 8
Hyperspectral Response of Dominant Plants in the Poyang Lake Wetlands to Heavy Metal Pollution

Caiying Ni, Dan Zhang, Pengfei Song, Siying Zhao and Wenjing Yang

8.1 Introduction

Wetland degradation is a global environmental problem. Among the various factors affecting wetland degradation, heavy metal pollution is a major threat. Strengthening the dynamic monitoring and identification of heavy metal pollutants in wetlands is an important part of preventing and controlling heavy metal pollution and subsequently wetland degradation [1]. The traditional method of "field sampling + lab analysis" has a high precision, but is very time and resources consuming. Besides, it is not able to obtain the continuous distribution and pollution degree of heavy metals in the studied area rapidly [2]. Remote sensing technology offers an opportunity to effectively overcome the deficiencies of the traditional methods by associating the pollution status of heavy metals in wetlands to the spectral parameters on large scale [3, 4].

The chemical composition of plants represents to a certain extent the composition of soil, bedrock, water bodies and the atmosphere surrounding them. Therefore, plants are often more preferred to be used for environmental monitoring than conventional soils, water bodies and the atmosphere [5]. Under stress by heavy metals, one of the most common physiological reactions of plants is the decrease of chlorophyll content. The amount of chlorophyll in a plant affects its spectral reflectance, which is the scientific base for the application of hyperspectral monitoring of heavy metal pollution [6, 7]. The advantages of using hyperspectral techniques to monitor environmental pollution by plants have been confirmed by many scholars. Studies by Wu [8] and Yang et al. [9] have shown that the direct monitoring of soil heavy metal pollution by hyperspectrum analysis is unsatisfactory, but the monitoring of heavy metal pollution in soils by the plants, which were growing on it, is a significantly more promising approach [10, 11]. For instance, Xu et al. [12] successfully

C. Ni (✉) · D. Zhang · P. Song · S. Zhao · W. Yang
Jiangxi Normal University, Nanchang, China
e-mail: ncy1919@126.com

T. Yue et al. (eds.), *Chinese Water Systems*, Terrestrial Environmental Sciences,
https://doi.org/10.1007/978-3-319-97725-6_8

established a hyperspectral estimation model to estimate cadmium content in reclaimed farmland soils in Liuxin mining area of Xuzhou using hyperspectral remote sensing technology. Jiang et al. [13] found that the hyperspectral estimation model of Geographically Weighted Regression (GWR) could predicted soil chromium of Fuzhou city precisely, and it was suitable for predicting spatial heterogeneity on the regional scale.

Poyang Lake wetland is one of the most important wetlands in the world. However, heavy metals from human activities such as mining and metallurgical processing have threatened the water quality of Poyang Lake since decades [14]. Consequently, the sediments of Poyang Lake and the wetland soils have accumulated heavy metals whereas the contamination with heavy metals is higher in the soil than in the sediments [15–17]. According to the literature, the most concerned metal ions in Poyang Lake area in previous years are Cu^{2+}, Zn^{2+} and Pb^{2+}, because of both their spreading over large areas and their heavy pollution intensity. Furthermore, they are not only easily accumulating in plants, but also show the tendency of compound pollution [18, 19].

Carex cinerascens Kükenth is the dominant plant with the largest distribution in the Poyang Lake wetlands. It has two growing seasons, which are interrelated with the annual rise and fall of Poyang Lake water level. It prefers contiguous growth and there are very few other plants in the cluster of *Carex cinerascens Kükenth*. These growth distribution characteristics of *Carex cinerascens Kükenth* make it the best target species for remote sensing technology to monitor the health of the Poyang Lake wetland.

Therefore, in this study, the *Carex cinerascens Kükenth* in the natural wetland of Nanjishan in Poyang Lake basin were studied. Pot experiments were conducted to study the relationship between the effects of heavy metal pollution by Cu, Pb on the values of chlorophyll content (SPAD) and spectral characteristic values of visible light bands. The characteristic spectral response of *Carex cinerascens Kükenth* was also discussed. Hence, this study provides reference and scientific base for using hyperspectral remote sensing technology to monitor heavy metal pollution in wetlands.

8.2 Materials and Methods

8.2.1 Experimental Equipment

Spectral data was obtained using the SVC HR-1024 Portable Spectra (Spectra Vista, USA). It has a wavelength range from 350 to 2500 nm whereas the range from 350 to 750 nm is assumed as representing the visible light bands in this study. The SPAD-502 chlorophyll meter (Konica Minolt, Inc.) was used to collect chlorophyll data. Heavy metals in plants and soil were determined using a graphite furnace atomic absorption spectrophotometer (VARIAN-AA240Z, Varian Technology, Inc., USA).

8.2.2 *Experimental Methods*

8.2.2.1 Plant Culture

Plants were collected from April 1 to May 25, 2012 in a natural wetland in the Nanji Montain area of Poyang Lake using a black watertight plastic basket (length/width/height: 48/38/15 cm). Black watertight plastic was used to reduce the interference of spectral acquisition to the *Carex cinerascens Kükenth.* The weight of soil in each basket equivalent was identical (25.87 kg/basket) and the thickness of the soil layer was the same as the height of the basket (Table 8.1). After *Carex cinerascens Kükenth* reached a stable growth in the baskets, the heavy metals copper and lead were injected into the soil as analytical grade pure chemical compounds $Cu(NO_3)_2$, $3H_2O$ and $Pb(NO_3)_2$. The concentration are shown in Table 8.2. After mixing and dissolving different concentrations of Cu, Pb and combinations of both, the solutions were used to evenly water the soil. The concentrations of N, P, and K nutrient elements, which are are integral for plant growth, were kept constant across treatments. Due to different Pb or Cu concentrations, the N nutrition imbalance was balanced with urea (ammonium nitrate, NH_4NO_3). Dihydrogen Phosphate

Table 8.1 The basic physical and chemical properties of soil

Indicators	Measured value	Indicators (mg/kg)	Measured value
Organic matter (g/kg)	22.83	Boron	0.07
Total nitrogen (g/kg)	1.20	Sulfur	38.69
Alkali-hydrolyzable nitrogen (mg/kg)	66.00	Silicon	38.30
Available phosphorus (mg/kg)	14.92	Zinc	5.35
Available potassium (mg/kg)	78.00	Iron	31.00
pH	5.50	Copper	5.37
Bulk density of soil (g/cm^3)	1.55	Manganese	39.83

Table 8.2 Concentration of metal ion species for each treatments

Ion/concentration	Cu	Pb	Cu + Pb	Cu + Pb + Zn
1	0	0	0	0
2	50	50	Cu100 + Pb400	Cu100 + Pb200 + Zn200
3	100	100	Cu100 + Pb800	Cu100 + Pb400 + Zn800
4	150	150	Cu200 + Pb400	Cu100 + Pb600 + Zn600
5	200	200	Cu200 + Pb600	Cu200 + Pb200 + Zn200
6	300	300		Cu200 + Pb200 + Zn600
7		500		Cu200 + Pb400 + Zn600

Potassium (chemical formula: KH_2PO_4) was used as P fertilizer and K fertilizer. The ratio of N:P:K was kept at 1:1:0.6. During the cultivation period the canopy was covered with double-layer plastic sheets during rainy days to maintain the same amount of water supply and to prevent elements of the box will be washed away by the rain.

8.2.2.2 Spectral Data Acquisition and Processing

The spectral characteristics of the plant is mainly obtained through its crown. In this experiment, the leaf spectrum of *Carex cinerascens Kükenth* was collected by SVC HR-1024 spectrometer (SVC Corporation, USA). The internal spectral resolution was $\leq$3.5 nm in the range of 350–1000 nm, $\leq$9.5 nm in the range of 1000–1850 nm and $\leq$6.5 nm in the range of 1850–2500 nm. Spectral measurements were taken with an angle of 25. Spectral data was collected at sunshine between 10:00 and 14:00 Beijing Time. During the collection process the sensor probe was put vertical downward in a distance of 15 cm from the targets. Each collection took place at 5 different spots with the same growth stage in each culture basket. 5 times were selected and the spectrum numbers were recorded. The measured raw spectral data was first screened excluding aberrant spectral curves and preprocessed using SVC HR-1024 PC Date Acquisition Software (Spectra Vista, USA) to extract partial spectral parameters for first-order differential spectroscopy. The spectral parameters were analysed and processed by the software ENVI (The Environment for Visualizing Images) to extract the relevant spectral parameters. Furthermore SPSS (Statistical Product and Service Solutions) was used to determine the correlation coefficient between the parameters.

8.2.2.3 Chlorophyll Data Collection and Processing

The relative content of leaf chlorophyll was measured using the SPAD-502 chlorophyll meter. Sunny weather was selected for chlorophyll determination. If possible, the same day as for the monitoring of the spectrum was used for the Chlorophyll measurement. 10 leaves with the similar growth were uniformly selected from each *Carex cinerascens Kükenth* box. In the upper, middle and lower position of each leaf the chlorophyll concentration was measured 3 times. The average value of the values retrieved from SPAD was recorded. The recorded chlorophyll data were used after the modified Grubbs rule and the mean SPAD values were averaged for each leaf.

8.2.2.4 Determination of Cu and Pb in *Carex Cinerascens Kükenth* and Soil

After the end of the cultivation of *Carex cinerascens Kükenth*, the leaves were stored in clean plastic bags and labeled to indicate the sample name and number. After washing in laboratory with tap water, the leaves were rinsed with deionized water three times and were cooled in a clean absorbent paper. The leaves were dried in the oven at 105 °C for 30 min. In the follow-up, oven temperature was reduced to 50 °C and the drying process continued for 48 h. Dried plant samples were grinded with an agate mortar grind over a nylon screen with 100 mesh. The grinded leaves were put into clean self-sealing bag, labelled and placed in a dryer spare till dissolution.

After the cultivation of plants, soil samples were collected in a depth between 5 and 15 cm within the culture baskets according to the quincunx method. In total, about 1 kg soil samples were collected. The soil samples were filled in labelled clean plastic bags. In the laboratory, the debris in the soil samples (grass/roots, etc.) was removed. After natural drying, the samples were crushed through a 100 mesh nylon screen. After mixing they were stored in the bottles.

The next step was the quantification of Cu and Pb concentration in the plants and the soils using the leaf and soil powder with a weight of 1 g and 0.5 g, respectively. The weight was determined by a high precision scale with an accuracy of 0.0001 g. Each sample was filled in a clean conical flask with perchloric acid and concentrated nitric acid. After digestion facilitated by an oven, the digests were filtrated and diluted, and the contents of Cu and Pb in leaves and soil were determined by graphite furnace atomic absorption spectrophotometer. Three blank samples were used for correction purposes. Each sample was treated with three replicates. The average of the three replicates was used to indicate the contents of Cu and Pb in the samples.

For the determination of available Cu and Pb in Soil, heavy metals in the soil samples were extracted by adding 4 ml of 0.1 mol/L HCl to 2 g of each of the sieved soil samples. The extraction took place in a horizontal reciprocating shaker at 25 °C for 2 h. Afterwards, the extract was filtered with filter paper. The first filtered 1ml of solution was discarded and the remaining filtrate was diluted to the appropriate level needed by the measuring instrument.

After the measurement, the Cu and Pb content was calculated by:

$$C = (\rho - \rho_o) * {}^{V}/_{M} \tag{8.1}$$

C-Cu, Pb content, mg/kg; ρ—measured heavy metal concentration, μg/mL; ρ_0—blank solution concentration, μg/mL; V—sample solution volume, mL; M—sample weighed, g.

8.3 Results and Discussion

8.3.1 Cu Contamination

Testing different Cu concentration, it was found that *Carex cinerascens Kükenth* died within a week for Cu concentration at a level of 300 mg/kg (Fig. 8.1).

Therefore, it can be confirmed that the upper limit of the tolerance concentration of *Carex cinerascens Kükenth* to Cu is within the range of 200 mg/kg, which is set as the existing concentration threshold. Hence, the Cu content, the SPAD value and the

Fig. 8.1 *Carex cinerascens Kükenth* at a Cu concentration of 300 mg/kg

extraction of spectral characteristic parameters were performed at Cu concentrations below 200 mg/kg.

With the increase of Cu concentration, the content of Cu in the leaves (leaf Cu) of *Carex cinerascens Kükenth* and the content of Cu in the soil (soil Cu) increased gradually. The leaf Cu was much higher than soil Cu. These results indicated that *Carex cinerascens Kükenth* could transfer Cu absorbed by the roots into the leaves, which is very beneficial for the spectral monitoring of its leaves. If Cu concentrations were in the range of 50–200 mg/kg, the contents of Cu in *Carex cinerascens Kükenth* and soil available Cu were significantly higher than for those plants, where no Cu was added. The content of Cu in leaves of *Carex cinerascens Kükenth* were sharply increased with added Cu concentration when Cu levels were higher than 100 mg/kg. As shown in the Fig. 8.2, adding Cu levels of less than or equal to 50 mg/kg, the increase of leaf Cu was 8.25 mg/kg, while when Cu was added at concentrations between 50 and 100 mg/kg, the increase of leaf Cu was 4.32 mg/kg. But when the added Cu concentration was higher than 100 mg/kg, the leaf Cu were relative high. When added Cu was between 100 and 150 mg/kg, the increase of leaf Cu was 28.85 mg/kg. Adding Cu at concentration between 150 and 200 mg/kg caused an increase of leaf Cu at 66.05 mg/kg.

As shown in Fig. 8.2, the available Cu content in soil increases most rapidly with a slope of 3.75 mg/kg when the concentration of Cu is increased from the CK (0 mg/kg) to the addition level of 50 mg/kg. For higher Cu concentration the related increase in Cu content in the soil continuously decreases with slopes of 2 mg/kg, 1.33 mg/kg and 1.12 mg/kg respectively.

Correlation analysis showed that there was a high positive correlation between added Cu concentration, leaf Cu and soil available Cu (Table 8.3). The correlation coefficient between added Cu concentration and leaf Cu was 0.893. The correlation coefficient between added Cu concentration and soil available Cu was 0.966. The correlation coefficient between leaf Cu and soil available Cu was 0.769. These

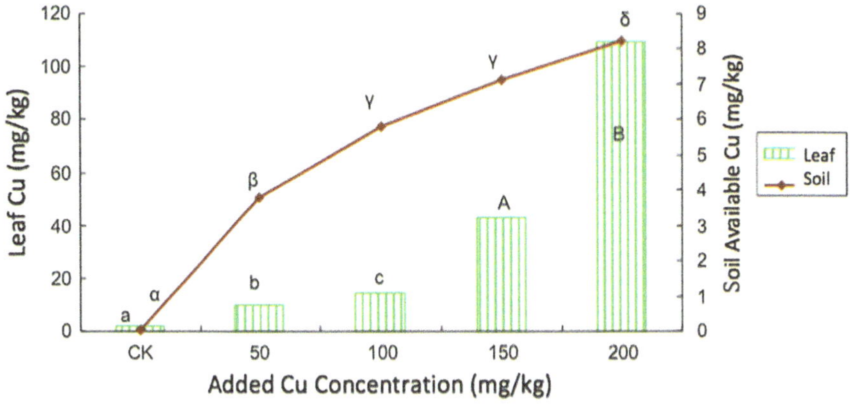

Fig. 8.2 Cu content of *Carex cinerascens Kükenth* leaves (green bars) and soil available Cu (dark red line) for different Cu level treatments. Different uppercase and lowercase letters indicate the significance levels of $p < 0.01$ or $p < 0.05$ respectively

Table 8.3 Correlation between Cu concentration, leaf Cu and soil available Cu

Type	Correlation coefficient
Added Cu concentration and leaf Cu	0.893
Added Cu concentration and soil available Cu	0.966
Leaf Cu and soil available Cu	0.769

significant correlation provided a scientific base for monitoring the Cu contamination of the soils by using the plant *Carex cinerascens Kükenth* as a proxy (Table 8.3).

The SPAD value of *Carex cinerascens Kükenth* decreased gradually with increasing Cu concentrations. That means the added Cu concentration affected the synthesis and decomposition of chlorophyll, which was beneficial to the spectral monitoring of Cu contamination by its leaves. The SPAD value of *Carex cinerascens Kükenth* was for all added Cu concentrations lower than for the control. There were also significant differences between derived leaf SPAD values for the different treatments (Fig. 8.3).

The slope of the decreasing graph of the leaf SPAD values became larger with the increase of Cu concentration, having an absolute increase in ratio from 0.74 to 1.15 to 1.36 to 1.46. The correlation analysis showed that there was a good negative correlation between added Cu concentration, SPAD value of *Carex cinerascens Kükenth and* and leaf Cu. The correlation coefficient between added Cu concentration and SPAD value of *Carex cinerascens Kükenth* was −0.994. The correlation coefficient between leaf Cu and SPAD value of *Carex cinerascens Kükenth* was −0.931. The SPAD value and the trend of Cu content in leaves of *Carex cinerascens Kükenth* negatively correlated with the added Cu concentration, which confirms the opportunity to determine the contamination with spectroscopic methods.

The spectral response of *Carex cinerascens Kükenth* to changes in added copper concentration depends on the growth stage (Table 8.4). Correlation analysis showed

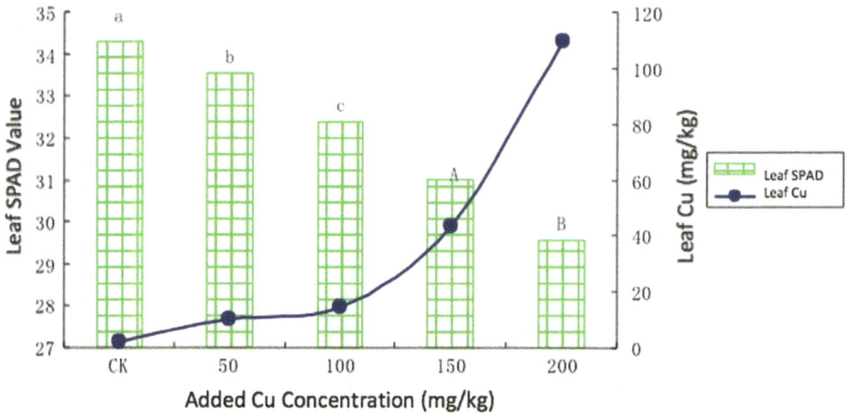

Fig. 8.3 SPAD value (green bars) and Cu content (blue graph) of *Carex cinerascens Kükenth* under different Cu concentrations. Different uppercase and lowercase letters indicate the significance levels of $p < 0.01$ or $p < 0.05$ respectively

Table 8.4 Correlation of leaf Cu content, SPAD value and spectral characteristic parameters

Growth period/characteristic parameters	Leaf Cu	Seedling SPAD	Elongating stage SPAD	Maturity stage SPAD
Green peak	−0.203	−0.538	−0.805*	0.348
Red valley depth	−0.631*	0.537	0.901*	0.534
Red edge position	0.271	0.791*	0.875*	−0.252
Red edge peak	−0.975**	0.614*	0.972**	0.967**
Red edge slope	−0.965**	0.666*	0.978**	0.977**
$NDVI_{705}$	0.222	0.806*	0.912*	−0.177
$R_{750-1250\,nm}$	−0.952*	0.661*	0.875*	0.989**

*Indicates significance level $p < 0.05$
**Indicates significance level $p < 0.01$

that the correlation coefficient between SPAD value and spectral eigenvalue in elongating stage was generally higher than that in the seedling and maturity stage, so it could be determined as the best monitoring period. The correlation analysis of copper and spectral eigenvalue shows that the red edge peak, red edge slope, red valley depth and $R_{750-1250\,nm}$ represents good spectral indexes to retrieve the copper content in soil.

8.3.2 *Pb Contamination*

Pb in leaves and Pb in soils increased at the beginning with the increase of Pb concentration, but reversed at very high added Pb concentrations (Fig. 8.4). Soil available Pb gradually increased if added concentrations were less than 150 mg/kg, significantly

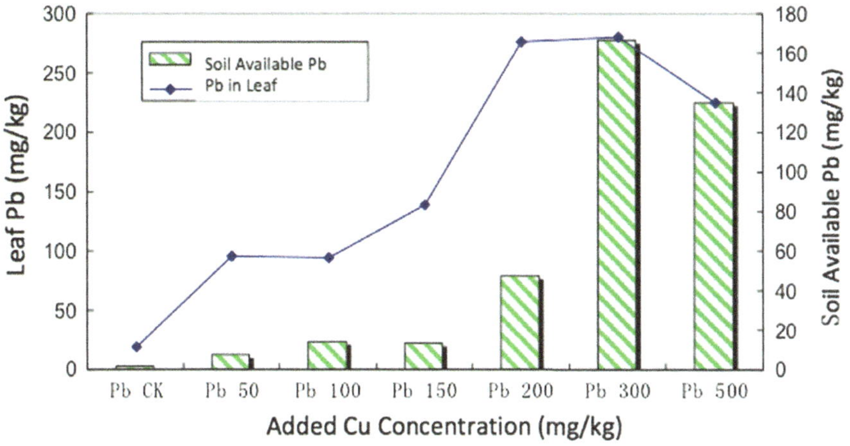

Fig. 8.4 Soil available Pb (blue graph) and leaf Pb (green bar) at different Pb concentrations

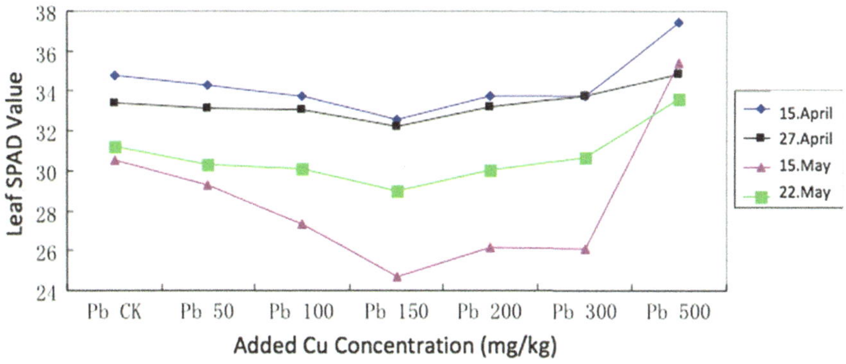

Fig. 8.5 Change of chlorophyll content (SPAD) value of *Carex cinerascens Kükenth* under different Pb concentrations and during different stages of the growth period (15 April–27 April seedling stage; 5 May elongating stage; 22 May maturity stage)

increased from 150–200–300 mg/kg, and decreased at even higher concentrations. However, leaf Pb gradually increased for increasing Pb concentrations up to 200 mg/kg. Leaf Pb remains almost unchanged for added lead concentrations between 200 and 300 mg/kg and subsequently decreased for the highest lead concentration of 500 mg/kg. In terms of *Carex cinerascens Kükenth* growth, 500 mg/kg Pb do not cause any harm which is may be caused by the chemical reduction of Pb in the leaves.

In contrast to the result for the copper pollution, the observed trends of measured SPAD values of *Carex cinerascens Kükenth* were not similar for each growth period (Fig. 8.5). The SPAD value of *Carex cinerascens Kükenth* decreased with the increase of Pb in leaf and in soil for lead concentration of less than 150 mg/kg. However, a positive correlation between the SPAD value and available Pb content in leaf and soil was found for lead concentration larger than 150 mg/kg. This phenomenon was

Table 8.5 Correlation coefficients between the spectral characteristic parameters and leaf Pb

Characteristic parameters	Correlation coefficient
Blue valley depth	−0.687*
Green peak	0.44
Red valley depth	0.559
Red edge peak	−0.866*
Red edge slope	−0.830*
$NDVI_{705}$	−0.733*
$R_{750-1250}$	−0.327

*Indicates significance level $p < 0.05$

observed for each stage of the growth period. Subsequently, spectral monitoring of lead-contaminated soils is only suitable for lead concentrations in the range between 0 and 150 mg/kg.

Table 8.5 showed the correlation between the lead concentration in the leaves and the spectral indices for added lead concentrations at a lower level from 0 to 150 mg/kg. These data showed that there was a significant negative correlation between the red edge peak, red edge slope, blue valley depth and $NDVI_{705}$ and leaf Pb, which both can be used to retrieve the lead content in soil of Poyang Lake wetland by using *Carex cinerascens Kükenth* as a proxy.

8.3.3 Combined Pollution by Cu and Pb

The relationship between the spectral eigenvalues and the combined addition of Cu + Pb was different for the three different growth stages of the plants (Table 8.6). At the seedling stage, the relationship between the concentration of added metals and the

Table 8.6 Correlation coefficients of spectral characteristic parameters and added concentrations of Cu + Pb at different growth stages

Characteristic parameters/growth stage	Seedling	Elongating stage	Maturity stage
Blue valley depth	−0.672*	−0.123	0.024
Green peak	0.333	0.352	−0.125
Red valley depth	−0.782*	−0.568	−0.365
Red edge Peak	0.452	−0.557	−0.760*
Red edge slope	0.353	−0.941**	−0.666*
$NDVI_{705}$	−0.689*	−0.404	−0.272
$R_{750-1250}$	−0.269	−0.887*	−0.750*

*Indicates significance level $p < 0.05$
**Indicates significance level $p < 0.01$

depth of the blue and red valley and $NDVI_{705}$ was significant, but these relationship were not significant at the elongating stage and the maturity stage. In the elongating stage phase, the slope of the red edge and $R_{750-1250}$ had better relationship than other parameters Pb + Cu with added concentration. At the maturity stage, the parameters were significantly related to the added metal concentrations and showed variations in the following order: red edge peak, red edge slope, and $R_{750-1250}$. According to the reproducibility of each growth period, the best monitoring period was the elongating stage (Table 8.6). Furthermore, it was recommended that the $R_{750-1250}$ and the red edge slope could be used as monitoring indexes for retrieving the combined content of Pb + Cu in the soil.

8.3.4 Discussion and Conclusion

From the above results, the different contaminated metal ions had different relations with the spectral characteristic values of *Carex cinerascens Kükenth*, such as copper and lead (Tables 8.4 and 8.5). The correlation between the spectral characteristic values for copper and *Carex cinerascens Kükenth* was higher than for lead.

In addition, the pollution of the plants with either copper or lead were significantly associated with at least four spectral eigenvalues (4 for lead and 7 for copper), while combined pollution with both heavy metals showed less significant correlations with spectral eigenvalues (even in the best monitoring period there were only 2 parameters having a good relationship with compound pollution of copper and lead). In addition to the above studies, a combined pollution with lead, copper and zinc was tested. However, hardly any significant correlation between leaf metal content and spectral characteristic parameters could be found for any stage of elongating (Table 8.7).

Summarizing, three main conclusion can be condensed from the study results. Firstly, using hyperspectral technology monitoring, a better corresponding relationship between certain spectral eigenvalues and the level of soil contamina-

Table 8.7 Correlation coefficients of spectral characteristic parameters and added concentrations of Cu + Pb + Zn at different growth stages

Characteristic parameters/growth stage	Seedling	Elongating stage	Maturity stage
Blue valley depth	−0.725*	0.009	0.515
Green peak	0.781*	0.071	−0.356
Red valley depth	−0.748	0.03	0.156
Red edge peak	−0.481	−0.031	−0.385
Red edge slope	−0.555	0.21	−0.344
$NDVI_{705}$	−0.789*	−0.592	−0.321
$R_{750-1250}$	−0.193	0.153	−0.287

*Indicates significance level $p < 0.05$

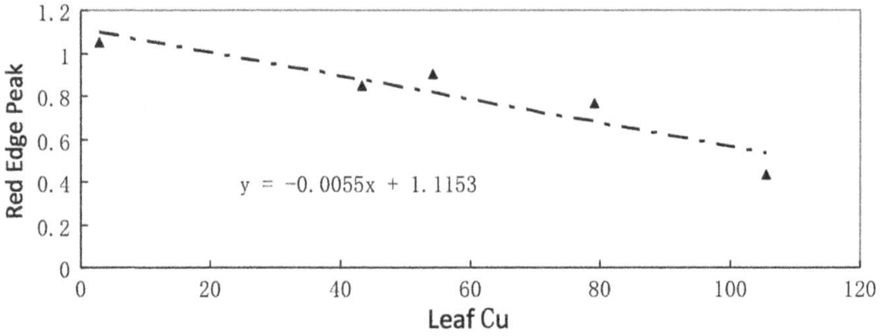

Fig. 8.6 Regression relationship between leaf Cu and red edge peak

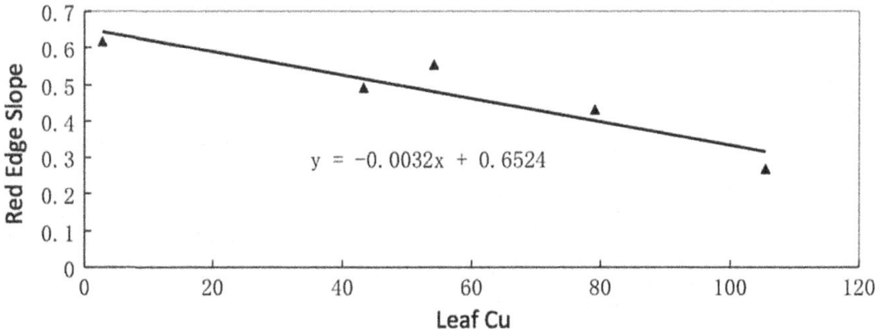

Fig. 8.7 Regression relationship between leaf Cu and red edge slope

tion by heavy metals can be obtained for contaminated sites. The two parameters of red edge peak and red edge slope were found to have extremely significant correlation and have a good linear relationship with the copper content in the plant, which can be expressed by a two linear regression equations.; $y = -0.0055x + 1.1153$ for the relationship between the Cu content in *Carex cinerascens Kükenth* and the Red Edge Peak (Fig. 8.6) and $y = -0.0032x + 0.6524$ for the relationship between the Cu content in *Carex cinerascens Kükenth* and the Red Edge Slope (Fig. 8.7). However, special attention should be paid to the use of spectral data for the retrieval of sites contaminated by multiple pollutants such as Copper, Lead and Zinc.

Secondly, all obtained data were from controlled test and the spectral data were collected from ground surface. In practice, when using spectral data from remote sensing, to monitor model and assess wetland pollution, special attention should be given to data selection, conversion, modification and processing [20, 21]. Different models use different patterns of data conversion and their correlation with heavy metals will differ, too [22]. At the same time, it is also necessary to eliminate the effects of weather and surrounding buildings to avoid erroneous conclusions [23].

Thirdly, to improve the application of spectral data to reveal the pollution status of wetlands, additional efforts are required to investigate the impact of different types of plants and vegetation. Considering that different plants have different heavy metals absorption and accumulation characteristics, the spectral responses of their vegetation will also be different [24–27].

Acknowledgements The study was funded by the project of National Natural Science Foundation of China (41061037) "Hyperspectral response of dominant plants to the heavy metal pollution in Poyang Lake wetlands" and "Poyang Lake biological integrity index ecological health assessment (41561097)".

References

1. Liu, S., X. Liu, J. Hou, G. Chi, and B. Cui. 2007. Application of plant spectroscopy in the stress response of brassica campestris (in Chinese). *Science in China Series E-Technological Sciences* 37 (5): 693–699.
2. Song, L., et al. 2014. Estimation of soil's heavy metal concentrations (As, Cd and Zn) in Wansheng mining area with geochemistry and field spectroscopy (in Chinese). *Spectroscopy and Spectral Analysis* 34 (3): 812–817.
3. Fu, X., et al. 2013. Research on hyper-spectral remote sensing in heavy metal pollution soil (in Chinese). *China Mining Magazine* 22 (1): 65–68.
4. Koponen, S., J. Pulliainen, K. Kallio, and M. Hallikainen. 2002. Lake water quality classification with airborne hyperspectral spectrometer and simulated MERIS data. *Remote Sensing of Environment* 79 (1): 51–59.
5. Dune, C.E. 1990. The biogeochemical survey of noble metal in West Canada. In *The proceedings of the 15th international geochemical exploration society*, vol. 245, no. 3. Prague: Geochemical Exploration Society.
6. Paetsikkae, E., M. Kairavuo, F. Sersen, E.-M. Aro, and E. Tyystjaervi. 2002. Excess copper predisposes photosystem II to photoinhibition in vivo by outcompeting iron and causing decrease in leaf chlorophyll. *Plant Physiology* 129 (3): 1359–1367.
7. Zhou, B. 2010. Hyperspectral fuzzy reasoning prediction model for lead pollution in rice (in Chinese). PhD thesis, China University of Geosciences.
8. Wu, J. 2005. Remote sensing geochemical fundamental studies on agricultural soil heavy metal pollution in suburbs (in Chinese). PhD thesis, Nanjing University.
9. Yang, L., Y. Gao, and Z. Hu. 2008. Estimating chlorophyll concentration of corn under copper stress with red edge (in Chinese). *Mining Research and Development* 5: 77–79.
10. Malley, D.F., and P.C. Williams. 1997. Use of near-infrared reflectance spectroscopy in prediction of heavy metals in freshwater sediment by their association with organic matter. *Environmental Science & Technology* 31 (12): 3461–3467.
11. Kemper, T., and S. Sommer. 2002. Estimate of heavy metal contamination in soils after a mining accident using reflectance spectroscopy. *Environmental Science & Technology* 36 (12): 2742–2747.
12. Xu, J., J. Dong, Y. Yang, K. Tan, and W. Cheng. 2013. Support vector machine model for predicting the cadmium concentration of soil-wheat system in mine reclamation farmland using hyperspectral data (in Chinese). *Acta Photonica Sinica* 43 (5): 530001.
13. Jiang, Z., Y. Yang, and J. Sha. 2016. Study on GWR model applied for hyperspectral prediction of soil chromium in Fuzhou City (in Chinese). *ACTA Ecologica Sinica* 37 (23): 8117–8127.
14. Lv, L. 1994. Investigation and analysis of heavy metal pollution in Poyang Lake (in Chinese). *Yangtze River* 25 (4): 32–38.

15. Jian, M., X. Gong, H. You, Z.Huang, and Zhujie. 2004. Assessment of pollution of heavy metals in water sedimints and aquatic plants in Poyang Lake basins (in Chinese). *Resources and Environment in the Yangtze Basin* 13 (6): 589–593.
16. Gong, X., C. Chen, W. Zhou, M. Jian, and Z. Zhang. 2006. Assessment on heavy metal pollution in the sediment of Poyang Lake (in Chinese). *Environmental Science* 27 (4): 732–736.
17. Chen, S., and Z. Huang. 2008. Apply geoaccumulation index and potential ecological risk index to assess heavy metal pollution in the wetland of Poyang Lake (in Chinese). *Jiangxi Chemical Industry* 1: 57–60.
18. Jian, M., H. You, and C. Ni. 2006. Characteristics of heavy metals contaminant status and migration in Raohe River of Lake Poyang (in Chinese). *Lake Science* 2.
19. Wan, J., W. Yan, and T. Xie. 2007. Research on heavy metals pollution status of Le'an River, Lake Poyang Basin (in Chinese). *Lake Science* 19 (4): 421–427.
20. Cheng, G., J. Li, and Z. Dai. 2017. Application of BP neural network in analysis of heavy metal pollution in soil (in Chinese). *Journal of Geology* 41 (3): 394–400.
21. Lu, J., et al. 2018. Hyperspectral estimation model of soil chrome content based on multivariate statistical analysis (in Chinese). *Journal of Shandong Agricultural University* 1: 1–3.
22. Zhang, L., Y. Wu, S. Liu, L. Zhou, and D. Cai. 2015. Research on monitoring of Pb content in the sewage irrigation area of Longkou city using hyperspectral data (in Chinese). *Environmental Engineering* 33 (8): 114–117.
23. Tu, Y., and T. Fei. 2016. From vegetation hyperspectral remote sensing to the diagnosis of soil heavy metal pollution (in Chinese). *Hubei Agricultural Sciences* 55 (6): 1361–1368.
24. Zhao, S. 2016. Hyperspectral response and relationship of cadmium pollution in paddy fields (in Chinese). PhD thesis, Jiangxi Normal University.
25. Zhang, W., K. Yang, T. Xie, C. Liu, and T. Sun. 2017. Correlation analysis on spectral fractional-order differential and the content of heavy metal copper in corn leaves (in Chinese). *Science Technology and Engineering* 17 (25): 33–38.
26. Song, P. 2015. Study on the spectral response of lead and its combined pollution to the stress of Carex cinerascens (in Chinese). PhD thesis, Jiangxi Normal University.
27. Liu, F., K. Yang, Y. Sun, H. Wei, and G. Shi. 2015. Harmonic analysis of spectrum red edge monitoring heavy metal pollution of corn (in Chinese). *Hubei Agricultural Sciences* 54 (15): 3747–3750.

Chapter 9
Assessment of Degradration Causes and Development of Protection Strategies for the Poyang Lake Wetlands

Xinghua Le

9.1 Characteristics and Trends of the Poyang Lake Wetland

9.1.1 Wetland Characteristics

9.1.1.1 Diversity and Internal Variability of the Ecosystem

Poyang Lake is a typical example for an seasonal fluctuating lake. Hence the lake's size depends on the season which leads to changes in the ecosystem. For example, the increasing amount of water during the rainy season is conducive to the growth of aquatic organisms in the ecosystem. The abundance of water leads to a dilution of contaminated parts of the water system, which has a positive effect on water quality [1]. In addition, the internal dynamics of the ecosystem makes it possible to conserve species and to provide valuable animal habitats in Poyang Lake [2].

Poyang Lake is a transition zone between land and water with abundant terrestrial and aquatic animal and plant resources. Aquatic organisms include mainly zooplankton, bottom benthic animals and fishes. Among them, there are 154 genus of phytoplankton, 8 phylums, 13 categories of bottom benthic animals and 122 species of fish [3]. In addition, animals and plants such as shellfish, shrimp, crabs, waterfowls, lotus and lake grass are abundant in the Poyang wetland ecosystem [4]. Plenty of different bird species exist in the lake area in different habitats distributed over the water surface, beach, grassland or the lakefront. 10 species of birds are listed as first class national protected animals. Among the mammals, more than 100 individuals of the freshwater finless porpoise population can be found in Poyang Lake.

X. Le (✉)
Jiangxi Mountain-River-Lake Development Office, Nanchang, China
e-mail: lexinghua@sina.com

T. Yue et al. (eds.), *Chinese Water Systems*, Terrestrial Environmental Sciences,
https://doi.org/10.1007/978-3-319-97725-6_9

9.1.1.2 Zonal Wetland Distribution

The area in Poyang Lake covered by wetland vegetation is about 2262 km^2 accounting for 80.8% of the total area of the lake. The distribution of vegetation is governed by the environmental gradients and the varying water depth. Under normal circumstances, the plant population can be divided according to their main habitats into four plant belts, namely the wet plant belt, the emergent plant zone, the floating plant zone and the submerged plant zone.

9.1.1.3 Variability of the Ecosystem

Variability due to external changes illustrates the fragility of the Poyang Lake wetland ecosystem. Factors such as hydrology, soil and climate are the major external elements for the wetland ecosystem of Poyang Lake. Changes of these factors affect the biomes and the ecosystem structure of the wetlands. For example, the start of the rainy season, the amount of rainfall, the temperature, the soil sedimentation/abrasion patterns and the variability in water turbidity all may significantly affect the food web of the ecosystem. Hence, if the ecosystem is fragile, a small change in the external factors may easily break the ecological balance, but the recovery is very difficult and may requires a long period of time.

9.1.2 Wetland Trends

9.1.2.1 Shrinking Lake Area

According to the statistics covering the years from 1950 to 2014, extraordinary floods in Poyang Lake occurred in 1954, 1983, 1995, 1998 and 1998–1999 [5, 6]. Among them, the largest surface of Poyang Lake was more than 5100 km^2 during the year 1954 flood. In contrast, on July 31, 1998, at the peak time of the last large Yangtze River flood, the flooded area of Poyang Lake was only 4,070 km^2. After 1998, the maximum annual surface area extension of Poyang Lake is gradually decreasing. In 1999, the maximum surface area of water was less than 4,000 km^2. Until 2013, the maximum water surface decreased to only about 2,800 km^2. Although some years show an increase in maximum water surface area in comparison to the previous year (e.g. the year 2012), in general, maximum annual water surface areas are decreasing.

Similarly, the minimum water surface area in Poyang Lake decreases every year. After 1998, the problem of longer seasonal low water levels appeared in the basin, which became even worse after 2003. The severe winter droughts in 2008 and 2009 led to difficulties in obtaining domestic water along some parts of the lake and caused drinking water problems. Although the situation slightly improved in 2012, the drought appeared again and water levels dropped to a historical minimum of less than 500 km^2 in March 2014 and many former water covered lake areas and rivers could be crossed without a boat. Summarizing, in the recent past, Poyang Lake shows a consecutive history of rare low water level and the water levels continue to decline, dry season continues to extend, which all together will severely challenge future water resources management.

9.1.2.2 Expansion of Dried-Out Plain Areas

The size of the temporally dry land areas of Poyang Lake depends on the water level. In 2013 these areas covered 1092.4 km^2 in the dry season and 91.38 km^2 in the wet season, which means that most of the floodplain was inundated in the wet season. The amount of dried-up land in winter increased slightly in recent years due to the generally declining water levels in Poyang Lake.

9.1.2.3 Decreasing of Desertified Land and Challenging Management

The proportion of desertified land in the natural wetland landscape of Poyang Lake is small and accounts for less than 2% of the total area. It mainly concentrates in Pengze and Duchang County. Under normal circumstances, sandy land will be exposed in the dry season. Strong winds during this season led to a gradually advances of the desertification to the periphery of the lake. However, after starting to actively manage the sandy land areas, the total area of sand desertification could be reduced in recent years. Hence, the desertification of the area is temporarily controlled, but the major factors are not completely eliminated and continue to exist. The current management strategy promotes planting of adapted vegetation on sandy land to achieve sand fixation. However, the vegetation fixed on the sand dunes is often extremely fragile and the external environment is harsh. Furthermore, these vegetation is difficult to maintain and can be destroyed easily.

9.1.2.4 Threatening of Biological Resources

In the early 1990s, the State Fisheries Department had carried out a survey of fish in Poyang Lake. At that time, Poyang Lake had 158 species of fish. However, in the census of 2008, there were only 122 species of fish left in Poyang Lake. In less than 20 years, 36 species of fish disappeared. Furthermore, species such as Lipotes vexillifer, Reeves shad and Chinese sucker are on the verge of extinction and the fish stocks of Black Carp, Grass Carp, Silver Carp and Bighead Carp are continuously decreasing. Nowadays, caught fishes are younger, smaller and have a worse quality than in previous decades. Some important migratory fish, such as Anchovies and Reeves Shad are rare or even near extinction in Poyang Lake.

Man-made projects damage the wetland and illegal hunting has reduced the abundance of birds tremendously. The degradation of wetland functions severely affects the migratory birds in terms of habitat, seeking food and temporal appearance.

According to the statistics by the Wildlife Conservation Administration for migratory birds, 107 species of waterfowl and more than 460,000 waterfowl were counted in the Poyang Lake area during the winter 2006/2007. These were about 20,000 migratory birds less than in the previous winter. In addition not only the number of birds but also number of species is decreasing. 326 species of birds were counted during the first Poyang Lake survey. However, according to the results of the second Poyang Lake survey in 2008, there were only 236 species of birds left.

In the past 50 years, the Poyang Lake wetland vegetation has been affected by both natural and man-made factors. The general trend shows that the wetland area is decreasing. The main features are as follows: the wetland area is continuously decreasing; the integrity of wetland vegetation belt is destroyed; the biomass of wetland vegetation declines year by year; the numbers of endemic species decrease and the numbers of alien invasive species increase.

Focusing on the aquatic system, the production of benthos, such as Alocinma longicornis and plankton is decreasing. Some benthos species such as Lamprotula scripta and Acuticosta chinensis became rare in Poyang Lake. Considering that the benthic environment is the bottom of the biological food chain and the most important producer, the entire wetland ecosystem is seriously threatened by a decrease in benthic production.

9.2 Causes of Wetland Degradation

The causes of wetland degradation are both natural and man-made [7]. At present, the research on wetland degradation mainly focuses on the status quo and driving forces of wetland degradation, as well as changes in soil structure and biological diversity in degraded wetlands [8].

9.2.1 Changes of the Climate

Poyang Lake is a subtropical humid monsoon climate zone. However, climate change is one of the factors leading to the degradation of Poyang Lake wetlands. Global warming provides the conditions for invasive alien species. It leads to changes in growth patterns and a deterioration of water quality. Another factor is the intensification of abnormal climate conditions. The extremely hot and dry conditions in 2003 and the drought in 2006–2007 exposed part of the lakes and lake beds of Poyang Lake. Additionally, the occurrence of snowstorms such as in 2008 causes large grassland freezing and damages the wetland ecosystem.

9.2.2 Drying up of the Lake

As mentioned above, the decreasing water levels of Poyang Lake, particularly in winter, affect the Poyang Lake Ecosystem. The prolonged withering of the aquatic vegetation causes damage to aquatic plants and disrupts the community structure. Large areas of turf dry up and are lost as habitats, especially for carps such as Carassius auratus.

The inundation time of natural lake embankments at the boundary between the main wetlands, rivers and small sub-lakes is shortened due to the prolonged time with low water level. Consequently, the number of fish, shrimp, mussels, snails and other small aquatic animals entering these sub-lake has dropped substantially. Considering their role in the food web, a significant reduction of these species impacts the migratory birds during winter.

9.2.3 Soil Degradation

The Poyang Lake wetland mainly consists of alluvial soils, meadow soils and swamp soils. The alluvial plain of the five main rivers and the dryland of Poyang Lake is covered by fluvo-aquic soil, Magan Soil and yellow soil, which generally have good soil properties. Paddy soil is the main cultivated soil in this area, which is fertile and has a good farming performance. However, compaction and erosion of soil is an increasing problem in Poyang Lake as well as the pollution of soils by fertilizer overuse.

9.2.4 Population Growth

The population of Poyang Lake Basin approximately tripled between 1950 and 2012 which led to a massive increase of human activities in the area of Poyang Lake. The growth rate of the population in the Poyang Lake Core Region is much faster than the population growth in the whole basin, with an average growth rate of 10%. In the next 20 years, this growth rate is assumed to remain constant [9] . Hence, more natural resources will be needed to meet the needs of an increasing population which will e.g. negatively impact the soil fertility of the area.

Since ancient times, land reclamation and utilization measures are applied in the area of Poyang Lake to meet the growing demand of the population and to improve the production and living environment. Reclamation has become the main form of development and utilization of land resources in the lake area and according to the statistics, the total reclamation area in Poyang Lake area reached 1210 km^2.

Poyang Lake wetland ecosystem degradation and overuse of resources are inseparable. Inefficient use of resources will inevitably lead to poverty. Poverty also restricts the improvement of living conditions at the lakeshore which again leads to a more unsustainable exploitation of resources. As a result of these interactions, the damage to the ecological environment can only aggravate.

9.2.5 *Agricultural Practices*

Inefficient single-crop based agricultural production increases the extent of diffusive source pollution in the Poyang Lake area. The planting of rice monocultures over a long period decreased the soil fertility significantly. In order to increase production, fertilizers are widely used in the area. Frequent crop failures due to large scale appearance of pest led to an increase of pesticide usage. Hence, both pesticides and mineral fertilizers demand is increasing. From 1990 to 1995, pesticide use increased by 3% annually. From 1995 to 1999, the annual increase rate reached 7.5% annually. Following the distribution of mineral fertilizers on the field , about 60–70% of the fertilizer enters the environment and pollute water and soil. Excess N, P from the fields and the soils flows into the river and eventually to the Poyang Lake wetlands.

9.2.6 *Exploitation of Wetland Resources*

Although cutting wood in the upper reaches is scarce, the development of the fruit industry has caused large losses of endemic vegetation, soil and clean water, a problem which has not been adequately handled yet.

Due to generally low levels of economic productivity in many rural areas along the lake, lakeside communities often plunder natural resources of the Poyang Lake that create economic value. As an example, the long-term ongoing predatory operations on several species led to the depletion of these resources. Despite the strengthening of fishery management, e.g. by establishing protection areas and defining periods with a fishing ban, the phenomenon of overfishing and catching fish during the protection periods and in protection areas remains due to the large area of the wetland system and limited management efforts. Furthermore, the degradation of wetland vegetation has caused the destruction of spawning grounds of main economic fish species like carp and crucian carp.

9.2.7 *Policy Impact*

After 1998, sand mining in the Poyang Lake area and river channels intensified. Mining activities, irregular waste disposal, navigation and coastal wading safety of the rivers as well as threatening the aquatic resources and the ecological safety.

Unreasonable land-use patterns are another important reason for the degradation of the ecological environment in Poyang Lake. Public unstructured planting of poplar caused the appearance of so-called "green deserts" in the basin. Since 2000, large-scale planting of poplar at the wetland banks in 11 counties (cities and districts) around Poyang Lake has not only affected flood discharge, but also resulted in a drying up of the soils in these wetland bank areas which resulted in a loss of biodiversity. Although rectification actions were carried out in 2007, they achieved limited success.

At the same time, the construction of cofferdam lakes also lead to a fragmentation of the wetland ecological environment. Investigations show that the number of cofferdam lakes is spreading in the Poyang Lake area. This behavior not only leads to a destruction of the ecological and fisheries production, but also lead to conflicts on ownership disputes in the Lake District.

In addition, the construction of the Three Gorges Dam has a certain impact on the hydrological regime of Poyang Lake. The water level is lengthened in the flood season and the reduction of water levels is more severe in winter. It leads to the aggravation of the vegetation degradation in the Poyang Lake wetland, the desertification of land, the shrinkage of lakes and the destruction the biodiversity.

9.2.8 *Urbanization*

Urbanization is another major cause of wetland loss and degradation. The expansion of urban areas is associated with an increasing discharge of industrial waste water, waste residue, domestic sewage and fertilizers, pesticides and other harmful substances into the wetlands [10]. These harmful substances not only cause serious damage to biodiversity, but also affect the surface water, groundwater and soil environment, deteriorating water quality, causing water shortage and parasitic epidemic. In fact, many wetland areas have been transformed into industrial, intense agricultural areas or into polluted discharge areas for domestic wastewater.

The construction of roads, buildings and parking lots resulted in soil sealing and compaction of many areas. Due to the increasing soil impermeability, precipitation is prevented from infiltrating into the soil and instead flows as urban runoff into the wetlands transporting municipal waste and pollutants with it. Subsequently, increasing salinity, turbidity, toxicity, as well as reduced dissolved oxygen levels affect the aquatic food web. In addition, excessive nutrients lead to eutrophication, algae outbreaks, and widespread death of aquatic communities.

9.3 Current and Recommended Wetland Protection Measures

Based on the principle of having a positive connection between environmental protection and economic development, the Poyang Lake ecological environment protection agency developed a comprehensive resource development and utilization plan based on the long-term observation and tracking data to positively combine the ecological environment protection with the resource utilization. This takes into account the richness and diversity of resources as well as the resilience of the ecosystem. In order to protect the wetlands, several suggestions are given in the following sections.

9.3.1 Improvement of Laws and Regulations

Main future improvements should include the laws and regulations itself but also adapt the lake and wetland management system in accordance with the law. It is necessary to revise and reformulate various types of the current lake-wetland protection regulations based on the results gained from practical experience. Concurrently, the introduction of new specific laws and regulations on water resources protection as well as the establishment of ecological compensation should be studied. Currently the administration dealing with the wetland management is fragmented and divided into many different authorities which limits the success of the current lake-wetland management. To overcome these shortcomings, the responsibilities of all involved departments need to be integrated and a more authoritative Poyang Lake protection and management system needs to be established. As a high-standard management body, the set-up of a higher-level coordination agency is suggested. Furthermore, the supervision and management of the wetland biodiversity should be strengthened.

Strengthen information campaigns, education and legal supervision on wetland conservation offers new opportunities to improve the wetland protection. Using public media such as radio, newspapers and newspapers can be actively used to educate the public on wetland conservation. Additional public reports and touring exhibitions will further raise public awareness of conservation and make the public aware of the structure of wetland ecosystems, the ecosystem services they provide and the serious situation of the current wetland degradation.

9.3.2 Application of Modern Science and Technology

The health of the Poyang Lake ecosystem maintains the ecological security of the middle and lower reaches of the Yangtze River. The structure of the Poyang Lake basin integrates mountains, rivers and lakes. The area of the basin and the administrative borders of Jiangxi Province are highly consistent making it an ideal study area for basin-wide ecology. Although research on the level of Poyang Lake Basin has reached a certain level, it is still not sufficient in terms of holistic system understanding and research depth.

Key scientific issues in the fields of hydrology, water environment disciplines and aquatic ecology should be applied to strengthen both the protection of the ecological environment in the Poyang Lake Basin and the comprehensive utilization of resources. This combined research activities should investigate the control mechanism of nitrogen and phosphorus loads, the status of Poyang Lake Basin ecosystems, the succession of the Poyang Lake water resources and hydrological processes to the succession of Poyang Lake wetlands and its influence on the middle and lower reaches of the Yangtze River. Further research topics may include the migratory birds, the risk assessment of habitats of rare species and an integrated management approach for the Poyang Lake Basin [11].

Strengthening the scientific and technological support of the Poyang Lake management system cannot solely focus on the single lake itself. Poyang Lake has a unique and irreplaceable important role in economic construction, social development and ecological environment protection in the middle and lower reaches of the Yangtze River. The existence of mountains, rivers and lakes is one of the main characteristics of Poyang Lake. Mountains, rivers and lakes are forming a comprehensive hydrological system, which needs to be studied together. In addition, interactions with the Yangtze River Basin need to be included as the hydrology of Poyang Lake partially controls the water resources on the middle and lower reaches of the Yangtze River and vice versa. Therefore, it is necessary to strengthen the scientific and technological support for the management of Poyang Lake wetlands. A sustainable development of the Poyang Lake system in accordance to science must always adhere to the development concept of "pollution first and post-treatment" and needs to look on the water resources from the perspective of the entire catchment. Hence, future management of the Poyang Lake basin should be guided by in-depth scientific proof and careful decision-making.

9.3.3 Establish and Improve the Poyang Lake Monitoring and Early Warning System

Hydrology is the most important ecological element of wetland ecosystems. The monitoring of hydrological parameters provides the foundation for the judgment of wetland ecological environment and the diagnosis of its health status. Hence, hydrological monitoring is indispensable for the maintenance of the wetland ecosystem structure and function.

Although systematic monitoring, analysis and evaluation activities of six major projects on the natural environment of Poyang Lake, the hydrological climate, water level and sediments, storm floods, water quality and water quantity and experimental research go back until 1959, much work has been added in more recent years. However, the current monitoring frequency still does not meet the requirements [12].

To better cope with the adaptive ecosystem management strategy of the Poyang Lake Basin, it is proposed to further improve the monitoring and early warning system for the hydrological and ecological systems in the five main rivers of Jiangxi Province as well as in Poyang Lake [13]. Improved monitoring networks will be established to monitor meteorological and hydrological data, the soil moisture content, the groundwater level as well as the water quality [14]. The establishment of automatic collection network, transmission and computer application system will form a comprehensive information database for the Poyang Lake Basin which allows to adjust the Poyang Lake ecosystem management measures and to adopt emergency protection measures.

9.3.4 Pollution Control

In order to protect the aquatic environment of Poyang Lake, the control and reduction of pollutants into the lake need to be improved. Firstly, the construction of modern sewage treatment facilities in Poyang Lake and the upper reaches of the city needs to be accelerated to achieve a better treatment of sewage water before entering the lake. The reclamation of urban sewage is an effective way to alleviate urban water shortage, control the total discharge of pollutants and to reduce pollution. It is proposed to actively develop and introduce zero waste and water saving technologies to the industry to further reduce the discharge of sewage.

For sediment and non-point source pollution appropriate measures need to be adapted on local conditions. In areas with sparse population, ecological restoration is a suitable method. In agricultural production areas where population is relatively concentrated, the channel protection project, soil and water conservation measures for forest, grass and soil conservation should be implemented. Besides, measures such as energy alternative construction and animal husbandry, adjusting agricultural planting structure, developing water-saving irrigation schemes and reducing fertilizer and pesticide use should be introduced.

9.3.5 Improving Ecological Environmental Ability of Regulation and Control

To improve the ecological environmental ability of regulation and control in Poyang Lake, most important is ecological regulation which is suggested to include the following three aspects:

1. Establish Poyang Lake water conservancy hub project: Establishing of the Poyang Lake water conservancy hub project can realize hydrological regulation and control. Proper regulation and control of the water level in Poyang Lake improves the conditions for the species and communities so that the structure and function of the Poyang Lake ecosystem can be maintained and biological integrity and ecosystem health can be protected.
2. Strengthening Ecological Hydraulic engineering: Water is the key factor of wetland ecology. Strengthening ecological water conservancy construction and Hydraulic engineering are key points to protect wetland resources and to prevent the degradation of the wetland ecosystem. Water resources management should

actively play a regulatory role on management. It is important to promote water conservation, water-saving irrigation technology research and promotion.

3. Coordination of the relationship between wetland resource protection and agricultural development: Wetland protection and agricultural development have interactions and mutual restraints. Regional agricultural production requires the support of land and water resources and a less frequent occurrence of disasters. Wetlands are important sources of regional water resources and can mitigate floods and regulate the regional microclimate. They are of great significance to regional agricultural production. Furthermore, efforts should be untertaken to rationally adjust the structure of the agricultural industry and promote high-yielding and stable-cropping techniques, which are conducive to the rational utilization of wetland resources and are conducive to the implementation of the key protection of agricultural wetlands.

9.3.6 Developing Communities with the Characteristics of Poyang Lake Wetland

1. Wetland products should be rationally developed and exploited: Poyang Lake wetland has abundant plant resources, with 476 species of higher plants. Among them, more than 200 species with economic use are known. If the rich plant resources in Poyang Lake can be properly utilized, for example als medicinal herbs, the economic benefits will be very significant.
2. Promoting the Development of Ecotourism: Poyang Lake is the most important international wetland that serves as a habitat for waterfowls in China. With beautiful natural scenery, beautiful ecological environment, rich tourism resources, the basic conditions for developing ecotourism are excellent.
3. Creating a process of regional ecologically friendly industry: Industrial access requirements for the Poyang Lake ecological economic zone should be strictly formulated. A number of high-tech industries, modern processing industries and related industrial projects should be established to achieve the development of an emerging industrial chain in the wetlands.
4. Establishing long-term wetland protection rewards and compensation mechanism for wetlands: It is suggested that factors should be included in the examination contents of local governments and relevant departments, such as assessment results of wetland ecological protection, discharge of major pollutants, environmental quality, environmental input and capacity. The governmental agencies and departments being most successful in environmental protection should be rewarded. Besides, a policy of paid use of natural resources and the environment should be formulated.

References

1. State Forestry Administration of the People's Republic of China. 2000. China wetland protection action plan (in Chinese). China Forestry Press.
2. State Forestry Administration of the People's Republic of China. 2001. Guide on implementing ramsar convention in China (in Chinese).
3. Wang, S., and H. Dou. 1998. Chinese Lakes (in Chinese). Beijing: Science Press.
4. Zhang, B., G. Lu, and H. Zhu. 1988. Poyang Lake research (in Chinese). Shanghai Science and Technology Press.
5. Jiangxi Provincial Department of Water Resources. 1995. Jiangxi water conservation (in Chinese). China Forestry Press.
6. Zhu, H., and B. Zhang. 1997. Poyang Lake (in Chinese). Hefei: China Forestry Press.
7. Min, Q., G. Tan, and J. Ye. 2009. Main issues and regulation measures of Poyang Lake ecosystem (in Chinese). *China Water Resources* 11: 44–47.
8. An, C., and Y. Zhong. 2008. The study on social and economic reasons to degradation of Poyang Lake wetland ecosystem (in Chinese). *Environmental Science and Management* 33 (7): 136–140.
9. An, C., and Y. Zhong. 2008. The study on social and economic reasons to degradation of Poyang Lake Wetland Ecosystem. *Environmental Science and Management* 33 (7): 136–140.
10. Jian, M., H. You, and C. Ni. 2006. Characteristics of heavy metals contaminant status and migration in Raohe River of Lake Poyang. *Journal of Lake Sciences* 18 (2): 127–133.
11. Jiang, X., et al. 2009. Recovery strategies of degradation wetland ecosystem in Poyang Lake (in Chinese). *Jiangxi Forestry Science and Technology* 12 (6): 50–53.
12. Shao, B., S. Zhou, T. Gu, and Q. Li. 2010. Spatio-temporal changes of NDVI and their relationships with runoff in the Poyang Lake basin (in Chinese). *Chinese Journal of Agrometeorology* 31: 288–294.
13. Yin, J., H. Wang, M. Zhan, P. Kong, and Z. Jiang. 2011. Analysis of the climate change law of Poyang Lake basin in the past 50 years (in Chinese). In *Proceedings of the first China Lake forum*, vol. 9.
14. Wang, F., D. Wu, and R. Li. 2008. Analysis on flood disaster characteristics in Lake Poyang region (in Chinese). *Lake Science* 20 (4): 500–506.

Chapter 10
Distribution of Aquatic Macrophytes in the Le'an River and Its Indicative Evaluation on Heavy Metal Pollution

Minfei Jian

10.1 Introduction

Macrophyte is an ecologically-diverse taxon that consists of a type of different groups of plants that adapt to the aquatic environment over a long period of time, such as aquatic vascular plants and higher algae (Chara, etc.). Macrophytes can be classified into four different types based on growth form: Emergent macrophytes, floating macrophytes, floating-leaved macrophytes and submergent macrophytes. Aquatic macrophytes exist between water-sediment, water-air and water-land interface. They are important components of the lake ecosystem and are important regulators of lake evolution and lake ecological balance. Besides, they improve water quality and increase the stability of water environment by inhibiting the growth of phytoplankton and thus reduce the stock of algae. So they are often used as indicators of changes in the lake environment and as a basis for establishing good lake ecology.

With the rapid economic development in various regions and the continuous utilization of lake resources, many lakes in China, especially the lakes in the middle and lower reaches of the Yangtze River, suffer serious eutrophication and algae blooms. Much of the lake aquatic vegetation has disappeared and the water quality has deteriorated. On the other hand, in some shallow weedy water lakes, the structure of the lake's aquatic plant community has changed due to the adjustment and change of fishery structures, such as the rapid development of aquaculture. The proportion of marsh plants has increased and lake bogginess has exacerbated. Therefore, the dynamic changes and distribution characteristics of aquatic macrophytes can accurately reflect the changes of lake water quality and the environmental characteristics of sediments. They can be used as indicator organisms for monitoring and evaluating the water quality of lakes.

M. Jian (✉)
Jiangxi Normal University, Nanchang, China
e-mail: jianminfei0914@163.com

T. Yue et al. (eds.), *Chinese Water Systems*, Terrestrial Environmental Sciences,
https://doi.org/10.1007/978-3-319-97725-6_10

As the largest freshwater lake in China, Poyang Lake is also the only of the four large Chinese lowland lakes without large eutrophication problems. In the Poyang Lake wetland ecosystem, aquatic macrophytes play a crucial role and are the major primary producers and important biological resources of the system. In the 1980s, 102 species of aquatic vascular plants divided into 38 families were found at Poyang Lake. These large aquatic plants are the main food source for many aquatic animals and birds in the Poyang Lake wetland ecosystem. Further, the rich diversity of macrophytes provides a variety of good habitat and spawning sites. They can be utilized by society as food, vegetables, herbs, fertilizers, weaving and papermaking raw materials. In addition, the macrophytes in Poyang Lake also play an indispensable role in maintaining water transparency, purification of pollutants and improving the oxygen saturation of the lake. However, the rapid economic development and continuous utilization of lakes in Poyang Lake area lead to a more and more serious loss of aquatic macrophytes in the Poyang Lake wetland area: The distribution area is shrinking gradually, more and more macrophyte species extinct, the total coverage area and the biomass of macrophytes decreases. This effects lead to the serious degradation of the lake ecosystem and the degradation of aquatic products, seriously affecting the yield and quality of aquatic products and threatening the sustainable development of aquatic products.

Le'an River is one of the five tributaries of the Rao River in Poyang Lake. Mineral resources such as copper, gold, silver, lead, zinc, manganese, coal and limestone are abundant in the river basin making it an important non-ferrous metal industrial base in northeast Jiangxi. Asia's largest porphyry open-pit copper mine—Dexing Copper Mine is located near the Le'an River, too. The deposit contains mainly copper sulfide, pyrite and polymetallic sulfide. At present, Dexing Copper Mine has a decisive influence on the economic development of Jiangxi and the proposed Poyang Lake Ecological Economic Zone.

In the early 90's, some studies have reported that the excessive discharge of sour and soda wastewater contaminated by heavy metals from the mining activities has destroyed the balance of river ecosystem in the Le'an River. In some river sections, heavy metals such as copper, lead and zinc seriously have exceeded which leads to recession and deterioration of the quality of the water and the underlying sediment. As heavy metals are difficult to degrade, accumulate easily and have a high toxicity, they present a severe harm to crop growth, yield and quality. When they are absorbed by biological enrichment, they will enter the food chain and groundwater bodies, people, livestock, birds and other life and health can be endangered. Therefore, the heavy metal pollution in the Le'an River Basin has been subject to great concern.

In this chapter, the community structure and distribution characteristics of aquatic macrophytes in the Le'an River-Poyang Lake are studied for three different water stages of the Le'an–Poyang Lake section. The survey data made a study on the plant composition, plant community structure and diversity of the Le'an River wetland. The physical and chemical analysis of the wetland water bodies, sediment and wetland topsoil were carried out. Through comparative analysis of background value and risk analysis, the status of heavy metal pollution, distribution characteristics, sources of pollution and environmental risk in Le'an River wetland water and soil environment were discussed. At the same time, based on the adaptation of Le'an River wetland ecosystem to the environment, response about aquatic macrophytes community to environmental factors were carried out.

10.2 Methodology

This study aims to investigate the status of heavy metal pollution and the structure and diversity of aquatic macrophytes community in the Poyang Lake—Le'an River Basin wetland which is located in the Rao River Basin (28°55′41 −29° 1′21″ N, 116° 29′35″ − 1117°48′24″ E. Le'an River originates from the southwestern foothills of Wulong Mountain at an altitude of 1496 m in the northeast of Wuyuan County on the border between Jiangxi and Anhui. It is 279 km long and has a basin area of 8456 km^2. The study identifies the major environmental factors that affect the aquatic macrophytes community in the Le'an River wetland and to analyze the response of the aquatic macrophytes community to the spatial and temporal distribution of heavy metals. In April, July and November 2011, representing average, high and low water level conditions, 13 sampling sites were set up in the section of Le'an River–Poyang Lake and its tributaries. All sampling points were georeferenced using the GPS positioning system. The sampling points are numbered in downstream direction and labeled as follows: Haikou Town (T1), Shanggukou village (T2), Xiagukou village (T3), Xiangtun hydrological station (T4), Jiedu Bridge (T5), Leping Bridge (T6), Qingjiawan village (T7), Hujia Factory (T8), Longkou village (T9), etc. Dawu River (D1), Dexing copper mine life area (D2), Dexing Copper mine Middle School (D3), Dexing copper mine production area (D4) (Fig. 10.1).

10.2.1 Collecting and Processing of the Samplings

The vegetation survey was conducted as a sample survey. Three $10 \times 10\,m^2$ standard sampling quadrat were repeated as sampling area of each sampling site. Using the so-called "Five point sampling methods" (Fig. 10.2), five quadrat with an area of $1\,m^2$ were selected within each $10 \times 10\,m^2$ standard sampling quadrat to investigate the species of large aquatic plants, plant cover, plant height and other indicators. In additional samples of the surface water, the underlying sediments and the surface soil were collected separately at each sampling point. The three samples of macrophytes were filled into polyethylene plastic bottles for the determination of heavy metal water samples, which were acidified with nitric acid to achieve a pH of less than 2. In each sampling site of the three $10 \times 10\,m^2$ standard sampling quadrats, the topsoil of the area was collected by the five point sampling methods. Firstly, the topsoil (10 cm thickness) in root zone was collected by using PVC sampling tube in each standard sampling quadrat, then the topsoil of the five sampling points were removed and mixed into one polyethylene bag followed by sealing, numbering and

Fig. 10.1 Location of the sampling sites in the study area

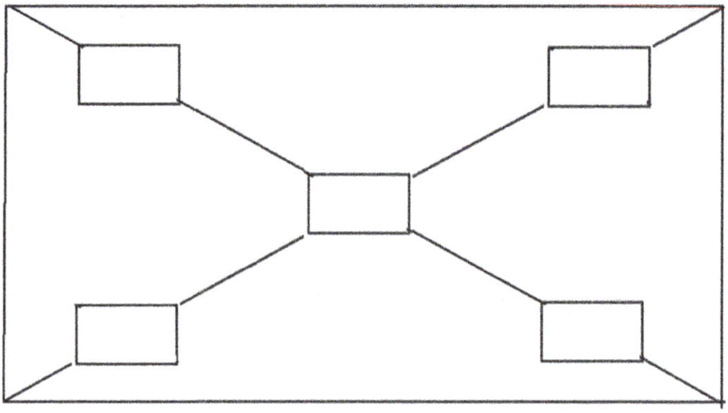

Fig. 10.2 Five point sampling method in each $10 \times 10\ m^2$ standard sampling quadrats

recording of the bag. The sediments were collected by using a grab sampler, which sampled sediment up to a depth of 15 cm below water level and repeated five points in each standard sampling quadrats, and then also mixed into one polyethylene bag too, and also sealing, numbering and recording. Each sampling site also had 3 repeated standard sampling quadrats.

In this study, 39 samples of the surface soil of the root zone, 39 sediments samples and 39 water samples were collected at the 13 sampling sites. All samples were taken to the laboratory for pretreatment and repeated testing. All water samples were stored in the refrigerator at 4 °C before analysis and treatment. Each sample was repeatedly tested 3 times. The topsoil of the the root zone and water sediments were dried at room temperature naturally, removing stones, plant debris and other foreign matter, and then oven dried to constant weight, with a mortar after grinding 60 mesh nylon sieve, respectively, shrink Sub-processed into polyethylene sealed bag for later use.

10.2.2 Analytical Methods

10.2.2.1 Determination of Physicochemical Indices of Overlying Water

The physicochemical indices of overlying water including Chemical oxygen demand (CODcr), total phosphorus (TP), total nitrogen (TN) and pH value were determined according to the requirements of the State Environmental Protection Administration of 2002 [1]. The pH value was determined by a precision pH meter. In order to reduce the error, each water sample was measured three times and the average value was used (Table 10.1).

10.2.2.2 Pretreatment and Determination of Heavy Metals in the Water and Soil Samples

Sample pretreatment of the water, soil and sediment was conducted in the MDS-8G multi-flux sealed microwave chemical workstation of Shanghai Xinyi. Chemical digestion using the HNO_3 digestion system was applied for the water samples and soil or sediment samples were digested by using the HNO_3-HCl-H_2O_2 digestion system (6 ml of 65% HNO_3, 3 ml of 36% HCl, 0.25 ml of 30% H_2O_2) [2]. After the chemical digestion and cooling, all samples were filtrated to a constant volume of ultrapure water (Nico, LT-RF20) was added. This processing steps were repeated done 3 times for each sample in order to reduce errors, and all treated samples were determined by using AA-240Z atomic graphite furnace spectrophotometer produced by the Varian company (USA). In the process of heavy metals analysis in soils or sediments, all used reagents are guarantee reagents (GR). In order to ensure the whole process quality control, the determination of heavy metals concentrations in GSS-3 (Identification value and uncertainty value) standard soil samples was repeated [3].

Table 10.1 The measuring methods of physicochemical indices of overlying water

Indicators	Determination methods
pH	By precision pH meter
COD_{cr}	By potassium dichromate Oxidation
TP	By potassium persulfate Oxidation-spectrophotometry
TN	By alkaline potassium persulfate Oxidation-Ultraviolet spectrophotometry

Table 10.2 The results of soil standard material samples by microwave-digestion

	Cu(mg/kg)	Pb(mg/kg)	Cd(mg/kg)
Determination value	12.01	24.69	0.059
GSS-3 (Identification value and uncertainty value)	11.4 ± 1.1	26 ± 3	0.06 ± 0.009

The determination results of GSS-3 standard soil samples are shown in Table 10.2. The determination results of the standard samples were within the agreed range. The relative standard deviations of heavy metals Cu, Cd and Pb were 3.89%, 1.90% and 2.57%, respectively, which means that the digestion method has high accuracy in comparison to GSS-3 standard method.

10.2.3 Data Processing Method

10.2.3.1 Methods for Plant Diversity Indices Estimation

Different diversity indices (Patrick species richness, Shannon-Wiener diversity index, Simpson dominance and Pielou evenness) of each sampling site were calculated by a field survey. The spatio-temporal distribution of vegetation diversity in Le'an River was analyzed and the software Spss18.0 was used to determine the Pearson correlation coefficient for all four different diversity indices. In addition, the TWINSPAN two-way indicator analysis method was used to cluster the aquatic macrophytes community in Le'an River wetland.

The formula for each diversity index is defined as following:

Patrick Species Richness Index (S):

$$S = S' \tag{10.1}$$

In Eq. (10.1), *S*' is the total number of species in the survey sample quadrat.

Shannon-Wiener Diversity index (H)

$$H = -\sum_{i=1}^{n}(n_i/N)log(n_i/N) \tag{10.2}$$

In Eq. (10.2), n_i is the number of individuals of the aquatic plant in the measured quadrat area and *N* is the total number of aquatic plants in the quadrat.

Simpson Index (D):

$$D = 1 - \sum(P_i)^2 \tag{10.3}$$

In Eq. (10.3) P_i is the important value of any species.
Pielou (J):

$$J = H/ln(S) \tag{10.4}$$

In Eq. (10.4) , H is the Shannon-Wiener diversity index of the sampling site and S is the total number of a species in the sampling site.

10.2.3.2 Analysis Methods of Spatial and Temporal Distribution Characteristics of Heavy Metals in the Soil and Water Environment

The concentrations of heavy metals Cu, Pb and Cd measured in soil and water environment of each sampling quadrat were determined and the content and distribution characteristics of heavy metals in the water and soil environment of Poyang Lake and Le'an River were analyzed. The Pearson correlation coefficients of heavy metals in the water and soil environment at the same water period and the same heavy metal elements in different water periods were both analyzed using the software SPSS 18.0.

10.2.3.3 Pollution Risk Assessment Method of Heavy Metal in Soil and Water Environment

Taking the Poyang Lake wetland soil environmental background and the GB15618-1995 soil environmental quality grade standard as evaluation criteria, 4 different methods including the single factor pollution index method, geoaccumulation index method, the Nemerow exponent method and the potential Ecological Risk Index method were used to evaluate the pollution status of heavy metals Cu, Pb and Cd in the wetland of Le'an River and Poyang Lake [3] (Table 10.3).

Table 10.3 The background values of heavy metals concentration in the wetland sediments of Poyang Lake and the grade standard for soil environment quality (mg kg^{-1})

Element		Cu	Pb	Cd
Background value of sediment in Poyang Lake		4.75	12.50	0.75
Grading by the soil environmental quality standard	Slightly polluted	35	35	0.3
	Medium polluted	100	250	0.75
	Heavily polluted	400	500	1
	Very heavily polluted	>400	>500	>1

Single Factor Pollution Index

The single contamination index method can be used to evaluate the pollution status of each pollutant [4]. The formula is given by:

$$P_i = C_i/S_i \tag{10.5}$$

In Eq. (10.5), P_i is the single factor pollution index of pollutant i. C_i is the measured value of pollutant i, (in mg kg^{-1}). S_i is the evaluation standard value of pollutant i in soil (mg kg^{-1}). In this study, the environmental quality grading system for heavy metals in wetland soils (Table 10.3) was selected as the evaluation criteria for the degree of heavy metal pollution in the study area. Depending on whether the ratio P_i of the measured concentration to the evaluation standard at each sample point is greater than 1, it is judged whether the corresponding standard is reached and the category of the evaluation index is determined. Finally, the worst category reached by each evaluation index is taken as the comprehensive evaluation result for the water sediment at each sampling site.

Geoaccumulation Index

The Geoaccumulation index method was proposed by the German academic Muller to study heavy metal pollution in sediments of the aquatic environment. This method directly reflects that the accumulation of exogenous heavy metals in the sediments of water bodies and the dataset is highly comparable. In recent years, it is widely used in domestic and international heavy metal pollution assessment of sediments [5]. The formula is:

$$I_{geo} = log_2 C_n/(K \cdot \beta_n) \tag{10.6}$$

In Eq. (10.6), C_n is the measured concentration (mg/kg) of some heavy metal in the sediment of the evaluation area,; βn is the reference value, which is the average geochemical background value of this heavy metal in the evaluation area. The background values of the average geochemical background value of the heavy metals Cu, Pb and Cd are 4.75, 12.50 and 0.75 mg kg^{-1}, respectively. K is a constant that is set in consideration of the fluctuation of background values caused by geological processes. In this study the K value of 1.5 was used. The index grading standards are given in Table 10.4.

Table 10.4 Standard for grading of different heavy metals pollution indices (i: Unpolluted. ii: Slightly polluted. iii: Moderately polluted. iv: Heavily polluted. v: Extremely polluted)

	Degree/class of pollution						
	i	ii	iii	iv	v	vi	vii
Nemero pollution index P_t	≤1	1–2	2–3	3–4	4–5	5–6	>6
Geoaccumulation index I_{geo}	≤0	0–1	1–2	2–3	3–4	4–5	>5
Potential ecological risk index RI	≤20	20–80		80–160		160–240	>240

Nemerow Pollution Index

The Nemerow Pollution index method was developed by the American Professor Nemerow and is typically applied to evaluate the average and extreme environmental pollution by multiple heavy metals at single points [6]. The Nemerow pollution index can be calculated as follows:

$$P_i = \sqrt{\frac{(\frac{C_i}{S_1})^2_{ave} + (\frac{C_i}{S_1})^2_{max}}{2}} \tag{10.7}$$

In Eq. (10.7), P_t is the comprehensive pollution index of heavy metal pollution in the sediments of the evaluation area; The definition of the values C_i and S_i is similar to the Single -factor pollution index method (Eq. (10.7)) [7]. $(C_i\ /S_i)_{max}$ is the maximum single pollution index ratio of heavy metal elements in the soil at the monitoring site. Subsequently, the ratio $(C_i\ /S_i)_{ave}$ is the average single pollution index ratio of heavy metal elements in the soil at monitoring sites. The relationship between the Nemerow Pollution Index P_t and the pollution level is shown in Table 10.4.

Potential Ecological Risk Index

The potential ecological risk index method is proposed by the Swedish scholar Hakanson in 1980 and can be used to quantitatively analyze the potential ecological risk of single pollutant and multi-factor integrated pollution in the assessment area in connection with toxicology, environmental effects and ecological effects [8]. The index can be calculated by a set of equations:

$$RI = \sum_{i=1}^{n} E_r^i \tag{10.8}$$

$$E_r^i = T_r^i \times C_f^i \tag{10.9}$$

$$C_f^i = C_s^i / C_n^i \tag{10.10}$$

In the Eqs. (10.8)–(10.10), *RI* is the comprehensive ecological risk index of the heavy metals in sediments. is the single coefficient of potential ecological risk of the heavy metal in sediments; is the toxicity coefficient of heavy metals (Cu = Pb = 5 < Cd = 30); is the enrichment coefficient of heavy metal i. is the measured value of heavy metal i in the sample and is the reference value of the sediment background Table 10.3. The levels of pollution corresponding to the Potential Ecological Risk Index are shown in Table 10.4.

10.2.3.4 Methodology to Analyse the Aquatic Macrophytes Community and the Environmental Impact Factor

The relationship between aquatic macrophytes communities and environmental factors was assessed using the Detrended Canonical Correspondence Analysis method (DCCA) [9]. DCCA can be combined with multiple environmental factors to better reflect the relationship between communities and their environment. Le'an River has long been affected by both, the exploitation of heavy metal minerals in the basin and the industrial and domestic sewage from a number of large and medium-sized cities. In order to surely identify the main environmental factors of the Le'an River wetland and the current status of the Le'an River pollution, the heavy metal elements Cu, Pb, Cd as well as the parameters pH, TP, TN, COD and additional 13 environmental factors were included in the analysis. The DCCA was conducted by the software Canoco [10].

10.2.3.5 Data Statistical Analysis Method

Analysis of variance (ANOVA), multiple comparisons and correlation analysis were performed by the statistical software SPSS (Version 18) [11]. The experimental data were expressed as mean ± standard deviation and the accuracy of the data was analyzed. The differences of heavy metal contents in topsoil, overlying water column and the wetland between different sampling sites in the same water phase were tested using Duncan's Multiple range test of Duncan's One-Way ANOVA with a significant level of $p = 0.05$. Pearson's bivariate correlation analysis was used to analyze the correlation between Patrick's species richness, Shannon-Wiener Index, Simpson dominance and Pielou evenness, as well as the correlations of heavy metal elements in overlying water, sediment and wetland in the same water period and in the different water periods. The significance of the derived correlation coefficients was analysed by twin-tailed test using significance levels of $p = 0.05$ and $p = 0.01$ to identify significant and very significant relationships.

Plant community classification was analysed by the statistical cluster analysis software TWINSPAN [12]. The aquatic macrophytes community of each sampling point in different water periods was classified based on the species important value. The species were classified by 5 classes using the following class boundaries as follows: 0–0.01,0.01–0.1,0.1–0.3,0.3–0.5,0.5–0.8. The class with the largest distribution is class 4.

The aquatic macrophytes communities and environmental factors were sorted by the ecological statistical analysis software CANOCO,. Detrended Canonical Correspondence Analysis (DCCA) was used to reveal the effect of different environmental factors on the large-scale aquatic plant community in the same water stage. The most important influencing factors and their correlation coefficients were sorted and ranked using significance levels of $p = 0.05$ and $p = 0.01$.

10.3 Dynamic Characteristics of the Aquatic Macrophytes Community in Poyang Lake-Le'an River Section

10.3.1 Floristic Composition Analysis of Plants Family, Genera and Species

According to the broad definition of aquatic plants, the aquatic plants in Le'an River wetland are divided into five life forms: hygrophyte, emerged plant, macrophyte, floating leaf formation and submerged plants.

According to the survey of rivers and lakes along the Le'an River during three different water periods (dry season, wet season and normal water period) in 2011, a total of 48 families, 124 genus and 166 species of aquatic macrophytes were found. Among them, there are 13 species of emerged plant, 1 species of floating leaf formation, 1 species of macrophyte, 8 kinds of submerged plants and 145 species of hygrophyte.

Among the 166 species of aquatic macrophytes, there are 1 familiy, 1 genera and 1 species fern: *Nephrolepis cordifolia*, which is *Nephrolepis* genera and Nephrolepidaceae familiy. The are 47 families, 123 genera and 165 species of angiosperms, that is accounting for 60.25% of the total families of angiosperms, 50.41% of the total genera and 35.79% of the total species in Poyang Lake wetlands. All the seed plants (Spermatophyta) in the study area are angiosperms, of which 113 species belong to 84 genera and 36 families are dicotyledons, while 52 species belong to 39 genera and 11 families to monocotyledons.

The composition of different families and genus of plants in the wetland of Le'an River is very different. In the 48 families, there are 6 families with more than 5 genus and there are 60 genus in the six families, that is accounting for 12.5% of the total number of families and 48.39% of the total genus respectively (Table 10.5). The vegetation families of Umbelliferae, Cyperaceae, Lamiaceae, Poaceae, Asteraceae, Cruciferae occupy an important position in the genus of wetland vegetation in the

Le'an River. Among them, the number of genus is largest for Asteraceae and Poaceae. However, there are 29 Monogeneric plant families in Le'an River wetland, accounting for 60.41% of the total number of families and 23.39% of the total number of genera.

The Le'an River wetland not only contains a huge difference in genus for each family, but also the number of included species for differs significantly for each genus. Polygonum contains the largest variety of in total 8 species but Carex, Cyperus, Artemisia, Stellaria, Cardamine and Ludwigia contain many species, too. The vegetation with more than three species are accounting for 5.64% of the total genus and 18.02% of the total species.

Asteraceae, Leguminosae and Poaceae are major plant families with more than 10,000 kinds of species each in the world. In addition, Rosaceae is the world's fifth largest family with about 5,000 species of *Cyperaceae*, 3,000 *Cruciferae*, 3000 *Umbelliferae*, 700 *Lamiaceae* and 200 *Polygonaceae*. These 9 families together account for only 18.36% of all families in Le'an River wetlands, but their genus and species account for 56.01% and 60.23% of all kinds, respectively. This demonstrates the dominance and importance these major families in the composition of flora in the region (Table 10.5).

However, the unispecific family and depauperate family account for 71.64% of the total vegetation in the Le'an River wetland indicating that the region is unstable in species composition. As the environment degrades and man-made damage worsens, some sparsely distributed species can easily disappear and species diversity can change very dynamically. For example, there was a large area of submerged plants a decade ago in the T4 sampling site. Submerged plants are rare in this area because they are sensitive to contamination by industrial and mining enterprises and domestic sewage. At the same time, decreasing levels of pollution in the study area formed a diversity of flora composition and the complexity of evolutionary succession.

10.3.2 Diversity of Aquatic Macrophytes

10.3.2.1 Diversity of Aquatic Macrophytes at Different Water Periods

The number of species in Patrick's species richness index varies from 3 to 57 during the investigating period,and the number of species in sampling site T1 was 57 in the wet season, which was the largest number of species, and the number of species in sampling site T3 was 3 in the Low-water period, which was the least number of species (Table 10.6).

The value of Shannon-Wiener index is between 0.23 and 2.84 during the whole year (Table 10.6). Sampling Site T1 had the highest value of 2.84 in the normal water period and sampling point D1 was associated with the lowest value of 0.23 which was obtained for the wet season. The range of the Simpson dominance in the different water period was from 0.1 to 0.97: The maximum appeared at D4 during the dry season and the lowest appeared at D1 in the normal water period, which was 0.1. The Pielou's evenness index varied from 0.21 to 0.93 during the three different water

Table 10.5 Statistics of vascular plant families in the wetland of Le'an River

Family	Genus (No.)	Species (No.)	Genus (%)	Species (%)
Umbelliferae	5	6	4.03	3.61
Rosaceae	3	4	2.42	2.41
Cyperaceae	6	15	4.84	9.04
Primulaceae	1	1	0.81	0.60
Lamiaceae	8	8	6.45	4.82
Euphorbiaceae	1	1	0.81	0.60
Polygonaceae	4	12	3.23	7.23
Violaceae	1	2	0.81	1.20
Poaceae	18	20	14.52	12.05
Geraniaceae	1	2	0.81	1.20
Amaranthaceae	3	5	2.42	3.01
Solanaceae	1	1	0.81	0.60
Papaveraceae	2	2	1.61	1.20
Juncaceae	1	1	0.81	0.60
Asteracea	18	22	14.52	13.25
Commelinaceae	1	1	0.81	0.60
Moraceae	1	1	0.81	0.60
Oxalidaceae	1	1	0.81	0.60
Scrophulariaceae	3	4	2.42	2.41
Saururaceae	1	1	0.81	0.60
Ranunculaceae	1	2	0.81	1.20
Asclepiadaceae	1	1	0.81	0.60
Boraginaceae	1	1	0.81	0.60
Dioscoreaceae	1	2	0.81	1.20
Plantaginaceae	1	1	0.81	0.60
Campanulaceae	1	1	0.81	0.60
Caryophyllaceae	3	5	2.42	3.01
Onagraceae	1	3	0.81	1.81
Cruciferae	5	9	4.03	5.42
Alismataceae	1	1	0.81	0.60
Urticaceae	2	2	1.61	1.20
Acanthaceae	1	1	0.81	0.60
Adoxaceae	1	1	0.81	0.60
Portulacaceae	1	1	0.81	0.60
Nephrolepidaceae	1	1	0.81	0.60
Pontederiaceae	1	1	0.81	0.60
Vitaceae	2	2	1.61	1.20
Lentibulariaceae	1	1	0.81	0.60
Araceae	3	3	2.42	1.81
Gentianeae	1	1	0.81	0.60

(continued)

Table 10.5 (continued)

Family	Genus (No.)	Species (No.)	Genus (%)	Species (%)
Leguminosae	4	4	3.23	2.41
Potamogetonaceae	1	2	0.81	1.20
Apocynaceae	1	1	0.81	0.60
Ceratophyllaceae	1	2	0.81	1.20
Hydrocharitaceae	2	2	1.61	1.20
Lemnaceae	1	1	0.81	0.60
Convolvulacea	2	2	1.61	1.20
Rubiaceae	2	2	1.61	1.20

Table 10.6 Comparison of species diversity indices for different water seasons of Le'an River

	Normal water period				Wet season				Dry season			
	S	H	D	J	S	H	D	J	S	H	D	J
T1	42	2.84	0.89	0.76	24	2.46	0.89	0.77	57	2.68	0.91	0.66
T2	30	2.36	0.88	0.69	6	1.17	0.65	0.65	9	1.61	0.77	0.73
T3	3	0.90	0.77	0.82	6	1.13	0.61	0.63	16	1.85	0.79	0.67
T4	40	2.61	0.96	0.71	6	1.24	0.60	0.69	12	2.30	0.89	0.93
T5	21	2.64	0.90	0.87	14	1.45	0.67	0.55	19	1.82	0.76	0.62
T6	18	2.58	0.91	0.89	10	1.44	0.70	0.63	28	2.31	0.84	0.69
T7	21	2.63	0.91	0.86	17	1.41	0.60	0.50	19	1.77	0.73	0.60
T8	23	2.82	0.93	0.90	13	1.76	0.75	0.69	17	1.64	0.70	0.58
T9	30	2.56	0.84	0.75	8	1.49	0.69	0.72	13	1.52	0.70	0.59
D1	9	1.45	0.28	0.66	3	0.23	0.10	0.21	7	0.59	0.27	0.31
D2	30	2.42	0.97	0.71	14	1.63	0.77	0.62	9	1.54	0.74	0.70
D3	37	2.39	0.96	0.66	4	0.65	0.34	0.47	7	1.29	0.67	0.66
D4	43	2.54	0.97	0.67	14	1.28	0.58	0.49	11	1.52	0.72	0.64
Average	26.7	2.37	0.86	0.77	10.7	1.33	0.61	0.59	17.2	1.73	0.73	0.64

periods. The maximum value was 0.93, which appeared at T5 and the lowest value was 0.21, which was at D1. Averaging these four kinds of diversity indices in Le'an River shows that all applied diversity indices approximately have the same trend. Values tend to be highest for the normal water period, followed by the dry and finally the wet period.

10.3.2.2 Distribution of the Aquatic Macrophytes Along the River Course

Patrick species richness fluctuates greatly for the different reaches of rivers. However, Patrick species richness is largest in the headwaters of the river (sampling site T1) during all three sampling periods. Species richness rapidly decreased in flow direction for the next sampling points T2 and T3. Species richness increased again in middle and lower reaches but resulting values were still lower than in the headwaters of the river (Fig. 10.3a).

The Shannon-Wiener index in the normal water, wet and dry periods is lowest at sampling point D1 with values of 0.9, 0.23 and 0.27 respectively. The highest index still appeared at T1, which were 2.84, 2.46, 2.68. The Shannon-Wiener index in the headwaters and middle and lower reaches of Le'an River are higher than in its upper reaches (T2 and T3) but also higher than in the Dawu River (Fig. 10.3b).

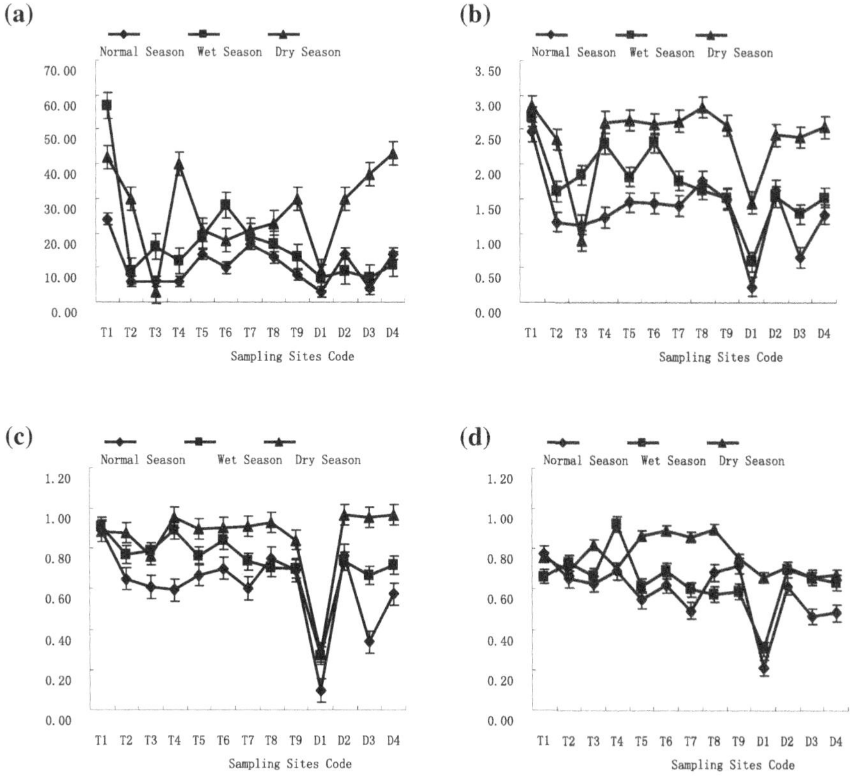

Fig. 10.3 Spatio-temporal distribution characteristics of different species diversity indices in the Wetland of Le'an River (A: Patrick species richness. B: Shannon-Wiener index. C: Simpson dominance. D: Pielou evenness)

The Simpson dominance at D1 was lowest with values of 0.28, 0.27 and 0.1 during the three water periods. The variation of this index values along the river course of Le'an River and the tributaries Dawu River are relatively low except for Sampling site D1, which causes a total range of values between 0.58 and 0.96 (Fig. 10.3c).

The Pielou Evenness index does not show a regular pattern along the river course of Le'an River, although station D1 has the lowest Pielou Evenness in each sampling period (Fig. 10.3d).

10.3.3 Correlation Analysis of Aquatic Macrophyte Communities

Pearson correlation analysis was conducted between each pair of the 4 indices (Table 10.7). The results show that there was a significant or very significant correlation between Patrick species richness, Shannon-Wiener index and Simpson dominance ($p < 0.05$, $p < 0.01$), while the correlation of Pielou evenness, Patrick species richness, Shannon-Wiener index and Simpson dominance was not significant ($p > 0.05$). This indicates that the evenness of species in the normal water period is low, which may form some dominant species.

The correlation analysis showed that there was a significant positive correlation ($p < 0.01$) between Shannon-Wiener index, Simpson dominance and Pielou evenness during the wet and dry periods, which means that all three indices can be used to parametrize plant diversity in Le'an River.

Table 10.7 Correlation analysis of four species diversity index for different water seasons

Water period		S	H	D	J
Normal	S	1			
	H	0.707**	1		
	D	0.594*	0.646*	1	
	J	−0.448	0.192	0.233	1
Wet	S	1			
	H	0.841**	1		
	D	0.686**	0.946**	1	
	J	0.385	0.813**	.896**	1
Dry	S	1			
	H	0.748**	1		
	D	0.503	0.916**	1	
	J	0.085	0.687**	0.852**	1

(*Note* *Correlation is significant at the 0.05 level (two-way test), **Correlation is significant at the 0.01 level (two-way test))

10.3.4 Community Structure of Aquatic Macrophytes

In order to better reveal the structure and classification of wetland plant communities, TWINSPAN was used to study the plant communities in the study area (Table 10.8). In 2011, 13 sampling plots in Le'an River were investigated for normal, wet season and dry season. Two-way Indicator Species Analysis (TWINSPAN) was used to cluster analysis of plant communities in the study area.

The results show that the 13 sampling sites were divided into 6 groups during the normal water period and divided into 5 groups during the wet and dry periods. According to the classification and nomenclature of "Chinese wetland vegetation" and the classification results of TWINSPAN, based on the indicator species and their combination of the habitat characteristics of river wetlands, the vegetation was divided into five different plant (Table 10.9).

Although, some differences in the plant community composition of the five plant groups, which are divided into three different water periods in the Le'an River wetland exist (Table 10.9), there are some common features between plant communities along the river flowpath. According to their distribution, they are mainly divided into the junction of Dawu River, the source and the confluence of Dawu River and Le'an River. Independently from the season, the main species in Dawu River are *Phalaris arundinacea and Cynodon dactylon* accompanied by *Oplismenus undulatifolius, Bidens pilosa, Cerastium fontanum, Hemarthria altissima* and *Viola mandsurica.*

Table 10.8 Two-way table species and quadrat classification by TWINSPAN

	Sampling sites code												
Investigation Period	D1	T5	T6	T7	T8	T9	T1	T2	T4	D2	D3	D4	T3
Normal	0	0	0	0	0	0	1	1	1	1	1	1	1
	0	1	1	1	1	1	0	0	0	0	0	0	1
		0	1	1	1	1	0	0	0	1	1	1	
	I	II	III				IV			V			VI
Sample Point	T1	T3	D2	D4	T5	T7	T8	D3	T4	T6	T9	T2	D1
Wet	0	0	0	0	0	0	0	0	0	0	0	1	1
	0	0	1	1	1	1	1	1	1	1	1		
			0	0	1	1	1	1	1	1	1		
					0	0	0	0	1	1	1		
	I		II		III				IV			V	
Sample point	T7	T8	T9	T2	T3	T5	T6	D1	T1	D4	T4	D2	D3
Dry	0	0	0	0	0	0	0	0	0	0	0	1	1
	0	0	0	1	1	1	1	1	1	1	1		
				0	0	0	0	0	0	0	1		
				0	0	0	0	0	1	1			
	I			II					III		IV	V	

Table 10.9 Communities structures of the main vegetation types at different water periods

Water period	Comm-unity number	Sampling sites code	Name of main river section	Type of vegetation community
Normal season	I	T3, D1	Merger of the Le'an River and Dawu River	*Miscanthus sinensis-Cynodon dactylon, Rhizoma Cyperi, Phalaris arundinacea*
	II	T5	Midstream of the Le'an River	*Rumex japonicus + Cynodon dactylon, Polygonum criopolitanum, Erigeron canadensis, Carex spp.*
	III	T6, T7, T8, T9	Down-stream of the Le'an River	*Carex cinerascens + Cynodon dactylon, Soliva anthemifolia, Alopecurus aequalis,Conyza canadensis, Polygonum hydropiper, Potamogeton crispus*
	IV	T1, T2, T4	Source and upper-stream of the Le'an River	*Miscanthus sinensis-Phalaris arundinacea + Glechoma longituba + Cynodon dactylon, Fagopyrum cymosum? Hemarthria altissima, Ranunculus sceleratus*
	V	D2, D3, D4	Dawu River Section	*Phalaris arundinacea + Cynodon dactylon + Malachium aquaticum, Viola philippica, Cerastium caespitosum, Galium aparin*
Wet Season	I	T1, T3	Source and upper-stream of the Le'an River	*Fagopyrum cymosum + Oenanthe javanica, Cyperu srotundus, Ludwigia prostrata, Alternanthera Philoxeroides, Centella asiatica*
	II	D2, D4	Dawu River Section	*Cynodon dactylon+ Phalaris arundinacea, Lophatherum gracile, Oplismentls undulatifolius, Bidens pilosa*
	III	T5, T7, T8, D3	Mid-stream and down-stream of the Le'an River	*Cynodon dactylon, Hemarthria altissima, Lobelia chinensis, Polygonum hydropiper, Cynanchum stauntonii*
	IV	T4, T6, T9	Mid-stream of the Le'an River	*Carex spp + Phalaris arundinacea + Hemarthria altissima, Polygonum hydropiper?Alternanthera philoxeroides? Vallisneria natans? Polygonum amphibium*
	V	D1, T2	Merger of the Le'an River and Dawu River	*Miscanthus sinensis-Cynodon dactylon + Miscanthus sinensis-Alternanthera philoxeroides,Rhizoma Cyperi, Discorea nipponica, Phalaris arundinacea*

(continued)

Table 10.9 (continued)

Water period	Comm-unity number	Sampling sites code	Name of main river section	Type of vegetation community
Dry Season	I	T7, T8, T9	Down-stream of the Le'an River	*Cynodon dactylon + Carex cinerascens, Gnaphalium affine, Soliva anthemifolia, Artemisia selengensis, Nymphoides peltatum*
	II	T2, T3, T5, T6, D1	Upper and middle reaches of the Le'an River	*Miscanthus sinensis-Cynodon dactylon + Celosia argentea, Rhizoma Cyperi, Hemarthria altissima, Polygonum orientale, Cyperus michelianus, Cyperus imbricatus*
	III	T1, D4	Source of the Le'an River	*Fagopyrum dibotrys + Oenanthe javanica + Hydrilla verticillata, Phalaris arundinacea, Cynodon dactylon, Alternanthera philoxeroides, Vallisneria natans, Ceratophyllum demersum*
	IV	T4	Midstream of the Le'an River	*Polygonum praetermissum + Polygonum hydropiper + Polygonum amphibium, Eleocharis yokoscensis, Marsilea quadrifolia , Phalaris arundinacea*
	V	D2, D3	Dawu River Section	*Cynodon dactylon + Phalaris arundinacea, Hemarthria altissima, Phragmites communis, Oplismentls undulatifolius*

Similar groups can be found in the headwaters of the river such as predominantly as *Fagopyrum tataricum, Oenanthe javanica, Glechoma longituba, Alternanthera philoxeroides* and *Centella asiatica.* Submerged plants such as *Hydrilla verticillata, Vallisneria natans, Ceratophyllum demersum* appear mostly in the dry season. The upper reaches downstream the confluence of Dawu River and Le'an River are dominated by *Celosia argentea, Cyperus rotundus* and *Cynodon dactylon*. In contrast, vegetation patterns in the middle part of the Le'an River varies significantly in dependence of the water stage. During the normal water period, species such as *Rumex japonicus* and *Cynodon dactylon* family with Polygonum criopolitanum and Erigeron canadensis, etc. can be found. In the wet season, *Cynodon dactylon, Phalaris arundinacea* and *Hemarthria altissima* with Polygonum hydropiper and Lobelia chinensis are dominant with this river section, followed by *Cynodon dactylon, and Polygonum sieboldii* with *Polygonum amphibium, Phalaris arundinacea* during the dry season.and so on. In the lower course *Carex cinerascens* and *Cynodon dactylon* with *Soliva anthemifolia, Gnaphalium affine* and *Erigeron canadensis* and so on appear.

10.4 Spatial and Temporal Distribution of Heavy Metal Pollution in the Soil and Water of Le'an River

10.4.1 Heavy Metal Content Analysis in the Soil and Water Environment

10.4.1.1 Heavy Metal Content Analysis in Surface Water

Table 10.10 compares the values of heavy metals in the overlying water of the Le'an River wetlands with the Chinese surface water environmental quality standards. The average concentration of Cu and Pb in overlying water of the Le'an River wetland over all sampling periods was 1013.47 and 3.54 g/L. For the normal water period, Cu content ranged over almost two orders of magnitude with values between 7.4–732.85 g/L, which was about two times higher variation than for Pb, which ranged between1.15 and 61.48 g/L. Comparing these concentrations with the other two periods shows that heavy metal concentrations vary greatly within the Le'an River vary greatly. In general, water pollution during wet season is more severe than for the normal water and dry period. According to the national surface water environmental quality standard, surface water along the sampling points of Le'an River achieves water qualities ranging from Class I to Class V, although the data shows that Copper pollution is more severe according to the standard.

Table 10.10 Concentrations of heavy metals (g/L) in the surface water of Le'an River wetlands

		Normal water period		Wet season		Dry season	
		Cu	Pb	Cu	Pb	Cu	Pb
Overlying Water	Mean	165.5	30.40	1013.47	80.76	220.54	3.54
	Max	732.85	61.47	3658.93	155.65	779.53	6.35
	Min	7.4	1.15	7.40	4.90	9.90	0.10
	Std.	227.03	17.36	1433.01	35.88	313.06	2.01
Quality standard for surface water	Class I	10	10				
	Class II	1000	10				
	Class III	1000	50				
	Class IV	1000	50				
	Class V	1000	100				

10.4.1.2 Analysis of Heavy Metals in Sediments of Water Bodies

Table 10.11 shows the heavy metal content in the sediments of Le'an River wetland and compares it with background value of heavy metals in Poyang Lake. The results showed that the average Cu concentration with 312.17 mg/kg in the sediments of Le'an River wetland was high during the normal water period, although Cd concentrations were lowest with 0.76 mg/kg. During the wet season, the Cu concentration in sediment of Le'an River wetland was very high with an average of 372.17 mg/kg and Cd concentration even tripled in contrast to the normal water water period. In contrast, average concentration of lead were with 6.01 mg/kg lowest during the wet season. During the dry season, average concentrations of Cu in the sediment of the Le'an River wetland decreased to 216.89 mg/kg, while the average concentrations of Pb and Cd were 6.94 mg/kg and 1.29 mg/kg, respectively. Compared with the background value of heavy metals inPoyang Lake, the average values of Cu, Pb and Cd were 65.7, 1.14, 1.01, 78.35, 0.48, 2.5, 45.67, 0.55 and 1.72 times of the background values of Poyang Lake for the normal, the wet and the dry period, respectively. This means that heavy metal elements are enriched significantly in the sediments, particularly Copper which average concentration for the wet season was 5 times more than the background value.

10.4.1.3 Analysis of Heavy Metal Content in the Surface Soil of the Wetland

As shown in Table 10.11, the Cu content of the three water periods in the wetland soil of Le'an River is with 591.86 mg/kg, which is 124.6 times of the background value of heavy metals in the Poyang Lake sediment, the highest of all samples. The average Pb content of each season ranged from 7.62 to 10.76 mg/kg, which was all not larger than the background values of Poyang Lake. This indicated that the Pb concentrations are not enriched at the surface of the wetland soils. Although absolute concentrations of Cd were not large, they exceeded the background values at each period. Cd average minimum was 1.04 mg/kg in wet season, which was 1.43 times more than the background value of heavy metals in Poyang Lake. In this study, the Cu and Pb concentrations were less compared with the study results written by Zeng Fanping [13], but the content of Cd increased rapidly. That means that the pollution control in Poyang Lake has achieved some improvements, but further supervision and control activities are needed.

Table 10.11 Concentration of heavy metals (mg /kg) in the sediments and soil of Le'an River during different seasons

		Normal water period			Wet season			Dry season		
		Cu	Pb	Cd	Cu	Pb	Cd	Cu	Pb	Cd
Sedi-ment	Mean	312.17	14.24	0.76	372.17	6.01	1.88	216.89	6.94	1.29
	Max	774.79	35.76	3.79	799.76	16.49	4.28	512.66	10.78	3.52
	Min	15.02	3.16	0.03	16.02	1.89	0.03	10.02	2.49	0.03
	Std.	305.26	11.50	0.99	298.33	4.49	1.51	158.58	2.95	1.04
Sur-face soil	Mean	329.83	10.76	2.22	591.86	8.12	1.07	261.71	7.62	1.49
	Max	985.58	32.67	9.84	2311.8	29.39	5.89	600.00	25.88	4.94
	Min	34.09	4.55	0.40	24.37	2.10	0.20	35.02	2.30	0.32
	Std.	319.47	7.30	2.61	677.47	7.03	1.54	208.05	6.00	1.36
Background value in Poyang		4.75	12.5	0.75						

10.4.2 *Spatial and Temporal Distribution of Heavy Metals in the Soil and Water Environment*

10.4.2.1 Spatial and Temporal Distribution of Heavy Metals of Overlying Water

Copper concentrations showed the same trend at each sampling site between the three sampling periods (normal water period, wet season, dry season) has roughly (Fig. 10.4a). At point T3 located in the mainstream of Le'an River the Cu content reached its peak. However, differences in concentrations between point T3 and D1 were not significant ($p > 0.05$) during the normal water period. This means that the concentrations of Cu in the confluence of the Dawu River and the Le'an River are similar. Cu concentrations reached a very high level in the upper reaches of the whole river system. In the Dawu River, Cu concentrations at each sampling section varied greatly between the sampling periods. Cu concentrations showed a decreasing tendency along the river course, but there was no significant difference ($p > 0.05$) between surface water concentrations contents in the upper reaches of Dawu River during the wet and dry periods, while differences were significant ($p < 0.05$) for the downstream location D1. The contents of Cu in the upper and lower reaches

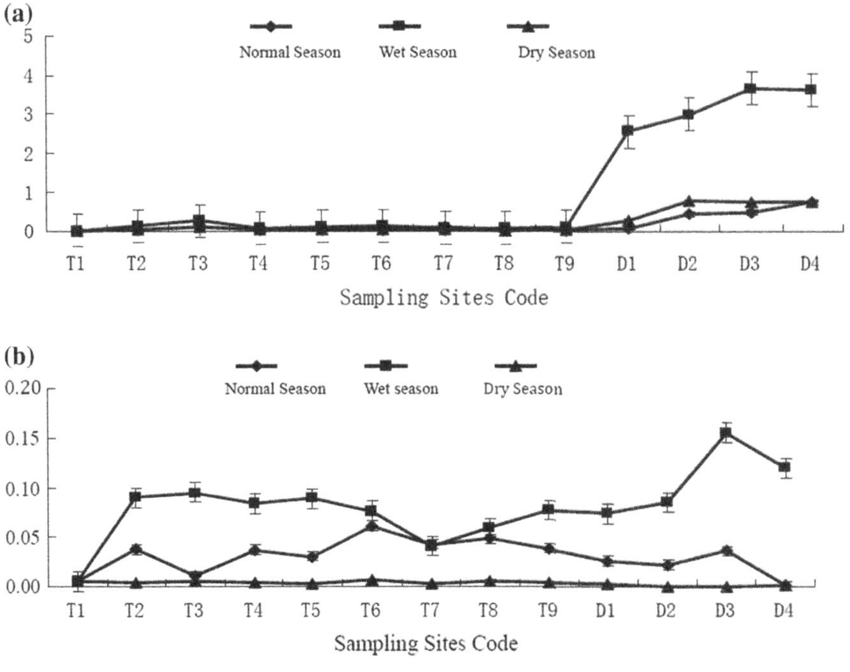

Fig. 10.4 Distribution characteristics of heavy metals contents (mg/L) in the overlying water of the wetland in Le'an River

of the Le'an River were relatively stable. Concentration differences between each season were significantly lower at location T3 in the middle and lower reaches of the Le'an River during the normal water period ($p < 0.05$). However, there was no significant difference in Cu content in surface water between the water and dry season ($p > 0.05$). This shows that in the middle and lower reaches of the Le'an River was not significantly polluted by Cu sources. Comparing the surface water Cu concentrations for each season, it was found that Cu concentrations in the wet season can reach 4–5 times higher values at some locations than during the other two water periods.

Pb concentrations in the whole river section fluctuated without revealing a simple spatial distribution relationship (Fig. 10.4b). However, Pb concentrations in the headwaters were significantly lower than in the other sampling areas. This indicates that the headwaters of Le'an River are less polluted by Pb. Comparing the Pb concentrations in the surface water during each period, it was found that Pb concentrations are largest during the wet season, followed by the normal water period and the dry season.

In summary, Cu concentrations in the headwaters of Le'an River wetland are low, while it increases in the upper reaches with location T3 being mainly affected. Further downstream, Cu concentrations in the surface water decreased gradually. Pb concentrations did now show an obvious spatial distribution. Cu and Pb concentrations have some common features during the three water periods: Concentrations hare higher during the wet season than during the other two periods. One explanation could be the mobilization of heavy metals from the pollution sources and their transport to the river system by surface runoff following precipitation events.

10.4.2.2 Spatial and Temporal Distribution of Heavy Metals in the Sediments

Copper concentrations in the river sediments of the sampling sites showed large spatio-temporal variations (Fig. 10.5a). Cu concentrations were largest at sampling point T3. There was a significant difference between T3 and other samples in Le'an River during the three water periods ($p < 0.05$). Sediment Cu concentrations gradually decreased in downstream direction of Le'an River and reached a stable level in the downstream and estuary. Summarizing, sediment Cu was relatively low at the source of Le'an River, high it in the upper reaches and low again in the middle and lower reaches. However, sediment Cu concentrations were very high in Dawu River but showed significant variations between the sampling sites was well which may be by the sedimentation processes in the river bed. Sediment Cu concentrations were significantly lower during the dry season is than that of during the normal water period and the wet season (Fig. 10.5a).

The concentrations of Pb in the sediments had different tendency for each of the three sampling periods (Fig. 10.5b). Concentrations at T7, T8 and T9 were significantly different from those at other locations. However, sediment Pb values in the upper and middle reaches of Le'an River and Dawu River were not significantly dif-

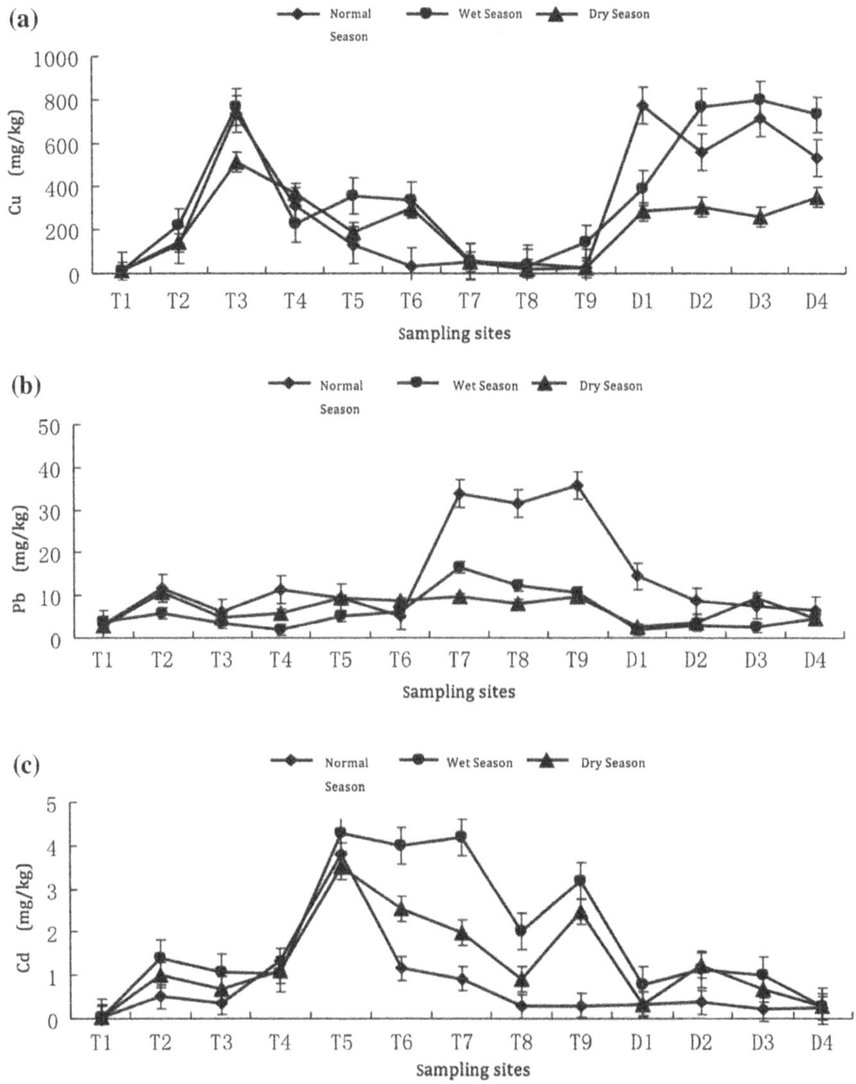

Fig. 10.5 Distribution characteristics of heavy metals contents (mg/L) in the sediments of the wetland in Le'an River (a: copper contents. b: lead contents. c: cadmium contents)

ferent from each other during the wet season. Sediment Pb values fluctuated largely during the dry season. Pb concentrations were relatively low in the source and upper reaches of Le'an River, while they were higher at further downstream sampling points. Surprisingly, sediment Pb concentrations were higher in the normal water period than it in wet and dry periods.

Cd concentrations in the sediments were quite different for different sampling points, but still showed some structural patterns (Fig. 10.5c). Cd was lowest near the source of Le'an River, which is significantly different from the other samplings. Then it increased slightly in the upper reaches and rapidly increased in the middle reaches at T4. Cd concentrations were highest at T5, but then decreased further downstream. At the estuary (T9) high Cd concentrations were measured during the wet and dry season but not for the normal water period. The significant difference of Cd concentrations at the middle and lower reaches of Le'an River and the Dawu River ($p < 0.05$) indicated the water was heavy polluted by Cd in the middle and lower reaches. Comparing over all sampling points for the three sampling periods, sediment Cd concentrations were highest in the wet season > dry season > wet season.

In summary, the Spatial–temporal distributions of Cu, Cd and Pb in the sediments of water bodies in Le'an River wetland were different, but all concentrations showed roughly the same trend for each water period. The Cu pollution in the middle and lower reaches is serious, while pollution by Cd is serious in the middle and lower reaches. The downstream and estuary of Le'an River are polluted by Pb. The large metal mining areas are located near the upstream section of Dawu River and middle reaches of Le'an River, which are affected by the acidic wastewater from mining and the industrial effluent from mineral processing over long periods. Therefore, the management of these two rivers should be strengthened.

10.4.2.3 Spatial and Temporal Distribution of Heavy Metals in Wetland Soils

The spatial and Temporal Distribution of Heavy Metals in wetland soils surface soil of the beach of wetland is shown in Fig. 10.6.

The spatial distribution of Cu has roughly the same trend along the sampling points for both the normal water period and the dry season. Soil Cu increased gradually from the wetlands at the source of Le'an River to the upper reaches. Maximum value appeared at T3 and then it gradually decreased with the river course. Difference were significant between T3 and the other sampling point in the upper and middle reaches ($p < 0.05$), but there was no significant difference ($p > 0.05$) between lower reaches and source area. Soil Cu concentrations were significantly higher in Dawu River than in the mainstream of the Le'an River during the normal water and dry periods. During the normal water period and dry season the Cu pollution in the surface soil is mainly concentrated in the confluence of the Le'an River and Dawu River, while the pollution is less in the estuary and the downstream reaches of Le'an River. However, Cu concentrations in the wetland soils fluctuated largely in the upstream part of

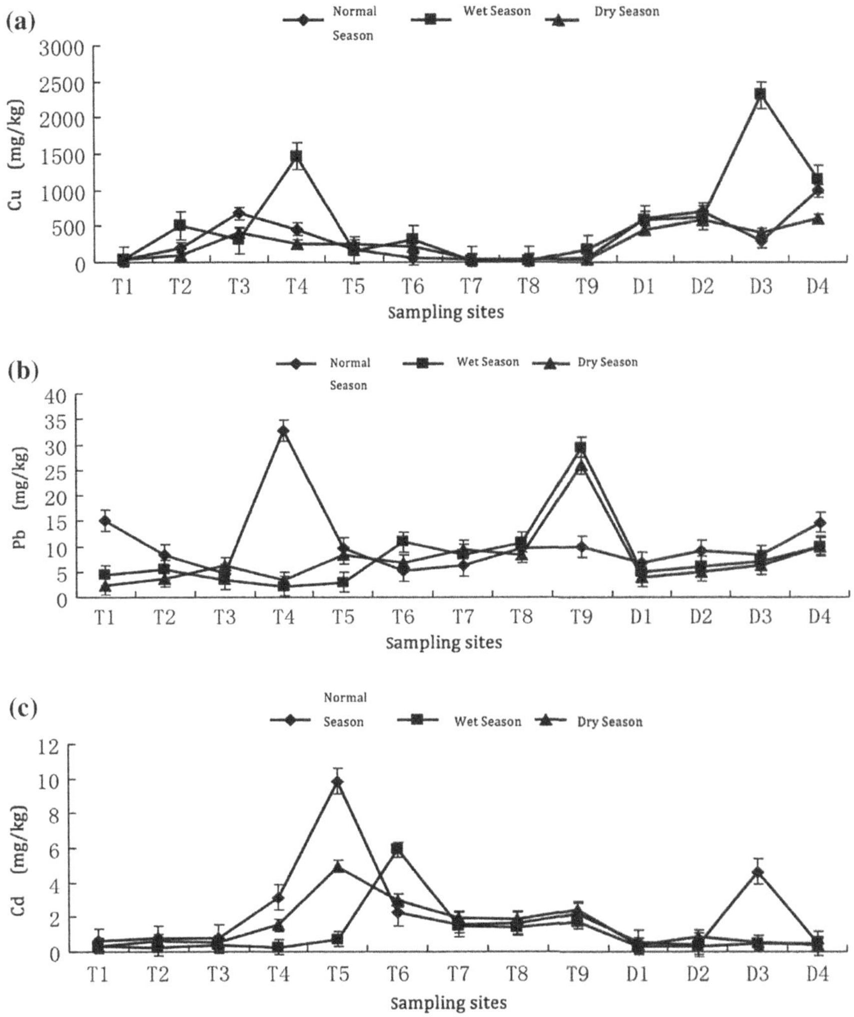

Fig. 10.6 Different contents of heavy metals (mg/kg) in the surface soils of Le'an River wetland

the system during the wet season, especially at T4 and D3, which was significantly different from other samples. At the lower reaches, cu concentrations were only slightly higher during the wet season than in the other two periods (Fig. 10.6a).

Soil Pb concentrations (Fig. 10.6b) were highest at sampling point T4 during the normal water period followed by elevated concentrations in the soil of the source area of Le'an River. However, soil Pb concentrations at T8 and T9 near the estuary were larger than at T5, T6 and T7 ($p < 0.05$). The results show that in the normal water period the Pb in the surface soil decreased gradually from the middle reaches to the lower reaches and then gradually increased towards the estuary. During the wet season, soil Pb concentrations were relatively constant along the river system. However, there was a large fluctuation in spatial variation during the dry season. During the dry season, Pb was relatively low at the source and upstream of the Le'an River, while the concentrations at T5, T6 and T7 were higher than in the source and the upstream section of the river.

Soil Cd concentrations were relative low (Fig. 10.6c) in the source and the upstream part of Le'an River as well as at Dawu River. Cd in the surface soil began to rise in downstream direction in the middle reaches with peak concentrations at T5 and T6. However the location of peak concentrations shifted between the different seasons. The maximum value was measured at T5 in the normal water period and dry season but at T6 during the wet season. There was no significant difference of soil Cd concentrations between the downstream and estuary areas of Le'an River, but these values were significantly higher than in the upstream reach and in Dawu River. Summarizing the soil Cd in Le'an River is highest in the middle reach followed by the downstream reach and the estuary of Le'an River and the source and upstream parts as well as Dawu River. Looking on the temporal dynamics, soil Cd was largest for normal water period but values in the wet season and dry season were roughly similar.

In summary, the spatio-temporal distribution characteristics of Cu, Cd and Pb concentrations in the soils of the wetlands are different. Cu pollution is concentrated in the upper reaches of Le'an River and Dawu River. Cd shows a significant increase trend in the middle reaches and Pb in the downstream reach of Le'an River. Soil concentrations of Cu were significantly larger during the wet season, while the concentrations of Cd had the maximum value in the normal water period. In contrast Pb values showed only slightly spatial differences comparing the three sampling seasons.

10.5 Environmental Risk Assessment of Heavy Metal Pollution in the Soil and Water of Le'an River

River water has a short residence time and is easily affected by various factors such as mining, industrial wastewater and domestic sewage. Therefore, the heavy metal pollution in the river is dynamic and can change rapidly. However, pollutants in the river sediment accumulate and can show stable concentrations over long periods. Research shows that heavy metals in river water can be deposited into the sediments leading to an extreme enrichment of concentrations there which can not be reversed easily. Therefore, the pollution status of sediments in the water body allows to improve the investigation of the pollution for longer time periods.

10.5.1 Assessment of Heavy Metal Pollution in Le'an River

Based on the present situation of heavy metal pollution in surface soil of rivers in April 2011, this study applied the One-factor index method, Geoaccumulation Index method, comprehensive Pollution Index method and Potential Ecological Risk Index (Eqs. (10.5))–(10.10)) to evaluate the pollution status of the heavy metals Cu, Pb and Cd in the sediment of Poyang Lake-Le'an River. The evaluation results are shown in Table 10.12.

1. The single factor index evaluation showed that 8 of 13 sampling sites were extreme polluted by heavy metals. 4 of 9 sampling sites in Le'an River are extreme polluted, while all 4 sampling sites in Dawu River are extreme polluted. Only one sampling sites in the upstream reach of Le'an River is slightly polluted and the rest of the samples showed moderate or severe pollution. However the evaluation result of the single factor index method is onesided. If a single heavy metal pollution factor exceeds the standard, it will be concluded that the entire sampling sites are polluted.
2. The results of the Geoaccumulation Index method show that the Geoaccumulation index of Cu in all samples in the basin is significantly higher than that of Pb and Cd. This means that the pollution by Cu is significantly higher than that by Pb and Cd. In the basin, only in the most upstream part of Le'an River has a slightly pollution with Cu according to the Geoaccumulation Index, while all other sampling sites are assigned with an index implying heavy pollution by Cu. Cu pollution in the Le'an River was at least partly caused by the inflow of heavy metals containing acid wastewater from Dawu River. Pollution levels gradually decrease after passing through the middle and lower reaches and decreased to moderate pollution near Longkou. The sediment of the entire basin are only considered by the index system to be uncontaminated or slightly polluted by Pb and Cd, which indicates that the associated pollution of Pb and Cd in heavy metal acid wastewater is relatively weak. The Geoaccumulation Index method comprehensively considers the background value of heavy metal elements in sediments and anthropogenic pollution factors and it can directly reflect the level of heavy metal pollution. On the other hand, the Geoaccumulation Index method also has certain weaknesses as it does not consider the bioavailability of various heavy metal elements and the contribution of various metal elements to different pollution points.
3. The evaluation using the Nemero Pollution index method showed that 10 of the 13 sampling sites are extreme polluted and two more are heavily polluted. Only Haikou sampling site in the upper reaches of the Le'an River is considered to be moderately polluted. The reason for this evaluation is that the single pollution index of Cu is higher, which increases the resulting Nemero Pollution Index significantly. The Nemero Pollution Index Method takes into account the highest pollution value and the average of each pollution values. It can reflect the combined effect of various pollutants in the sediment of the water body. But it also stresses the role of the highest pollution items in environmental quality

Table 10.12 The evaluation results of different evaluation methods in different sampling sites of Le'an River (i:Unpolluted. ii: Slightly polluted. iii: Moderately polluted. iv: Heavily polluted. v: Extremely polluted)

Site	Single factor index	Cu			Pb			Cd			Cu+ Pb+ Cd		
		I_{geo}	Pi	E_r^i	I_{geo}	Pi	E_r^i	I_{geo}	Pi	E_r^i	P_t	$I_{geo}max$	RI
T1	ii	1.08	3.16	15.81	−2.57	0.25	1.26	−5.42	0.04	1.05	iii (2.38)	ii (1.08)	i (18.12)
T2	iv	4.16	26.85	134.25	−0.68	0.94	4.68	−1.13	0.69	20.63	v (23.68)	iv (4.16)	iii (159.56)
T3	v	6.69	154.83	774.14	−1.64	0.48	2.41	−1.61	0.49	14.76	v (115.48)	v (6.69)	v (791.31)
T4	v	5.45	65.64	328.21	−0.72	0.91	4.54	0.25	1.79	53.68	v (47.73)	v (5.45)	v (386.44)
T5	v	4.2	27.49	137.45	−0.98	0.76	3.79	1.75	5.06	151.66	v (20.96)	iv (4.20)	v (292.90)
T6	v	2.11	6.47	32.33	−1.89	0.4	2.02	0.04	1.54	46.25	iv (4.98)	iii (2.11)	iii (80.60)
T7	iv	2.93	11.45	57.24	0.85	2.7	13.49	−0.28	1.23	37.01	v (8.87)	iii (2.93)	iii (107.73)
T8	iii	2.63	9.26	46.28	0.75	2.52	12.61	−2.02	0.37	11.08	v (7.14)	iii (2.63)	ii (69.97)
T9	iii	2.02	6.09	30.44	0.93	2.86	14.31	−1.94	0.39	11.75	iv (4.83)	iii (2.02)	ii (56.49)
D1	v	6.77	163.11	815.57	−0.37	1.16	5.8	−1.74	0.45	13.44	v (121.70)	v (6.77)	v (834.80)
D2	v	6.29	117.26	586.29	−1.11	0.69	3.47	−1.58	0.5	15.03	v (87.49)	v (6.29)	v (604.78)
D3	v	6.65	150.97	754.87	−1.31	0.61	3.03	−2.33	0.3	8.96	v (112.60)	v (6.65)	v (766.86)
D4	v	6.22	112.00	560.00	−1.50	0.53	2.66	−2.18	0.33	9.96	v (83.54)	v (6.22)	v (572.62)

assessment. Even if only one indicator is high in the evaluation items, the comprehensive rating value will be high. At the same time, the Nemero Pollution Index does not consider the difference between the poisoning of various heavy metal pollutants. It can only reflect the integrated degree of pollution and it is difficult to reflect the true poisonous effect on the living organisms.

4. The evaluation of the sampling points with the Potential Ecological Risk Index shows that only the Haikou sampling point is in a non-polluting (low-risk) state but 7 sampling sites are extremely polluted. These areas are mainly affected by the heavy metal acid wastewater. The potential ecological risk index was highest at T3. Heavy metal pollution in sediment decreased in flow direction of Le'an River. The potential ecological risk index takes into account not only the heavy metal content, but also the ecological effects of heavy metals. Environmental effects and toxicology are combined to consider not only the risk status of a single heavy metal element, but also consider the synthesis of all heavy metal pollutants in the substrate. But at the same time, the potential ecological risk index has weaknesses as well, e.g. by introducing subjectivity in the weighting scheme of the integrated ecological risk.

In summary, the evaluation results of the four evaluation methods are both consistent and show slight differences only. In particular, for areas with heavy metal pollution, the evaluation results of the four methods are highly consistent. For example, the six sites T3, T4, D1, D2, D3 and D4 are evaluated as extremely polluted by each method. However, for the evaluation of other areas with different degrees of pollution, although the results of the four methods have small differences, the overall evaluation results for the reaches of Le'an River are still in agreement.

10.5.2 Comparison of Different Evaluation Methods Evaluation Results

The main reasons for obtaining different evaluation results at the sampling sites using different evaluation methods are:

1. The single-factor index method uses the category of the worst index as the final level of pollution.
2. The Geoaccumulation Index method emphasizes individual pollution levels without regarding bioavailability and the joint effects of various heavy metals.
3. The Nemero Pollution index excessively stresses the contribution of the maximum of all single heavy metal pollution.
4. The Potential Ecological Risk Index considers the biological validity and can reflect the comprehensive effects of various heavy metal pollution.

The Geoaccumulation Index method and Nemerow index method rates pollution levels generally higher than the potential ecological risk index. The Cu concentrations in Longkou Village (T9) was lower than the national first-grade soil standard.

The content of Pb was near the first-level national soil standard value, while Cd was slightly higher than the national standard of soil level one. Additionally considering both the background value of lake sediment and the potential ecological risk assessment making a rating as slightly polluted more reasonable. Cu concentrations in Shangkuankou (T2) were near the secondary standard soil of the country and the contents of Pb and Cd are lower than that of Poyang Lake sediments, which is the considered background value. Hence, using the extremely pollution level result of the Nemerow comprehensive index method is not reasonable and the rating as moderately polluted by the potential ecological risk assessment seems to be more reasonable. Therefore, for the assessment of heavy metal pollution in the bottom of Le'an River-Poyang Lake section, the potential ecological risk method is more accurate and objective than the single-factor index method, the cumulative index method and the comprehensive pollution index method and should be preferred.

10.6 Responses of Large Aquatic Plant Communities to Environmental Factors in Le'an River

The DCCA method was used to analyze the influence of environmental factors on the aquatic macrophytes community in the Le'an River wetland and to analyse the main environmental factors, which affect the aquatic macrophytes community. 14 environmental factors that were mainly interfered by human were considered in the analysis:

W1: Cu concentration in surface water
W2: Pb concentration in surface water
W3: pH value in surface water
W4: COD concentration in surface water
W5: TN concentration in surface water
W6: TP concentration in surface water
S1: Cu concentration in sediments
S2: Pb concentration in sediments
S3: Cd concentration in sediments
S4: pH value in sediments
A1: Cu concentration in wetland soils
A2: Pb concentration in wetland soils
A3: Cd concentration in wetland soils
A4: pH value in wetland soils.

10.6.1 Responsive Relation Between Aquatic Macrophytes Community and Environmental Factors During the Normal Water Period

As shown in Table 10.13., the eigenvalues of the first three axes of the DCCA during the normal water period are 0.731, 0.424 and 0.328, respectively. The interpretation rates of species data were 16.9%, 9.9% and 7.6%, respectively, and the cumulative explanation rate of the first three axes of DCCA was 34.4%. The first two sorting axes have larger eigenvalues and contain more ecological information. Two-dimensional scatterplots are made according to the first two axes, as shown in Fig. 10.7.

The DCCA ordinations show that Cu in surface water body and Cu in the wetland soil and sediments are the most important environmental factors in the direction of the first axis, and all of them have a significant positive correlation with the first axis ($r = 0.755$, $P < 0.01$; $r = 0.632$, $P < 0.01$). However, pH value as well as Pb concentrations in the sediments and the topsoil were the main factors in the opposite direction of the first axis, showing a significant negative correlation. Studies have shown that the pH of the soil substrate has a strong negative correlation with the heavy metal concentration in the soil. Correlation analysis showed that there was a significant negative correlation between sediment Cu and sediment pH ($r = -0.865$, $P < 0.01$). At the same time, there was a significant negative correlation between soil Cu and soil pH ($r = -0.844$, $P < 0.01$). Summarizing, the first axis shows that the Le'an River wetland is affected by Cu pollution.

Figure 10.7 shows that the sampling points T2, T3, T4, D1, D3 and D4 are mainly distributed in the positive direction of the first axis, indicating that Cu is the dominant factor. However, T5, T6, T7, T8 and T9 are mainly distributed in the negative direction of the first axis, indicating that the middle and lower reaches of the Le'an River are mainly affected by Pb concentrations. The most important environmental element on the second axis is the Cu in the surface water, which is significantly and positively correlated with the second axis ($r = 0.622$, $p < 0.01$). It can be seen from Fig. 10.7 that the sampling points T4, D2, D3 and D4 are distributed along the second axis, reflecting that the main influencing factors of the tributaries are the Cu concentrations in the surface water.

10.6.2 Responsive Relation Between Aquatic Macrophytes Communities and Environmental Factors During the Wet Season

As can be seen from Table 10.14, the eigenvalues of the first two axes before the DCCA sorting in the wet season are 0.804, 0.669 and 0.237 respectively. Their interpretation rates for species data are 16.3%, 13.5% and 4.8% respectively, and a two-dimensional scatter plot is made based on the first two axes, as shown in Fig. 10.8.

Table 10.13 Ordination axis eigenvalues,c orrelation coefficients of species, environmental factors and the axis in normal season

	Axis 1	Axis 2	Axis 3	Inertia
Characteristics	0.731	0.424	0.328	4.313
Cumulative percentage variance of species	16.9	26.8	34.4	
Cumulative percentage variance of species-environment relation	16.9	26.8	34.4	
Correlation coefficient between each environmental factor and ordinal axis	r			
Cu concentration in surface water	0.1536	0.6217*	−0.0031	
Pb concentration in surface water	−0.6141*	0.1134	0.3251	
Cu concentration in sediments	0.755**	0.4899	−0.027	
Pb concentration in sediments	−0.5188	0.1543	−0.5079	
Cd concentration in sediments	−0.362	−0.2569	−0.0534	
Cu concentration in wetland soils	0.6315*	0.4631	−0.0004	
Pb concentration in wetland soils	−0.0218	−0.35	−0.1199	
Cd concentration in wetland soils	−0.3879	−0.1864	−0.2073	
COD	-0.2321	0.1694	−0.3659	
TN	-0.6028*	−0.1313	0.2182	
TP	0.3712	0.2376	−0.1115	
pH in sediments	−0.7108**	−0.4702	0.1097	
pH in wetland soils	−0.6418*	−0.4983	0.0332	

Note *Correlation is significant at the 0.05 level (2-tailed), **Correlation is significant at the 0.01 level (2-tailed)

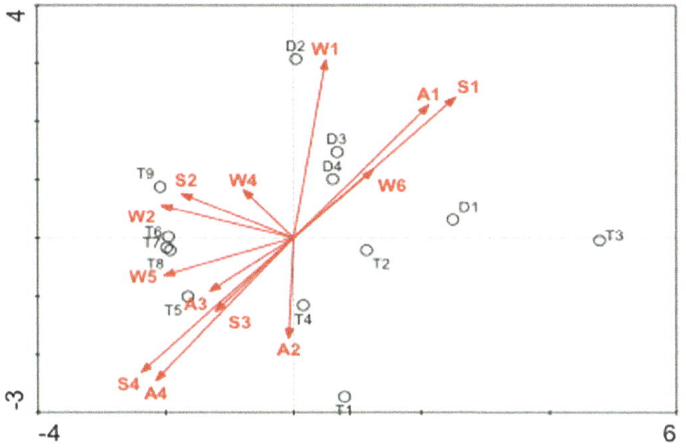

Fig. 10.7 DCCA ordination diagram of macrophyte communities and environmental factors in normal season

The DCCA ordinations Fig. 10.8 show, the first axis has a significant correlation with the Cu concentration in the sediment and the pH in the surface soil.

It has a positive correlation with Cu in sediment of water body and negative correlation with the pH in the surface soil. Related. In the sequence diagram, the sampling points T2, T3 and D1 are mainly distributed in the positive direction of the first axis. It indicates that the Cu in the sediment of water body is the main influencing factor, that mainly affects the confluence of Dawu River and Le'an River. However, the sampling points T4, D2 and D4 are distributed in the opposite direction of the first axis. It means that the major aquatic macrophyte community in the tributaries is mainly affected by pH. On the second axis, Pb concentration in overlying water, Cd concentration in sediment of the water body and Pb concentration in the surface soil of the wetland beach were the most important environmental elements. And Cd concentration in sediment was the most important factor, which negatively correlated with the second axis ($r = -0.603$, $P < 0.05$). In the positive direction of the second axis, only in the vicinity the sample points T1 there is no obvious interrelated anthropogenic environmental factor. It indicates that the aquatic macrophytes at is rarely influenced by human disturbance but mainly affected by natural environmental factors. The sampling points T5, T6, T7, T8 and T9 are mainly distributed in the negative direction of the second axis, where the aquatic macrophytes communities are mainly contaminated by heavy metals Pb and Cd in soil and water environment.

Table 10.14 Ordination axis eigenvalues, correlation coefficients of species, environmental factors and the axis in wet season

	Axis 1	Axis 2	Axis 3	Inertia
Characteristics	0.804	0.669	0.237	4.938
Cumulative percentage variance of species	16.3	29.8	34.6	
Cumulative percentage variance of species-environment relation	16.3	29.8	34.6	
Correlation coefficient between each environmental factor and ordinal axis	r			
Cu concentration in overlying water	−0.0493	−0.0401	0.2712	
Pb concentration in overlying water	0.1934	−0.5448	0.3897	
Cu concentration in sediment of the water body	0.4047	−0.1548	0.2768	
Pb concentration in sediment of the water body	−0.197	−0.3863	−0.3648	
Cd concentration in sediment of the water body	−0.2325	−0.6027*	−0.2611	
Cu concentration in wetland soils	−0.0592	−0.1078	0.4973	
Pb concentration in wetland soils	−0.2505	−0.5214	0.1297	
Cd concentration in wetland soils	−0.2153	−0.3082	0.393	
COD	0.2705	−0.1801	−0.1632	
TN	−0.0973	−0.077	0.1412	
TP	−0.0561	−0.1721	−0.2264	
pH in sediment of the water body	−0.2923	0.2171	0.0024	
pH in wetland soils	−0.456	0.1931	−0.1521	

Note *Correlation is significant at the 0.05 level (2-tailed), **Correlation is significant at the 0.01 level (2-tailed)

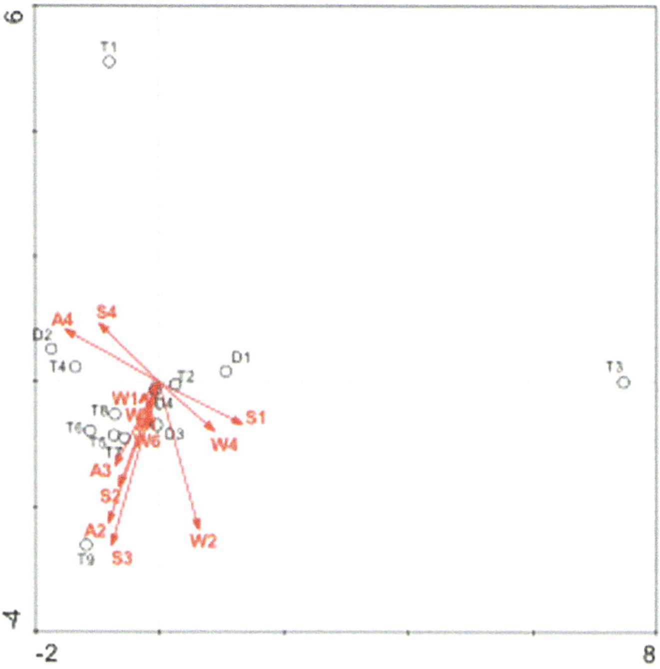

Fig. 10.8 DCCA ordinations of macrophyte communities and environmental factors in wet season

10.6.3 Responsive Relation Between Aquatic Macrophytes Communities and Environmental Factors During the Dry Season

As can be seen from Table 10.15, the eigenvalues of the first two axes in DCCA during the dry season are 0.822, 0.775 and 0.191, respectively. The interpretation rates of species data are 15.1%, 14.2% and 3.5% respectively. According to the first two Axis to make two-dimensional scatter plot, as shown in Fig. 10.9.

DCCA ordinations shows that the Cu sediment and Pb concentration in the the surface waterin the positive direction of the first axis are the most important environmental elements, while the pH value in the opposite direction of the first axis is the most important. There is a strong correlation between the Cu concentration in sediment and the first axis (Fig. 10.9). The sampling points T2, T3 and D1 are mainly distributed in the positive direction of the first axis, indicating that Cu in the sediment of water body in the confluence of Dawu River and Le'an River is the dominant factor. In the positive direction of the second axis, Pb concentration in the surface and TN are the main influencing factors, but the Pb is dominant. It has a significant positive correlation with the first axis ($r = 0.754$, $p < 0.01$). In the opposite direction of the first axis, pH of the surface soil in the wetland beach is the most important factor.

Table 10.15 Ordination axis eigenvalues,correlation coefficients of species, environmental factors and the axis in dry season

	Axis 1	Axis 2	Axis 3	Inertia
Characteristics	0.822	0.775	0.191	5.452
Cumulative percentage variance of species	15.1	29.3	32.8	
Cumulative percentage variance of species-environment relation	15.1	29.3	32.8	
Correlation coefficient between each environmental factor and ordinal axis	r			
Cu concentration in overlying water	−0.267	0.1408	−0.172	
Pb concentration in overlying water	0.3543	−0.1513	−0.1511	
Cu concentration in sediment of the water body	0.497	0.0998	−0.1462	
Pb concentration in sediment of the water body	−0.1001	0.3445	0.1783	
Cd concentration in sediment of the water body	−0.117	0.3604	0.0055	
Cu concentration in wetland soils	0.1326	0.1284	−0.2495	
Pb concentration in wetland soils	−0.1146	0.7536**	−0.0307	
Cd concentration in wetland soils	−0.1288	0.196	−0.1899	
COD	0.1793	0.2543	0.4267	
TN	−0.2789	0.4353	0.2901	
TP	−0.1227	0.3719	0.664	
pH in sediment of the water body	−0.2806	−0.3864	−0.0459	
pH in wetland soils	−0.3714	−0.4056	0.1036	
Cu concentration in overlying water	−0.0298	−0.3324	−0.2511	

Note *Correlation is significant at the 0.05 level (2-tailed), **Correlation is significant at the 0.01 level (2-tailed)

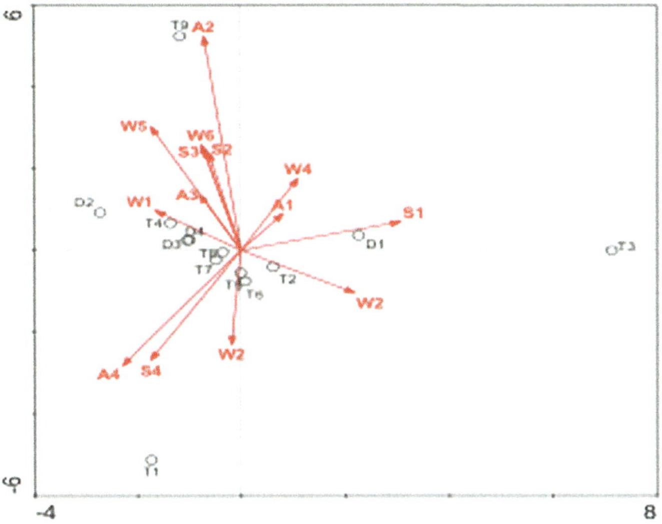

Fig. 10.9 DCCA ordinations of macrophyte communities and environmental factors in dry season

The sampling sites D2, D3 and D4 are mainly distributed in the positive direction of the second axis (Fig. 10.9), indicating that Pb in the surface in the wetland beach and TN are the main environmental factors, which affect the aquatic macrophytes community in the dry season. The aquatic macrophytes communities in sampling sites of the downstream (T6, T7, T8 and T9) were mainly affected by the pH of the surface soil in the wetland.

After comprehensively analyzing the responsive relationship between water and soil environmental factors and aquatic macrophyte communities in the above three different water periods, the following different relationships are obtained,

1. The water and soil environment of the source of Le'an River is relatively good during the normal water period and is less affected by various man-made pollution. Cu in the sediment in the upper reaches of the Le'an River and the Dawu River is the dominant factor affecting the community structure of large macrophyte, while the middle and lower reaches are mainly affected by Pb in the the surface waterand Pb in the sediment of water body.
2. During the wet season, the source of Le'an River is little affected by the anthropogenic influence factors. The aquatic macrophyte communities in the upper reaches of Le'an River and the Dawu River was mainly affected by Cu in the sediments of the water body and pH, while the lower reaches were polluted by the heavy metal Pb and Cd.
3. During the dry season, The aquatic macrophyte communities in Dawu River was mainly affected by Cu in the sediments of the water body, TN and Pb in the surface of wetlands. The confluence of the Le'an River and the Dawu River

was mainly affected by Cu in the sediment. However, the pH value in the soil and water environment in the middle and lower reaches affects the structure of aquatic macrophyte communities.

10.7 Results and Conclusions

1. A total of 48 families, 124 genera and 166 species of aquatic macrophytes are found in the contaminated area of Le'an River. Among them, there are 1 species of 1 genera of 1 species fern plant, 113 species of 84 genera of 36 families dicotyledon, 52 species of 11 genera of 11 families monocotyledons. Due to the heavy metal pollution in the upstream, the number of aquatic macrophytes in the typical polluted area of the Le An River wetland has declined. It affects also the composition and structure of the flora in the polluted area. The main advantage of the plant species in the Le'an River polluted area are: *Poaceae, Asteraceae, Cyperaceae, Umbelliferae, Cruciferae and Polygonaceae*. The main advantage of the plant genera in the Le'an River polluted area are: *Polygonum, Carex, Cyperus, Artemisia, Stellaria, Cardamine and Ludwigia.*
2. 4 plant diversity indices showed roughly the same changes in the three water periods of the Le'an River and the Dawu River: the plant diversity indices in normal water period > dry season > wet season. And plant diversity indices in the source, the upper reaches and the middle > lower reaches and Dawu River. The distribution of plants at each sampling site was uneven. Some dominant species were formed.
3. Two-way indicator species analysis (TWINSPAN) was used to cluster the wetland vegetation in Le'an River. The vegetation types were divided into 5 plant groups for the normal water period, wet season and dry season, respectively. According to their distribution, the location is mainly divided into Dawu River section, source area of Le'an River, the confluence of Dawu River and Le'an River, the upstream section, the middle section, the downstream section and the estuary section. In Dawu River *Cynodon dactylon* and *Phalaris arundinacea* are the dominant species. More *Fagopyrum dibotrys*, *Oenanthe javanica*, *Hydrilla verticillata* and *Vallisneria natans* are found in the headwaters of Le'an River. At the junction of Le'an River and Dawu River, the dominant species are *Miscanthus sinensis*, *Cynodon dactylon*, *Alternanthera philoxeroides* and *Celosia argentea*. In the downstream reach of Le'an River, *Carex*, *Cynodon dactylon*, *Potamogeton crispus* and *Nymphoides peltatum* are the dominant species.
4. The Cu concentration of heavy metal elements is the highest in the water and soil environment of Le'an River wetland, with the highest average of 591.86 mg/kg. Compared with the background value of heavy metals in the soil of Poyang Lake, the enrichment of Cu was the most abundant, the maximum is 124.6 times. The concentration of Cu in the surface watercan exceed the national Class V surface water quality standards. And Pb concentration in the surface wateris relatively

low, which are below the national Class III surface water quality standards during the three water periods.

5. The spatio-temporal distribution of heavy metals Cu, Pb and Cd in the water and soil environment of Le'an River wetland is different. The heavy metal Cu is in wet season > normal water period > dry season. The difference of heavy metals Pb during three water period is not big. Cd is in the normal water period > wet season ≈ dry season. The same heavy metal element showed the same trend of the same coastal distribution in all three water periods. Cu pollution in the middle and upper reaches of the Le'an River and the Dawu River was more serious in the sediments of the water, while Cd pollution was also heavy in the middle and lower reaches. The Pb in the Le'an and the Dawu River tributaries did not change much. The Cu pollution in the surface of wetland beach mainly concentrated in the Dawu River and the upper reaches of the Le'an River. The Cd concentration increased obviously in the middle reaches and the enrichment phenomenon of Pb appeared in the lower reaches. Correlation analysis showed that the heavy metal element Cu in the water and soil environment of Le'an River wetland have the same source, that is mail from the acidic wastewater of Dexing Copper Mine in the upstream. At the same time, Pb and Cd in sediments of water body also have certain homology in wet season and dry season.
6. Four kinds of evaluation methods were used to evaluate the heavy metal pollution in the Le'an River wetland. The evaluation results were both consistent and slight differences. It is shown that Haikou is the least heavily polluted or slightly polluted by heavy metals, while the area close to Dawu River is severely polluted by heavy metals. According to the water flow, the heavy metal pollution in the bottom sediments of all regions showed a decreasing trend. Compared the four evaluation methods, the potential ecological risk method is more accurate and objective. Because of its contact with heavy metals toxicology, environmental effects and ecological effects, heavy metal pollution status of the Le'an River wetland can be a comprehensively reflected.
7. The aquatic macrophytes community in the Le'an River wetland can respond well to the impact of environmental factors. The source of Le'an River is less affected by human activities, where the environment is good, the species is rich and the diversity of life is high. The sensitive species of Poyang Lake—Le'an are distributed in this region; while the aquatic macrophytes after the source are affected by the heavy metals and acid waste water pollution from mines. The upper reaches of Le'an River and the Dawu River are mainly affected by the Cu in the sediments of the water body and pH values. The structure of the aquatic macrophytes community is degraded. There are less species, less biodiversity, less sensitive species and more pollutant-resistant species The influence of heavy metal Cu in the middle and lower reaches greatly reduced, while the combined pollution of heavy metals Pb and Cd in soil and water environment became one of the main factor.

References

1. State Environmental Protection Administration of China. 2002. In *Water and wastewater monitoring and analysis method*. China Environmental Science Press.
2. Liu, L., et al. 2008. Determination of heavy metals in soils and plants with microwavedigestion and ICP-AES (in Chinese). *Environmental Chemistry* 27 (4): 511–514.
3. Poyang Lake research Editorial board. 1988. In *Research on Poyang Lake (in Chinese)*.
4. Xia, J. 1996. In *Detailed explanation of soil environmental quality standard (in Chinese)*, 84–86.
5. Ma, T., et al. 2011. Potential ecological risk assessment of heavy metal pollutants in surface sediments of the lakes in Nanjing (in Chinese). *Journal of Ecology and Rural Environment* 6: 37–42.
6. Li, Z. J. Ding, and L. Peng. 2004. *The principle and method of environmental quality evaluation*, Chemistry IndustryPublishing House, Beijing.
7. Jian, M., L. Li, H. Yu, J. Xiong, and G. Yu. 2015. Heavy metals pollution on the water and sediments and its influence on the submerged macrophyte community in the wetland of Poyang lake (in Chinese). *Ecology and Environmental Sciences* 24: 96–105.
8. Xu, Z., S. Ni, X. Tuo, and C. Zhang. 2008. Calculation of heavy metals' toxicity coefficient in the evaluation of potential ecological risk index (in Chinese). *Environmental Science & Technology* 2 (8): 112–115.
9. Jian, M., M. Jian, L. Li, and Other. 2015. Distribution pattern of submerged plants in typical wetlands of Poyang Lake and its influence factors of water environment (in Chinese). In *Resources & environment in the Yangtze Basin* 24.5, 765–772.
10. Li, K., Y. Hu, Y. Fan, and Other. 2007. Influence of environmental factors on distribution of plant communities and composition of species in alpine grassland (in Chinese). *Chinese Journal of Agrometeorology* 28.4: 378–382.
11. Yang, Q., and J. Zhang. 2000. The ordination axes clustering based on detrended canonical correspondence analysis ordination and its application to theanalysis of the ecological gradients of plant communities (in Chinese). *Acta Ecologica Sinica* 20 (2): 199–206.
12. Feng, Y., K. Ma, Y. Zhang, and J. Qi. 2008. DCCA analysis of plant species distributions in different strata of oak (quercus liaotungensis) forest along an altitudinal gradient in Dongling mounta in china (in Chinese). *Journal of Plant Ecology* 3: 568–573.
13. Zeng, F., H. Xiao, and W. Zhou. 2009. Distribution and retention of heavy metals in floodplain topsoils of Le'an River. *Environmental Science & Technology* 2: 96–101.

Part IV
Ecology

Chapter 11
Structure and Diversity Analysis of the Microbial Community in the Surface Waters of Poyang Lake Basin

Xiaolan Huang

11.1 Introduction

The river-lake ecotone is a buffer transition zone between the river ecosystem and the lake ecosystem [1]. The river-lake ecotones are affected by human activities, natural disasters, and the interactions of the biogeochemical cycle with rivers. Compared with the adjacent river ecosystem or lake ecosystem, the river-lake ecotone is a uniquely mixed ecological area, with relatively high levels of primary and secondary production which is also an important water purification area and pollution buffer zone. Thus, the river-lake ecotone forms a region where the energy and matter flow through the ecosystem is unique. Compared with the ecosystem of the open water body, the river-lake ecotone has a strong environmental gradient, biological heterogeneity, biodiversity, high levels of productivity and so on [2, 3]. Being influenced by hydro-geological effects, the extensions of the river-lake ecotone vary with the gradient of the elevation. The plants and biological surroundings changes along with the different soil textures such as sand, clay and silt or bedrock [4, 5]. From the land to the water, the species distributions in the river-lake ecotone changes from terrestrial, to wetland and aquatic species, then to benthic, free living and plankton species. This study aimed to identify the microbial diversity in the lake-river ecotones of Poyang Lake by using DNA sequencing techniques. Information on the microbial distribution and the microbial gene pool are provided and key species are delineated. Hence, this work will provide new knowledge on the link between the microbes population and environmental variations, which will be beneficial for preserving the ecosystem.

X. Huang (✉)
Jiangxi Normal University, Nanchang, China
e-mail: huangxiaolanlan@163.com

T. Yue et al. (eds.), *Chinese Water Systems*, Terrestrial Environmental Sciences,
https://doi.org/10.1007/978-3-319-97725-6_11

Table 11.1 Location of the sampling points in the Poyang Lake basin

Name	Description
N-H1	Main lake of Poyang Lake in Nanji Mountain
N-H2	Butterfly Lake in Wanjiaqiao section
N-H3	Between the Butterfly Lake and Chang Lake
N-H4	Gan River 200 m from Nanjiang Bridge
W-H5	Downstream the north branch of Gan River
W-H6	Wucheng at the confluence of Gan River and Xiushui River
W-H7	Dahu Pond in Wucheng
R-H9	Confluence Changjiang section of Rao River and Le'an River
H-H11	Hukou near the mouth of Poyang Lake
G-H8	Southern Gan River
Y-H10	Yao Lake in Nanchang

11.2 Selection of Sampling Points

This study selected eleven sampling points (Table 11.1) in different areas of Poyang Lake area (Fig. 11.1).

The collection of water samples took place in September 2016. The samples were collected with a one liter plexiglass water sampler in a water depth of 0.5 m. Three samples (two bottles of one liter, one bottle of 500 ml) were collected from each sampling point. The water samples were autoclaved with high temperature autoclave. Water temperature, pH and conductivity of the samples were measured parallel to the sampling. After they returning the samples to the laboratory, total nitrogen (TN), total phosphorus (TP) and other metal elements were determined.

11.3 Principles and Procedures of High-Throughput Sequencing

11.3.1 Principle of High-Throughput Sequencing

The principle of high-throughput sequencing is to hybridize the Polymerase Chain Reaction (PCR)-amplified single-strands with primers and incubate with enzymes such as DNA polymerase, ATP sulfatase, luciferase and 5′-phosphosulfate, so that the deoxyribonucleotide triphosphate (dNTP) will be connected to the primer. At the time of the sequencing reaction, each of the dNTPs added was paired with a to be determined template. Concurrently, an equimolar amount of pyrophosphate

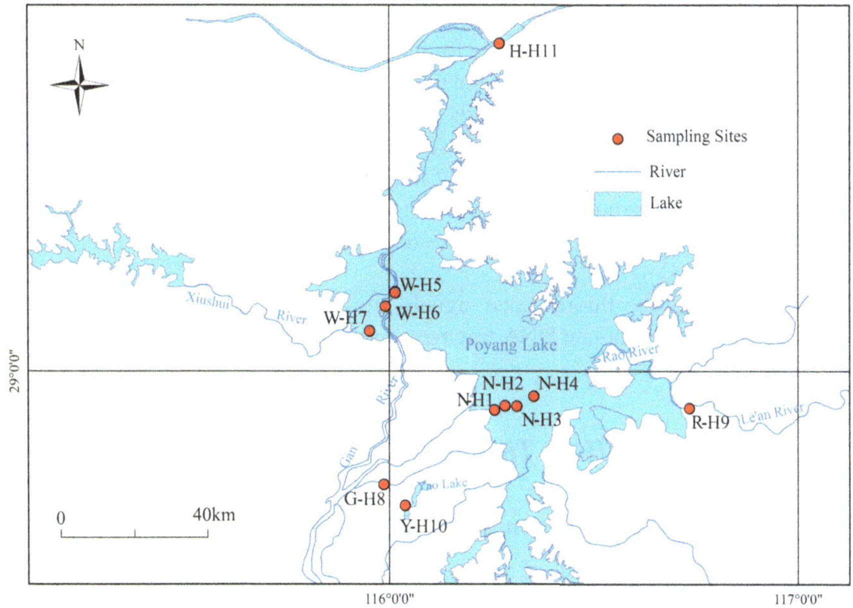

Fig. 11.1 Location of sampling sites (Note: There is slight overlap due to the distance between W-H5 selection and the W-H6)

groups was released, which formed ATP under catalysis of ATP sulfatase. Driven by ATP and catalyzed by Luciferase, Luciferin is transformed into oxyluciferin, simultaneously emitting a visible light signal that is proportional to the amount of ATP. ATP and unincorporated dNTP are degraded, the light signal is quenched and the initial reaction system is re-established, the next type of dNTP is added and the cycle begins again.

11.3.2 High Throughput Sequencing Procedures

The sample was first filtered with a 0.80 μm filter to remove the sediments and other impurities. It was then filtered with a 0.45 μm filter to obtain the microorganisms in the water [6]. High throughput sequencing is mainly done on an Illumina sequencing platform [7, 8], and its specific operations include the following:

1. The DNA fragment is complementary to the base of the primer and then immobilized on the chip
2. A section of the DNA fragment is attached to a DNA primer through complementary base paring and is immobilized on the chip
3. The other end of the DNA fragment is also complementary to another primer in the vicinity, they become attached thereby forming a "bridge"

4. Polymerase Chain Reaction (PCR) amplification, results in DNA clusters
5. The resulting DNA amplicons are formed linearly into single strands
6. The modified DNA polymerase and the dNTP with four fluorescent markers are added, with only one base synthesis per cycle.
7. The surface of the reaction plate is scanned with a laser, reading the sequence for each template in the first round of nucleotide polymerization.
8. The "fluorescent group" and "terminating group" are chemically incised, the 3' cohesive end is restored, and the procedure to polymerize the second nucleotide continues.
9. The results of the fluorescence signals are collected in each round and the sequence of the template DNA fragments are known.

11.3.3 Analysis of High Throughput Results

11.3.3.1 Effective Sequence Analysis

In this study, eleven samples were analyzed by high-throughput sequencing, and 333794 effective base sequences were obtained. The average sequence of each sample was 30345, and the average length of each sequence was 438.74bp. Among them, the most valid base sequence in all the samples was N2, and the effective base sequence was 36164, and the average sequence length was 434.70bp. The number of effective bases found in all samples was R9, only 21525 valid primers were found, and the average length of the sequence was 443.30bp. By analyzing the effective base sequence, it is helpful to better evaluate the bacterial community structure and diversity level in all samples. The length distribution of each sample is shown in Fig. 11.2. It can be seen from the figure that most of the effective bases in all the samples are between 421 and 460 bp, which is in good agreement with high throughput requirements and standards.

11.3.3.2 Operational Taxonomic Units (OTU) Generation and Comparison

In the process of bioinformatics, each sequence obtained from the sequencing is derived from a bacterium. To better understand the microbial composition in a sample, we need to categorize all the valid sequences according to the similarities between them, each category is an OTU, usually on the basis of a 96%–98% similarity for an appropriate OTU division [9, 10]. In this article, a similarity of 97% or more is required in order to be classified as an OTU. In this paper, 1137 OTUs were obtained by classifying the 333794 valid sequences obtained from the 11 samples collected in the Poyang Lake and Huhejie Strip. The number of OTUs in each sample was H-H11(854) > W-H6(647) > N-H2(579)> N-H1(560)> N-H3(554)> W-H5(515)> W-H7(496)> Y-H10(488)> G-H8(478)> R-H9(475)> N-H4(444).

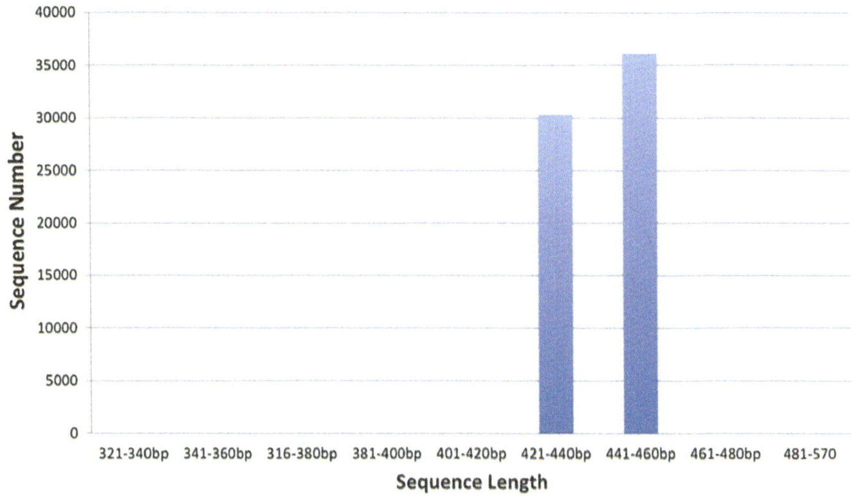

Fig. 11.2 Length distribution of optimized sequence

11.3.3.3 Dilution Curve

The dilution curve is built with the number of individuals and the number of species. It can be used to compare the abundance of species in different samples of sequencing data, and can also be used to reflect the validity of the sample sequencing. When the curve plateaus, that means the sequencing of the sample is reasonable, that is to say that the data obtained via the sequencing can encompass the study area. If the dilution curve shows an upward trend, that means there are hidden species in the sample that have not yet been found. In this study, the dilution curve was plotted and the diversity index was carried out at a similarity level of 97%. The dilution curve is shown in Fig. 11.3. It can be seen from Fig. 11.3, when all the samples achieve 17384, all sample dilution curves tend to be flat. This means that all microbial species in all samples had been covered and no hidden species were detected.

11.3.3.4 Microbial Community Diversity Analysis

The diversity of microbial communities in the surface water of the lake is illustrated by the Shannon-Wiener index. At the same time, the abundance-based coverage estimator (ACE), index and Chao1index (Community abundance index) are used to define the abundance of the samples while analyzing the saturation level of the sequencing results. The sequencing depth for all samples are illustrated using the dilution curve while the depth of sequencing coverage is described using the coverage index (Coverage). When the coverage is 100%, it indicates that the microbes

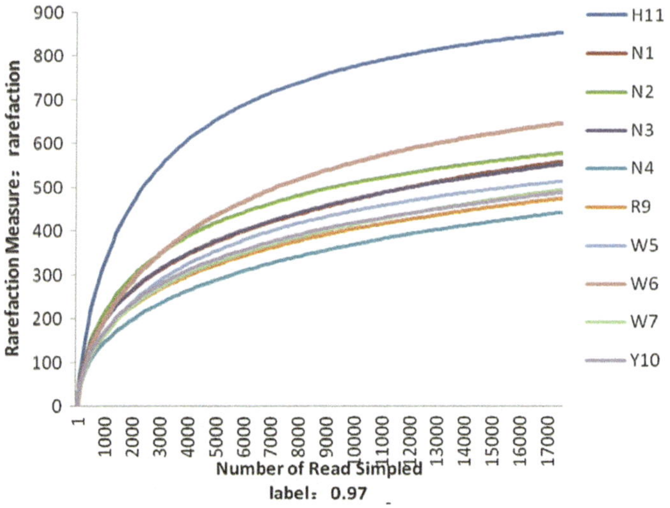

Fig. 11.3 Rarefaction curve

Table 11.2 Diversity index analysis results

Simple ID	Reads	Label: 0.97					
		OUT	Ace	Chao1	Coverage	Shannon	Simpson
N-H1	17384	560	731	711	0.9908	4.6	0.024
N-H2	17384	579	658	672	0.9933	4.46	0.0397
N-H3	17384	554	687	693	0.9917	4.62	0.022
N-H4	17384	444	588	610	0.9920	3.9	0.0639
W-H5	17384	515	591	611	0.9934	4.06	0.0412
W-H6	17384	647	750	754	0.9916	4.42	0.035
W-H7	17384	496	743	673	0.9913	4.12	0.0385
G-H8	17384	478	612	623	0.9926	4.54	0.0233
R-H9	17384	475	601	604	0.9925	4.32	0.0371
Y-H10	17384	488	610	656	0.9925	4.25	0.0352
H-H11	17384	854	916	914	0.9933	5.44	0.0233

detected in the sequencing contains all the microbes present in the test sample. For the microbial diversity indices for all samples in this study, see Table 11.2.

The coverage indices in all samples reached more than 99% (Table 11.2). It indicates that the microbial community measured in this sequencing could represent 99% of all microbial communities in the sample. In terms of abundance, the ACE index of H-H11, W-H6, W-H7 and N-H1 is relatively high, while the ACE index of N-H4 and W-H5 is maintained at a relatively low level. Through the analysis of the chao1 index can also draw the same conclusion. It also shows, indirectly, that the calculation of these two indices are basically similar. The Shannon index represents the microbial diversity. The higher the fragrant Warner index, the higher the diversity of microbes in the sample. As can be seen from Table 11.2, the Shannon index of H-H11 samples in the eleven samples were the highest and the N-H4 samples were the lowest.

11.3.3.5 Phylum Distribution Analysis of the Bacterial Community

This study classified the eleven samples based on their phylums. The specific results are shown in Fig. 11.4.

As shown in Fig. 11.4, the microbes detected in the eleven samples belonged to 34 phylums. Among them, Proteobacteria (41.2%) had the highest abundance, and was followed by Bacteroidetes (20.7%). The proportion of these two bacteria in the total gene sequence reached 61.9%. Within the phylum of Proteobacteria, Betaproteobacteria (20.3%) had the highest abundance, followed by Gammaproteobacteria (11.1%), Alphaproteobacteria (6.4%), and δ—Proteus (Deltaproteobacteria, 1.3%). The phylum Actinobacteria (8.9%) had the third highest relative abundance of microbes, where Cyanobacteria with a relative abundance of 7.78% ranked fourth and Verruvermicrobia (4.01%) ranked fifth. The relative abundance of the other phylums were less than 4%. At the same time, it can be seen from Fig. 11.4 that the proportion of Bacteroidetes between the different samples is quite different. The results showed that

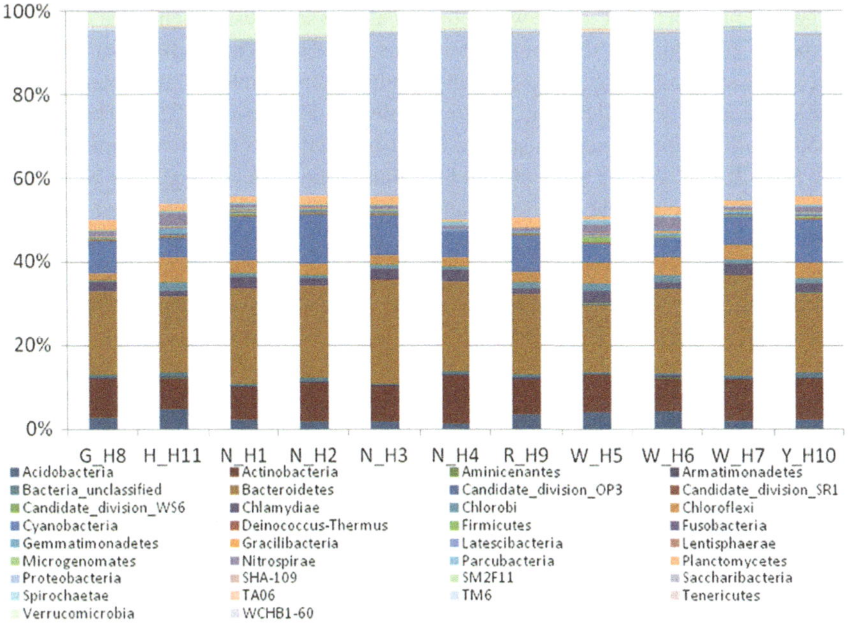

Fig. 11.4 Analysis of bacteria community on phylum distribution

dynamic changes were greatest with Acidobacteria, Spirochaetae and Chlorobi, and the variation coefficients of the dominant bacteria ranked as follows Cyanobacteria > Verrucomicrobia > Actinobacteria > Bacteroidetes > Proteobacteria. Univariate analysis of the relative abundance of bacteria at the phylum level found that there was a significant difference in the relative abundance of the classified bacteria at almost all phylum levels ($p < 0.05$).

11.3.3.6 Genus Distribution Analysis of the Bacterial Community

Through the structural analysis of the bacterial community genus distribution, the results indicated that the bacterial community structure in the 11 samples could be classified into 414 genera (For the convience of mapping and maintaining a neat figure, any bacterial genus with a relative abundance of less than 1% is grouped as [other]). The histogram of bacteria community on genus distribution was shown in Fig. 11.5.

As can be seen from Fig. 11.5, 414 genera were detected, and those with the highest relative abundance was hgcI_clade (20.9%), CL500-29_marine_group (6.9%), Rheinheimera (4.6%), Sporichthyaceae_unclassified (3.6%) and Flavobacterium (3.1%). The relative abundance of other bacterial genera was less than 3%. The five bacteria genera above are the dominant bacterias. The large differences ($p < 0.05$) in the community structure of the microbial community between the different samples were also apparent with the univariate analysis of the relative abundance of the genera present. HgcI_clade was found in all the samples, which were tested for all species belonging to the dominant species. But between the different samples, hgcI_clade still had a large difference in abundance: in sample G8, the abundance of hgcI_clade was 8%, while in the W-H6 sample, the proportion is 28%. The relative abundance of Rheinheimera in sample N-H4 reached 22.6%, but was not detected in Y-H10. It can also be seen in Fig. 11.5 that certain bacterial genera were only present in a single sample or several samples, but are not detected in other samples. Acidovorax has a relative abundance of 15% in sample R-H9%, but it is not found in the sample H-H11. At the same time, the sample Y-H10 has the highest degree of eutrophication of all eleven samples with a comparable level in the bacteria genera present. The proportion of Cyanobacteria_norank and Synechococcus in Y-H10 is much higher than in other samples. It can be seen that these two species of bacteria are more adaptable to eutrophic water bodies, and can become indicators of water eutrophication.

11.3.3.7 Species Distribution Analysis of the Bacterial Community

The results showed that the bacterial community structure in the eleven samples belonged to 678 species (In order to facilitate the mapping and graphics, any species with a relative content of less than 1% is grouped as [other]). The 678 species were

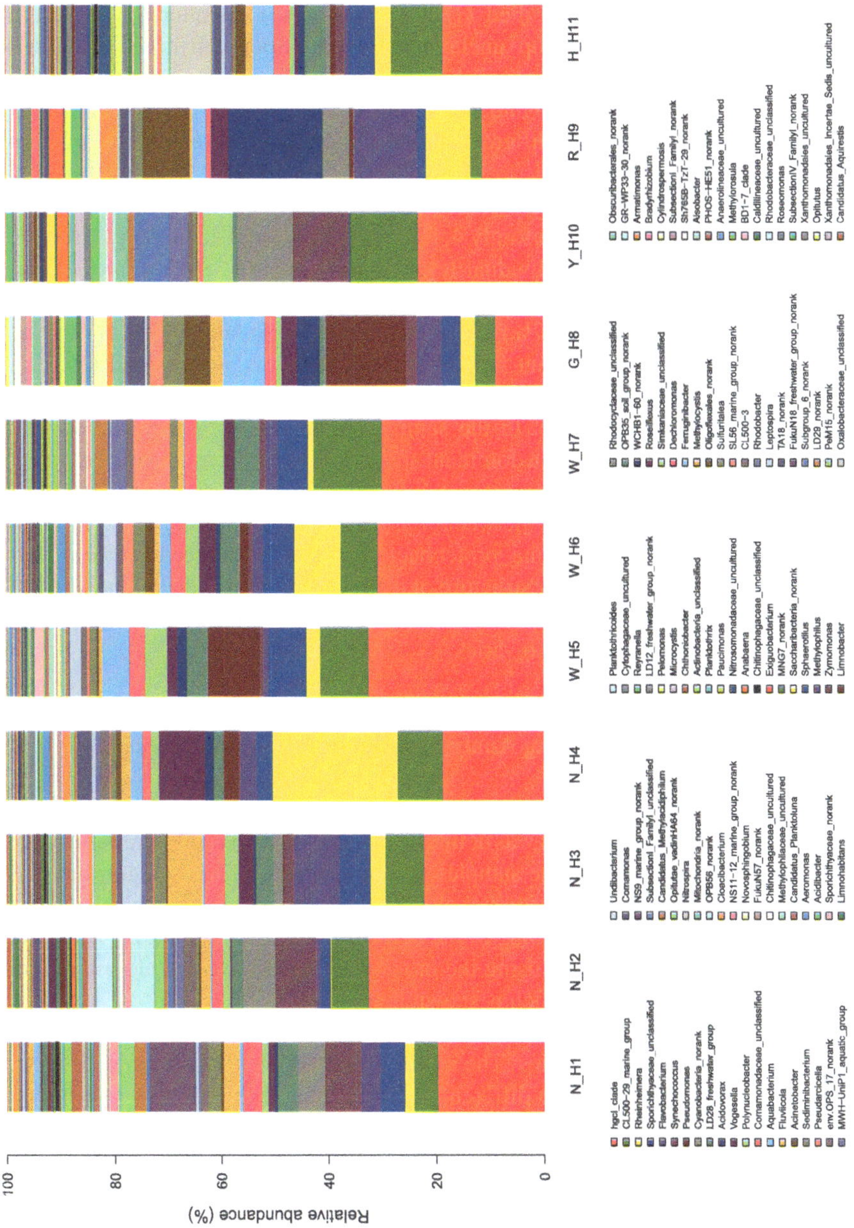

Fig. 11.5 Analysis of bacteria community on genus distribution

arranged according to their relative abundance, where only the top 100 species was included in the bacterial community structure histogram plot, see Fig. 11.6.

Those with the highest abundance according to species distribution are bacterial communitieshgcI_clade_unclassified (12.3%), hgcI_clade_uncultured_ bacterium (8.6%), CL500-29_marine_group_unclassified (6.1%), Sporichthyaceae_ unclassified (4%) and Rheinheimera_uncultured_gamma_proteobacterium (3.6 %). The univariate analysis of the relative abundance of the species present reveals that there is a great difference between the different strains in the bacterial community structure at the species level.

The relative abundance of hgcI_clade_unclassified in sample N-H2 is 18%, but the relative abundance in G-H8 is only 4.4%. For the bacterial community structure of bacteria measured at the individual level, in the sample G-H8, the most abundant species were Aquabacterium_unclassified (8.6%), Pseudomonas_unclassified (8.5%) Pseudomonas_peli (7.7%) and hgcI_clade_unclassified (5.3%). In the sample R-H9, the most abundant species were Acidovorax_unclassified (19.1%), Flavobacterium_unclassified (7.4%), hgcI_clade_uncultured_bacterium (6.5%) and hgcI_clade_unclassified (5.6%). In the sample N-H4, Rheinheimera_uncultured_ gamma_proteobacterium was the dominant species with a relative abundance was of 21%. In other samples, the content of Rheinheimera_uncultured_gamma_ proteobacterium was very low, between 1% and 6%. In the sample Y- In H10, Rheinheimera_uncultured_gamma_proteobacterium was not detected. The difference in the bacterial community structure of different samples is due to the impacts of human activity which varies with the different samples. Furthermore, the source of surface water pollutants is also varies.

11.3.3.8 Similarity Clustering Analysis of Bacterial Communities

The similarities and differences between the relative abundance of the bacterial communities in the multiple samples was delineated and contrasted using a tree structure. Firstly, the distance between the samples is calculated by using an algorithm that describes the relationship and structure of the community. The hierarchical clustering analysis is carried out according to the beta diversity distance matrix. Then, the tree structure is constructed by using the unweighted group average algorithm and the similarity clustering tree which is shown in Fig. 11.7.

As it can be seen from Fig. 11.7, two major clusters could be obtained from the 11 samples bacterial communities. The samples R-H9 and G-H8 are quite different from the other samples is quite different, which is consistent with the conclusion drawn from the microbial community structure study. The samples Y-H10 and N-H2 are also different from the other samples.

The effects of human activities on the samples Y-H10 and N-H2 were significant, which leads to a higher eutrophication degree of samples Y-H10 and N-H2 than the other samples. Seven samples with similar variability differences can be seen from the similarity tree, and the degree of similarity clustering of these seven samples is almost the same as the spatial distribution of the sampling points.

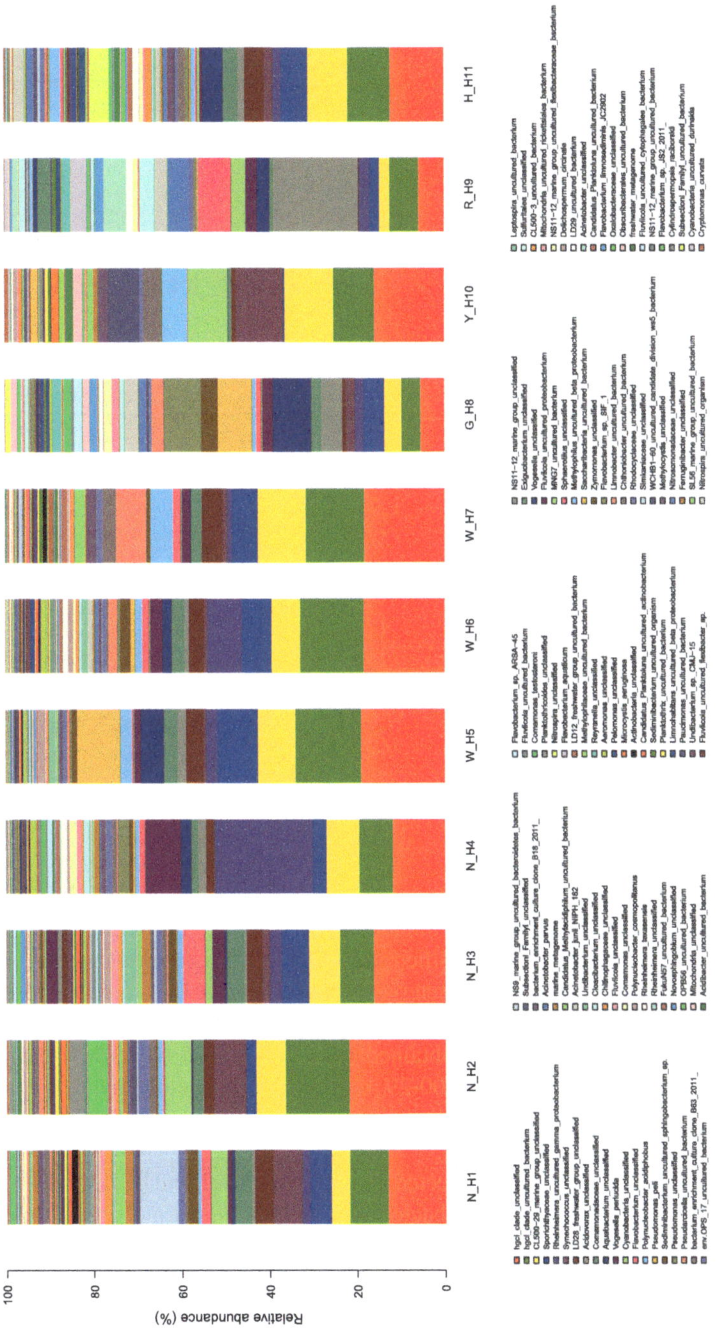

Fig. 11.6 Analysis of bacteria community on species distribution

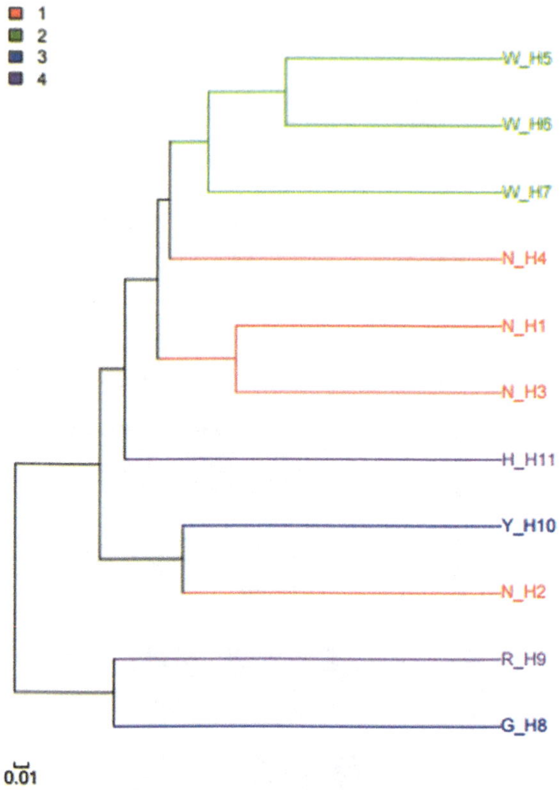

Fig. 11.7 Shared tree

11.4 Summary

1. The microbial Shannon-Wiener index of Poyang Lake and river-lake ecotone are between 3.9 and 5.44. The microbial population can be termed as rich.
2. The microorganismcan be classified into 34 phylums distribution groups. Among them, Proteobacteria, Actinobacteria, Baceroidetes and Cyanobacteria are dominant.
3. A total of 414 genera were obtained from the studied bacteria at the genus. Among them, hgcI_clade, CL500-29_marine_group, Rheinheimera, Sporichthyaceae_unclassified and Flavobacterium dominate the microbial community structure.
4. A total of 678 species were detected at the species level, and hgcI_clade_ unclassifie, hgcI_clade_uncultured_bacterium, CL500-29_marine_group_ unclassified, Sporichthyaceae_unclassified and Rheinheimera_uncultured _gamma_ proteobacteriumwere the dominant species of bacteria in the investigated surfacewater bodies under high water level.

Acknowledgements This work was supported by National Natural Science Foundation of China (41561002) "Study on temporal and spatial variation of bacteria and its water environmental response mechanism in River-lake ecotone of Poyang Lake".

References

1. Zeng, N., and S. Wen. 2008. Conservation and rational use of wetlands in Poyang Lake Area (in Chinese). *Jiangxi Forestry Science and Technology* 6: 48–51.
2. Hansel, C.M., S. Fendorf, P.M. Jardine, and C.A. Francis. 2008. Changes in bacterial and archaeal community structure and functional diversity along a geochemically variable soil profile. *Applied and Environmental Microbiology* 74 (5): 1620–1633.
3. Huang, X.L., et al. 2012. Vertical microbiological variation of a coastal aquifer in southern China. *Water Science and Technology* 65 (4): 703–712.
4. Jiang, F., et al. 2015. Hydrological and sediment effects from sand mining in Poyang Lake during 2001–2010 (in Chinese). *Acta Geographica Sinica* 70 (5): 837–845.
5. Sun, X., et al. 2009. Diversity and phylogenetic analysis of planktonic bacteria in eutrophic zone of Lake Wuliangisuhai (in Chinese). *Biodiversity Science* 17 (5): 490–498.
6. Huang, X., J. Chen, S. Zhou, L. Xie, and C. Fu. 2010. Prokaryote diversity in water environment of land-ocean ecotone of Zhuhai City (in Chinese). *Chinese Journal of Applied Ecology* 21 (2): 452–457.
7. Mardis, E.R. 2008. The impact of next-generation sequencing technology on genetics. *Trends in Genetics* 24 (3): 133–141.
8. Lou, J., Y. Liu, and Y. Li. 2014. Review of high-throughput sequencing techniques in studies of soil microbial diversity (in Chinese). *Chinese Agricultural Science Bulletin* 30 (15): 256–260.
9. Zhu, G. 2008. Study on 16SrDNA diversity of bacteria in south of Poyang Lake (in Chinese). PhD thesis. Nanchang University.
10. Liu, L., et al. 2012. Bacterioplankton diversity in Xiamen Houxi River watershed along an urban-rural gradient and its relation to environmental factors (in Chinese). *Chinese Journal of Applied and Environmental Biology* 18 (4): 591–598.

Chapter 12
Trends of Vegetation Ecosystem Distribution in Jiangxi Province

Zemeng Fan, Zhengping Du and Tianxiang Yue

12.1 Introduction

Climate is one of the most important factors in determining vegetation types and their distribution on Earth, while vegetation is the most direct reflection of Earth's climate change. In 1807, Humboldt and Bonpland studied the distribution of climate and species in their book "Essai sur la geographie des plantes" [1]. In 1823, Schouw explored the relationship between plant flora and vegetation type distribution and multiple climatic factors [2]. In 1872 Griesebach formed the first map of the vegetation distribution according to the vegetation appearance and its climatic factors. In 1892 Merriam proposed a zoning system for zonal distribution of North American fauna and flora based on temperature [3]. In 1916, the representative of the Anglo-American scholar studied the relationship between vegetation and climate in North America with dynamic succession, and put forward the theory of vegetation succession. To improve the drawback of the categories of the prior systems that are too coarse and inapplicable to the whole world, Holdridge (1947) devised a classification. The classification unit is called "life zone". According to the climatic indexes including annual average biological temperature (MAB), average annual precipitation (TAP) and potential evapotranspiration (PER), Holdridge life zone model (HLZ model) was established [4].

In many ecosystem models and vegetation classification systems, the Holdridge life zone classification system can well reflect the interrelationship between climate and vegetation. Thus it is widely used to environmental assessment, ecological zoning and prediction of global impacts on ecosystems [5–9]. Yang compared four commonly used vegetation classification models, namely Penman model, Holdridge life zone system model, Kira model and Thornthwaite model [10]. He pointed out that the HLZ model has a clear advantage in simulating temperate coniferous and broad-

Z. Fan (✉) · Z. Du · T. Yue
Institute of Geographic Sciences and Natural Resources Research (CAS), Beijing, China
e-mail: fanzm@lreis.ac.cn

T. Yue et al. (eds.), *Chinese Water Systems*, Terrestrial Environmental Sciences,
https://doi.org/10.1007/978-3-319-97725-6_12

leaved mixed forest, warm-temperate deciduous broad-leaved forest and subtropical evergreen broad-leaved forest. In early studies on vegetation-climate relationship using the HLZ model, the input parameters are only the point data of meteorological observation [11]. With the emergence of high-precision and high-speed climatic surface modeling tools, grid data of climatic factor are used as input parameters. It is possible to study the spatial distribution of vegetation by HLZ model [6, 7]. This chapter aims to use self-developed surface modeling tools HASM-OC at the regional scale. HLZ ecosystem was simulated and analysed by using the results of climate surface modeling of Jiangxi Province on different time scales. The significant correlation between climate change and the evolution of HLZ vegetation ecosystem on different time scales provides a reference for predicting the possible impact of climate change on vegetation ecosystems [12].

12.2 Holdridge Life Zone Ecosystem Model

Each HLZ ecosystem type shows a certain type of vegetation, but also reflects the numerical range of temperature and precipitation. In this sense, the life zone is a complex result of the surface water and heat system of the earth and it determins the vegetation community. It can also say that it is the product of the interaction between the biological effects of the climate system and the climate effects of the vegetation system [6–9]. Therefore, to a certain extent, the simulation of HLZ ecosystem effectively reflects the potential vegetation distribution under objective temperature and precipitation conditions in the area.

Firstly, the annual average temperature, annual rainfall and potential evapotranspiration of each grid is calculated as logarithmised values [6–8]:

$$M(x, y, t) = InMAB(x, y, t) = In[\frac{1}{365}\sum_{j=1}^{365} TEM_{>0}(j, x, y, t)] \tag{12.1}$$

$$T(x, y, t) = InTAP(x, y, t) = In\sum_{j=1}^{365} p(j, x, y, t) \tag{12.2}$$

$$P(x, y, t) = InPER(x, y, t) = In\frac{58.93MAB(x, y, t)}{TAP(x, y, t)} \tag{12.3}$$

$$d(x, y, t) = \sqrt{(M(x, y, t) - M_{i0})^2 + (T(x, y, t) - T_{i0})^2 + (P(x, y, t) - P_{i0})^2} \tag{12.4}$$

In the HLZ model, the annual average biological temperature (MAB) refers to the daily average temperature (TEM), which is accumulated for all temperatures more than 0 °C. P is the daily precipitation. TAP is the annual precipitation. PER is the annual evapotranspiration rate. M_{i0}, T_{i0}, P_{i0} are the annual average biological temperature, annual precipitation and annual potential evapotranspiration of the ith kind of life zone center coordinates in the hexagonal coordinate system of HLZ. If $d_k(x, y, t) = \min d_i(x, y, t)$, point (x, y) are the kth kind of Ecosystem type.

12.3 Analysis of HLZ Ecosystem Changes

The HLZ ecosystem was simulated in terms of the interpolation results of temperature and precipitation during the periods from 1960s to 2000s. The simulated results show that HLZ ecosystem types in Jiangxi Province include cold temperate rainforests, cold temperate moist forests, cold temperate rainforests, warm temperate humid forests, warm temperate moist forests, subtropical humid forests and subtropical moist forests. Cold temperate rainforests only appeared in the 1980s, and subtropical moist forests appeared only in the 1990s and 2000s. All other HLZ ecosystem types exist in all periods. Comparing the distribution of the HLZ ecosystems in Fig. 12.1, cold temperate moist forests and cold temperate rainforests turn to warm temperate moist forest in the high mountain areas along the border between Hunan and Jiangxi and in the Jiuling Mountains region in the south of Xiu Shui basin. In the central hills of central Jiangxi Province, a large area of warm temperate humid forests transformed into subtropical humid forests.

According to Table 12.1, warm temperate moist forests covered about 30% of the area in the 1960s–1980s, but which reduced sharply between 1990s and 2000s. The change of warm temperate moist forests is similar to that of cold temperate rain forest, which increased the area between the 1960s to 1990s, but drastically reduced in 2000s. The main HLZ ecosystem type in Jiangxi Province is subtropical humid forest, which area reduced between 1960s and 1970s. In the 2000s subtropical humid forest was up to about 77% of the all forest. From Table 12.1, the area of cold temperate moist forests and cold temperate rainforests decreased by less than 0.2% in the past three decades, while the warm temperate humid forests decreased by more than 12%. The area of warm temperate moist forests and subtropical humid forest has increased. In addition, the subtropical humid forest area has increased by more than 10%.

Table 12.2 shows the total proportion of changes in the HLZ ecosystem type and the climatic changes in the HLZ ecosystem types. For example, compared to 1960s, in 1970s a total of 14.80% of the area changed its HLZ ecosystem type. 14.80% of the pattern of HLZ ecosystem changes due to climate-driven factors, of which 4.35% was due to climate and warming. In contrast 69.47% of the area changes were due to cold climate. 0.17% were due to dry climate, 25.17% of the area changes due to the climate becomes wet and 0.74% because climate becomes cold and wet.Compared to 1960s there are about 21% of the area of HLZ ecosystem type changed in 1970s, except in

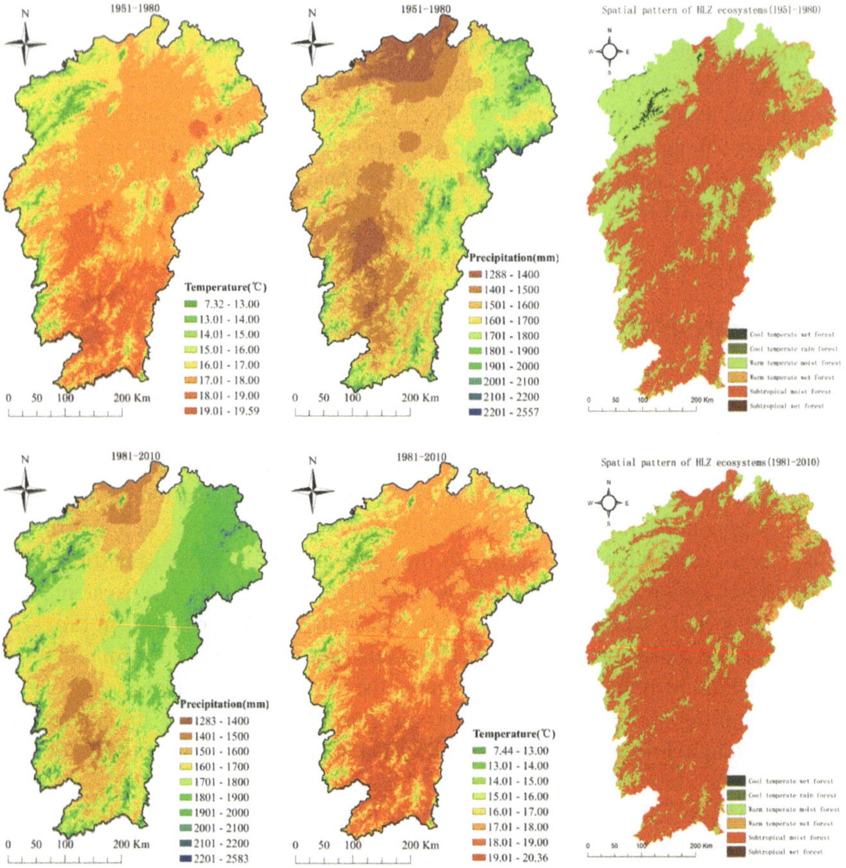

Fig. 12.1 HLZ ecosystem distribution in Jiangxi Province

Table 12.1 Simulation of HLZ vegetation

Year	Cold termperate rainforests	Cold temperate moist forests	Cold temperate rainforests	Warm temperate humid forests	Warm temperate moist forests	Subtropical humid forests	Subtropical moist forests
1960s	0	27390	1482	776951	61837	1803734	0
1970s	0	19060	3735	943938	167851	1536810	0
1980s	14	31979	6265	781283	176495	1675358	0
1990s	0	59	7963	367250	292965	1866362	136795
2000s	0	10193	1021	491539	107138	2060966	537
First 30 yrs	63	64245	7278	928879	102682	1568247	0
Last 30 yrs	0	1357	2514	598890	221194	1846985	454

Table 12.2 Climate factors of HLZ vegetation change (%)

Year	Total share	Warm	Cold	Dry	Wet	Warm and dry	Warm and wet	Cold and wet
1970s–1960s	14.80	4.35	69.47	0.17	25.17	0	0	0.74
1980s–1970s	20.68	52.43	30.88	4.72	10.27	1.10	0	0.13
1990s–1980s	21.06	57.31	0	1.36	34.97	1.58e-2	6.33	0
2000s–1990s	20.95	34.85	3.02	58.25	0.12	3.65	0.11	0
30 Years	22.99	68.68	13.09	3.57	11.88	0.77	0.07	0.94

1970s. Compared with the 1960s, the 1970s mainly caused by the climate change, which was with a proportion of 69.47%. Compared with the 1970s and 1980s, the main reason for the change of the HLZ ecosystem type are mainly climate warming in the 1980s and 1990s. The impact of the proportion was 50–60%. Compared to the 1990s, the reason of HLZ ecosystem type change was mainly less precipitation in the 2000s. It means that the reasons of the change of HLZ ecosystem type in different years are different.

On the whole, compared with 1960s, the main reason for the change of HLZ ecosystem type was caused by the change of temperature in the 1970s. After the 1980s, the influence of precipitation change on the transformation of HLZ ecosystem was intensified. In the 2000s, the precipitation exceeded the temperature becoming the main reason for the change of HLZ ecosystem type. The results of the three-year time scale show that the change of temperature is the main reason to the change of HLZ ecosystem type, which leads to nearly 70% of the pattern of HLZ ecosystem changes. It is much higher than the simulation results with ten years scale.

Compared to the first three decades, the temperature increases by 0.61 °C, the temperature increase to 3.67%, the precipitation increases by 112 mm and the precipitation increase to 7.02% in the last three decades (Table 12.1).

In order to further explain the change of main vegetation type, the transformation probability between the different HLZ ecosystem types during the two 30 years periods (Table 12.3). The horizontal axis represents the HLZ ecosystem type from 1951 to 1980 and the vertical axis represents the HLZ ecosystem type from 1981 to 2010. Table 12.3 shows that the original map of the warm temperate moist forests had only 80862 spots, while the new map has been more spots than the original map. The new spots from warm temperate wet forest are the largest, followed by cold temperate moist forests. That is, the formation of a new warm temperate moist forest is more affected by the increase of local precipitation and secondly by the impact of local climate warming. Similarly, the formation of new warm temperate humid forests is first from subtropical humid forests, which are transformed by the local

Table 12.3 Conversion matrix of HLZ vegetation type for 1951–2010

	Last 30 years	Cold termper-ate rain-forests	Cold temper-ate moist forests	Cold temperate rainforests	Warm temperate humid forests	Warm temperate moist forests	Subtropical humid forests
First 30 years		63	64245	7278	928879	102682	1568247
Cold temperate moist forests	1357	0	1239	115	0	3	0
Cold temperate rainforests	2514	63	1	2450	0	0	0
Warm temperate humid forests	598890	0	6130	0	490560	21817	80383
Warm temperate moist forests	221194	0	56875	4713	72948	80862	5796
Subtropical humid forests	1846985	0	0	0	364920	0	1482065
Subtropical moist forests	454	0	0	0	451	0	3

temperature, followed from warm temperate moist forests, which are transformed by local precipitation. The formation of subtropical humid forests are transformed from the warm temperate humid forests, which are local warm and humid. As can be seen from Table 12.3, the largest conversion area is the subtropical humid forests, which are from warm temperate humid forests.

12.4 Discussion

Using the HLZ ecosystem model, this study simulates and analyzes the trends in potential ecosystem types in Jiangxi Province. From 60 years in the different historical periods, there were wetland temperate rainforests, cold temperate moist forests, cold temperate rainforests, warm temperate moist forests, warm temperate moist forests, subtropical humid forests and subtropical humid forests in Jiangxi Province. Therefore the main forests with a large area are subtropical humid forests, warm temperate moist forests and warm temperate moist forests. Numerical simulations show that the small changes of climate may have a greater impact on the potential ecosystem ecosystems in Jiangxi Province. At the time scale of 30 years, about 20% of the HLZ ecosystem in Jiangxi Province change due to climate change.

Although the average temperature in the three decades from 1981 to 2010 is only 4% higher than the average temperature of 1951–1980, and the average annual precipitation from 1981 to 2010 is only 7% higher than that of 1951–1980, but the effect of small temperature change on the potential ecosystem type of Jiangxi Province is much larger than it of precipitation change. The main response of the Jiangxi Province ecosystem to the humidification of the climate is that the large area of warm temperate humid forests migrate to subtropical humid forests, and the temperate forests and cold temperate forests are disappearing or drastically reduced.

References

1. Humboldt, V.A., and A. Bonpland. 2012. *Essai sur la geographie desplantes*. Paris: Editorial Maxtor.
2. Schouw, J.F. 1823. *Grundzuge einer allgemeinen Pflanzengeographie*: Mit 4 Taf. Reimer.
3. Merriam, C.H. 1892. The geographic distribution of life in North America: With special reference to the mammalia, vol. 7. Biological Society of Washington.
4. Holdridge, L.R. 1947. Determination of world plant formations from simple climatic data. *Science* 105 (2727): 367–368.
5. Fan, Z. 2005. Design and application of resources and environment modelbase system: Spatial trend and scenarios modeling of terrestrial ecosystems in China (in Chinese). PhD thesis. Institute of Geographical and Natural Resources Research, CAS (In Chinese).
6. Yue, T., Z. Fan, and J. Liu. 2005. Changes of major terrestrial ecosystems in China since 1960. *Global and Planetary Change* 48 (4): 287–302.
7. Yue, T., Z. Fan, and J. Liu. 2007. Scenarios of land cover in China. *Global and Planetary Change* 55 (4): 317–342.
8. Yue, T. 2011. Surface modeling: High accuracy and high speed methods, vol. 2. New York: CRC Press. ISBN: 1439817588.
9. Fan, Z., J. Li, and T. Yue. 2013. Land-cover changes of biome transition zones in Loess Plateau of China. *Ecological Modelling* 252: 129–140.
10. Yang, Z., G. Zhou, and D. Yang. 2003. Comparison of simulated vegetation distribution in China produced by four popular climate-vegetation classification models. *Acta Phytoecological Sinica* 27 (5): 587–593. (in Chinese).
11. Nalder, I.A., and R.W. Wein. 1998. Spatial interpolation of climatic normals: Test of a new method in the Canadian boreal forest. *Agricultural and Forest Meteorology* 92 (4): 211–225.
12. Yue, T., Z. Fan, C. Chen, X. Sun, and B. Li. 2011. Surface modelling of global terrestrial ecosystems under three climate change scenarios. *Ecological Modelling* 222 (14): 2342–2361.

Chapter 13
Poyang Lake Basin and Its Ecosystem Evolution

Liu Musheng

13.1 Spatial Distribution Characteristics of the Poyang Lake Basin Ecosystem

Investigating the spatial distribution of various ecosystems is an important factor that determines the overall status of ecosystem services and their spatial differences. It is also an important basis for the protection and utilization of ecosystem services [1, 2]. In order to comprehensively and objectively understand the spatial structure of the ecosystem in Poyang Lake Basin, the results of remote sensing interpretation of land cover based on MSS, TM/ETM images in 1985, 2000 and 2010 were collected. The datasets were provided by the Institute of Geographic Sciences and Natural Resources Research, Chinese Academy of Sciences. Spatial pattern and dynamic changes of the ecosystem in the Poyang Lake Basin were assessed by spatial structure analysis, data analysis and data processing using the ARCGIS 10.0 software. According to the classification system of land cover, remote sensing [3] was combined with the identification, determination and analysis of ecosystem types to classify the five major types of ecosystems in Poyang Lake Basin as follows (Table 13.1):

1. Farmland ecosystems, including paddy fields and dry land in the land cover classification system;
2. Forest ecosystems, including forest land, shrubbery, woodland and other woodland in the land cover remote sensing classification system;
3. The grassland ecosystem which includes high-coverage grassland, middle-coverage grassland and low-coverage grassland in the land cover remote sensing classification system;

L. Musheng (✉)
Jiangxi Mountain-River-Lake Development Office, Nanchang, China
e-mail: 23114860@qq.com

T. Yue et al. (eds.), *Chinese Water Systems*, Terrestrial Environmental Sciences,
https://doi.org/10.1007/978-3-319-97725-6_13

Table 13.1 Land cover classification system

First class		Second class		Definition
No.	Name	No.	Name	
1	Arable land	–	–	Land for planting crops, including cultivated land, new wasteland, fallow land and crop rotation land. Garden land; beach land and tidal flats cultivated for more than three years
		11	Paddy field	Irrigated fields to grow aquatic crops such as rice and lotus root, including cultivated land for rice and dryland crop rotation
		12	Rainfed cropland	Cultivated land where plants grow by natural precipitation. Dry cultivated land with water sources and facilities which can be irrigated under normal years. Cultivated land that is mainly planted with vegetables and rotationcrops
2	Wood land	–	–	Trees, shrubs, bamboos, coastal mangroves and other forestry sites
		21	Closed forest land	Natural wood and plantation with a crown density >30%, such as timber forests, economic forests and shelter forests
		22	Shrubbery	Short forest and thickets below 2 m crown height and a crown density >40%
		23	Sparse wood land	Sparse wood land with a crown density between 10% and 30%
		24	Other	Young afforested land, cut-over land and gardens (such as mulberry field, tea garden)
3		31	High coverage grassland	Natural grassland, improved grassland and cut grass land with more than 50% coverage
		32	Moderate coverage grasslands	Natural grassland and improved grassland with 20–50% coverage
		33	Low coverage grassland	Natural grassland with 5%–20% coverage
4	Waters	–	–	Natural terrestrial waters and water conservancy facilities
		41	Waterway	Natural or artificial river, artificial channel and embankment
		42	Lakes	Land surface that is below perennial water level in natural water area
		43	Reservoirs	Land surface that is below perennial water level in man-made water area

(continued)

Table 13.1 (continued)

First class		Second class		Definition
No.	Name	No.	Name	
		44	Permanent glacier	Land surface that is covered by glaciers and snow cover
		45	Shoals	The tidal zone between the high and low tide levels
		46	Floodlands	Land surface between the flat water level and flood level of rivers and lake waters
5	Artificial areas	–	–	Urban and rural residential areas, mining areas and areas for transportational infrastructure
		51	Urban land	Land surface covered by large, medium and small cities and counties
		52	Rural settlements	Land surface covered by rural settlements
		53	Other	Factories, mines, large industrial areas, oil fields, salt fields, quarries and other sites, roads, airports and special sites
6	Unused land	61	Sand dune	Surface that is covered with sand and the vegetation coverage is below 5% of the land, including the desert but not including the beaches along shorelines
		62	Gobi	Land surface that is dominated by gravel with less than 5% vegetation cover
		63	Saline soil	Land surface covered by saline soils, where vegetation is scarce and only salinity-tolerant plant can grow
		64	Marshes	Flat and long-term wetted land surface
		65	Bareland	Vegetation cover of less than 5%
		66	Bare rock	Surface consisting of rock and gravel
		67	Other	Other types of unused land, including alpine deserts, tundra and so on

4. The wetland ecosystem includes the canals, reservoirs and ponds, beaches, and wetlands in the land cover remote sensing classification system;
5. Urban ecosystems, including urban land, rural settlements and other construction land in the land-cover classification system.

13.1.1 Characteristics of Spatial Distribution of Ecosystem

According to the classification result of remote sensing images (Fig. 13.1) and the field investigation, the spatial distribution characteristics of the ecosystem in Poyang Lake Basin are analyzed as follows (Table 13.2):

1. Farmland: The area of the farmland ecosystem is 42,716.39 km^2, accounting for 27.13% of the total basin area with major distributions in Gan River, Poyang Lake and Fu River Basin. To be more precisely, farmland ecosystems concentrate in the Poyang Lake Plain, Gan Fu Plain, Jitai Basin but also in some of the hilly and mountainous subbasins.
2. Forest: This kind of ecosystem has the largest area, with an area of 97,608.1 km^2, accounting for 62% of the total basin area, which mainly distributes in the upper and middle reaches of Gan River, Fu River, Xin River, Rao River and Xiu Shui River. The largest distribution area is in Gan River Basin, accounting for 54.39%. The amount of secondary forest in the basin is growing with increasing rates due to the launch of man-made reforestation programs.
3. Grassland: This kind of ecosystem has a small area of 6,452.14 km^2, accounting for 4.1% of the total basin area. It is located mainly in the Gan River basin.
4. Wetlands: The total area of wetland ecosystems is relatively small with only 7165.55 km^2, accounting for 4.55% of the total basin area. It mainly distributes around Poyang Lake, accounting for 58.39% of the total wetland area.
5. Urban: Urban ecosystems have the smallest area of 3501.14 km^2, accounting for 2.22% of the total area of the river basin and are mainly distributed in the plains of the catchment.

13.1.2 The Relationship Between Ecosystem Distribution and Elevation

Altitude elevation is closely related to topography. Poyang Lake Basin has many different landform types. Referring to China's topography of mountain classification table method and the specific circumstances of the Poyang Lake Basin, the elevation can be divided into five levels (Table 13.3).

The proportion of area belonging to each altitude class is different for the different ecosystems (Table 13.4). The farmland ecosystems are mainly distributed in plains, hillocks and terraces, which have altitudes of <100 m. The area belonging to this

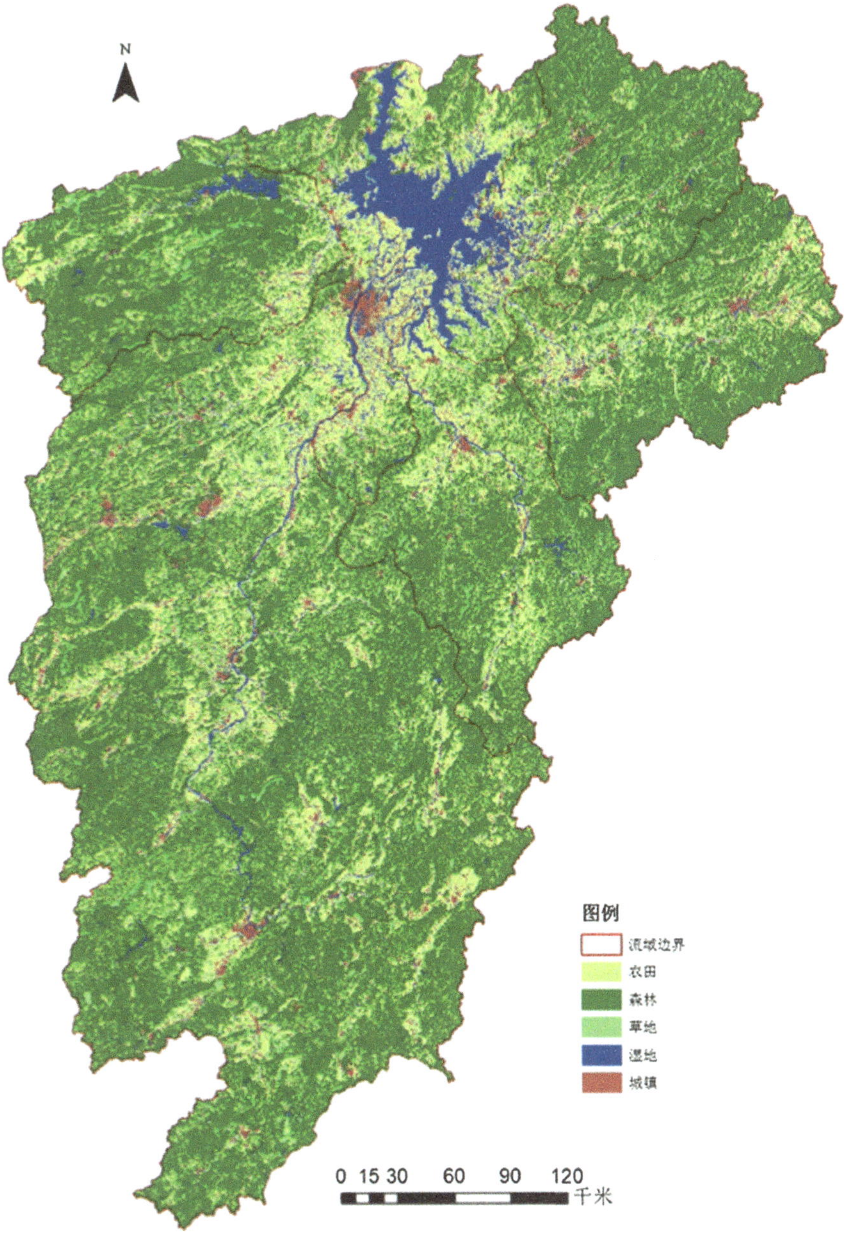

Fig. 13.1 Ecosystem distribution in Poyang Lake basin in 2010 (Yellow: Farmland; Green: Forest; Light Green; Glassland; Blue: Wetland; Red: Urban)

Table 13.2 Distribution of Poyang Lake Basin ecosystem in 2010 by area (km^2)

Basin	Farmland ecosystem	Forest ecosystem	Grassland ecosystem	Wetland ecosystem	Urban ecosystem	Total
Rao River	3199.38	8696.94	346.24	332.71	338.49	12913.76
Poyang Lake	7378.49	4477.4	535.13	4183.61	681.45	17256.08
Xiu River	3151.2	10349.74	545.12	445.45	159.59	14651.1
Xin River	4044.75	10794.72	513.31	274.47	308.71	15935.96
Gan River	19954.41	53085.65	4087.97	1422.68	1686.92	80237.63
Fu River	4988.16	10203.65	424.37	506.63	325.98	16448.79
Total	42716.39	97608.1	6452.14	7165.55	3501.14	157443.32

altitude class covers 26,769.98 km^2 which accounts for 62.67% of the total area of farmland ecosystems. They are distributed also in hilly areas with altitudes between 100–300 m, accounting for 30.3% of their total area. The forest ecosystems are mainly distributed in low and high hilly areas of 100–500 m elevation above sea level, with areas of 38,562.85 km^2 and 24,399.56 km^2, accounting for 39.51% and 25.00% of the total area, respectively.

Grassland ecosystems are mainly distributed in low hilly areas of 100–300 m, with an area of 2601.34 km^2, accounting for 40.32% of the total area. The second largest part of the grassland ecosystem is location in the plains with altitudes less than 100 m covering an area of 1817.53 km^2 or 28.17% of the total grassland ecosystem area. Wetland ecosystems are mainly distributed in plains and terraces with an altitude of less than 100 m. The total area of wetland ecosystems in this low altitude area is 6518.30 km^2, accounting for 90.97% of the entire wetland ecosystem.

The distribution of urban ecosystems is similar to that of farmland ecosystems, accounting for 76.50% of all urban ecosystems being in plains, hillocks and terraces at altitudes <100 m, and 21.76% in low hilly areas of 100–300 m. This indicates that farmland ecosystems are mainly distributed peripheral to the urban ecosystems meaning that the distribution pattern of both ecosystems have a strong correlation. In terms of the proportion of ecosystems in different elevation zones, forest ecosystem are dominant from low altitude to high altitude accounting for 62% of the total basin area.

13.1.3 Relationship Between Ecosystem Distribution and Slope

According to the Second National Land Survey Technical Regulation (TD/T1014-2007), the slope of the topography is divided into five grades (Table 13.5).

Table 13.3 Classification of elevation

Class of elevation	Elevation (m)
5	>1000
4	500–1000
3	300–500
2	100–300
1	<100

In addition to the influence of elevation, the distribution of ecosystems is also closely related to slope (Table 13.6). Farmlands, wetlands and urban ecosystems are mainly distributed in areas with a slope of less than or equal to 2°, with areas of 30018.70 km^2, 6636.09 km^2 and 2839.94 km^2, accounting for 70.2%, 92.61% and 81.11% of the total area, respectively. In general, the area of ecosystems decreased with an increase of slope.

The forest ecosystems are mainly distributed in the zone with slopes of 6–15° with an area of 36552.16 km^2, accounting for 37.45% of the total area, followed by the class with 15–25° with an area of 20807.54 km^2, accounting for 21.32% of the total area of forest ecosystems. Grassland ecosystems are mostly distributed in the area of $\leq$2° with an area of 2421.57 km^2, accounting for 37.53% of the total area, followed by areas with slopes of 6–15° and 2–6° with a total area of 25.43% and 17.85% of all grassland ecosystem areas, respectively. Forest ecosystem occupy the relative majority of area for each individual slope class except of areas with a slope of less than 2°, where the distribution of farmland ecosystem is the largest, accounting for 49.49% of the total area which belongs to this slope category. With increasing slope, the distribution of farmland ecosystems decrease. At areas with high slopes (15–25° and >25° class) forest ecosystems are dominant sharing 91.45 and 92.44% of the area belong to each slope class, respectively.

13.2 Analysis of Dynamic Changes in the Ecosystem Structures of Poyang Lake Basin

Ecosystem composition refers to the area and proportion of different types in the study area such as forests, grasslands, wetlands, farmlands and urban ecosystems. The natural and socio-economic conditions of the region play important roles for the formation, development and evolution of the ecosystem structure. The relationship between them is inextricably linked. The actual ecosystem structure is the result of the joint action of land use and land transformation by humans and the process of constant changes of the natural environment and socio-economic conditions. The purpose of analyzing the dynamic changes in ecosystem structure is to understand the spatio-temporal distribution and allocation of different ecosystem types, the quantitative ratio and the mutual transformation of various ecosystems, and to provide the basic data for revealing the current ecological status and its resilience.

Table 13.4 Ecosystems distribution in different elevation zones. The area is given in km^2 and the proportion in percentage

Ecosystem type		Elevation (m)				
		<100	100–300	300–500	500–1000	>1000
Farmland	Area	26769.98	12942.61	2161.50	797.40	44.89
Farmland	Proportion	62.67	30.30	5.06	1.87	0.11
Forest	Area	16798.60	38562.85	24399.56	16216.13	1630.96
Forest	Proportion	17.21	39.51	25.00	16.61	1.67
Grassland	Area	1817.53	2601.34	945.89	692.13	395.25
Grassland	Proportion	28.17	40.32	14.66	10.73	6.13
Wetlands	Area	6518.30	610.89	28.61	7.62	0.13
Wetlands	Proportion	90.97	8.53	0.40	0.11	0.00
Urban	Area	2678.25	761.95	50.70	9.38	0.85
Urban	Proportion	76.50	21.76	1.45	0.27	0.02

13.2.1 Changes of Ecosystem Composition

The ecosystem composition is calculated by:

$$P_{ij} = S_{ij}/TS \tag{13.1}$$

P_{ij} is the area ratio of the ith type of ecosystem in the jth year, and S_{ij} is the area of the ith type of ecosystem in the jth year. TS is the total area. The change rate of ecosystem type refers to the quantitive change of ecosystem area in the study area within a certain period of time. The calculation method is:

$$K_i = (U_b - U_a)/U_a * 100\% \tag{13.2}$$

A positive value K_i indicates an increase in the area of the ith type of ecosystem at the end of the study period. Subsequently a negative value indicates a decrease in area covered by a certain ecosystem type. U_a/U_b is the ratio between the area of an ecosystem type at the beginning of the study period and the end of the study period. Using the software ARCGIS10.0, the results of the three-phase remote sensing interpretation of land cover were summed up to obtain the characteristics of the ecosystem structure in Poyang Lake Basin (Table 13.7).

Table 13.5 Classification of slope

Class of elevation	Elevation (m)
1	≤2°
2	2°–6°
3	6°–15°
4	15°–25°
5	>25°

The statistical analysis shows that the total area of Poyang Lake Basin is 157,443 km^2. It could be shown (Fig. 13.2) that, in the past 25 years, Poyang Lake Basin is dominated by forests and farmland, accounting for 89.13% of the total basin area. However, the data shows that the ecosystem composition in the region changed which reflected the trend of land use in this area.

Table 13.7 shows that in the mid-1980s, the structural proportions of farmland, forest, grassland, wetland and urban ecosystem in Poyang Lake Basin were 27.35, 61.94, 4.54, 4.59 and 1.59% of the total area. For the year 2000, the ratio of area composition in Poyang Basin was 27.21, 62.13, 4.36, 4.62, and 1.68%. From mid 80s to 2000 (Table 13.8), the area of farmland ecosystems decreased by 222 km^2 (−0.52% of their previous area), the area of forest ecosystems increased by 305.06 km^2 (+0.31%), and the area of grassland ecosystems decreased by 278.85 km^2 (−3.9%), while the area of wetlands ecosystem increased by 53.15 km^2 (+0.74%) and the area of urban ecosystem increased by 142.61 km^2 (+5.7%) as presented in Table 13.8.

From 2000 to 2010, the areas belonging to farmland ecosystem and grassland ecosystems further decreased by 122.53 km^2 (−0.29% area loss in comparison to the year 2000) and 414.12 km^2 (−6.03%), but also the area covered by forest ecosystem and wetland ecosystem decreased by 209.38 km^2 (0.29%) and 112.53 km^2 (−1.55%). This loses were compensated by an increase in urban ecosystem area by 858.59 km^2 (32.49%). Focusing on the entire investigation period from mid-1980s to 2010, the absolute value of farmland, grassland and wetland ecosystem decreased and the absolute value of forest and urban ecosystem increased. Among them, the largest decrease was referred to grassland ecosystems with 692.97 km^2 (−9.7% less area than in the 1980s) and the farmland ecosystem which decreased by 344.53 km^2 (−0.8%). The area of the wetland ecosystems slightly decreased by 59.38 km^2, the area of grassland ecosystems decreased by 692.97 km^2 and the area of wetland ecosystem decreased by 59.38 km^2 (−0.82%). On the contrary, the area of urban and forest ecosystems increased by 95.68 and 1001.2 km^2, whereas the change of both, total change and relative change in area was largest for urban ecosystem among all ecosystems. Further interesting is that both forest and wetland ecosystems experienced a process of first increasing, then decreasing area whereas the trends of the remaining three ecosystems remain constant.

Table 13.6 Ecosystems distribution in different slope zones. The area is given in km^2 and the proportion in percentage

Ecosystem type		≤2	2°–6°	6°–15°	15°–25°	>25°
Farmland	Area	30018.70	7550.70	4093.55	948.65	104.8
Farmland	proportion	70.27	17.68	9.58	2.22	0.25
Forest	Area	18742.29	16599.59	36552.16	20807.54	4906.5
Forest	proportion	19.20	17.01	37.45	21.32	5.03
Grassland	Area	2421.57	1151.96	1640.74	950.26	287.6
Grassland	Proportion	37.53	17.85	25.43	14.73	4.46
Wetlands	Area	6636.09	374.21	127.87	23.43	3.94
Wetlands	Proportion	92.61	5.22	1.78	0.33	0.05
Urban	Area	2839.94	521.66	112.53	22.31	4.69
Urban	Proportion	81.11	14.90	3.21	0.64	0.13

Table 13.7 Temporal changes of the Poyang Lake Basin ecosystem composition. The area is given in km^2 and the proportion in percentage

Year	Statistics parameters	Farmland ecosystem	Forest ecosystem	Grassland ecosystem	Wetlands ecosystem	Urban ecosystem
Mid 80s	Area	43060.81	97512.48	7144.96	7224.95	2500.01
Mid 80s	Proportion	27.35	61.94	4.54	4.59	1.59
2000	Area	42838.78	97817.5	6866.12	7278.11	2642.61
2000	Proportion	27.21	62.13	4.36	4.62	1.68
2010	Area	42716.39	97608.1	6452.14	7165.55	3501.14
2010	Proportion	27.13	62	4.1	4.55	2.22

13.2.2 *Conversion of Ecosystem Types*

Using the spatial analysis function of ARCGIS software, the conversion of ecosystems from one type to a specific other one can be obtained for all three periods. The results are visualized by the specific transition matrix (Table 13.9). Rows refer to the earlier and column to the later period of measurement. Bolded diagonal elements represent proportions of each land use/land cover class that were static (persisted) between 1973 and 2015. The loss column and gain row indicate the proportion of the landscape that experienced gross loss and gain in each class, respectively.

The Ecosystem transition matrix is a good expression for the changes of the Poyang Lake Basin ecosystem composition as it is able to show the source, direction and result of ecosystem conversion. Using the ecosystem transition matrix, the proportions of different types of ecosystems between different investigation periods can be calculated (Tables 13.9 and 13.10).

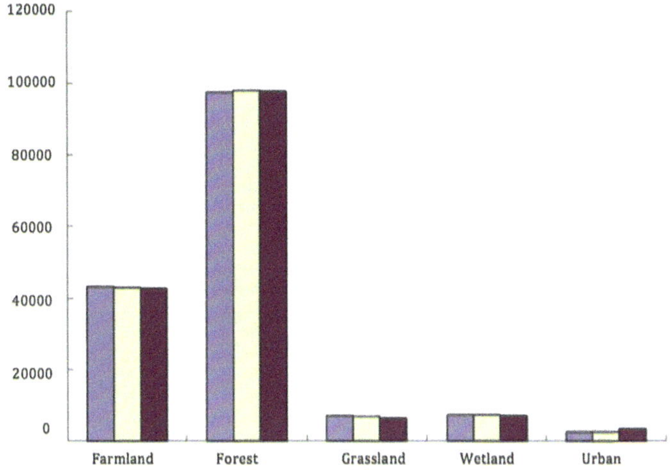

Fig. 13.2 Temporal changes ecosystem areas in km^2 (Blue:1985. Yellow: 2000. Purple: 2010)

Table 13.8 Absolute and relative change of ecosystem in Poyang Lake Basin. The area is given in km^2 and the rate change in percentage

Year	Statistics parameters	Farmland ecosystem	Forest ecosystem	Grassland ecosystem	Wetlands ecosystem	Urban ecosystem
Mid 80s–2010	Changed area	−344.53	95.68	−692.97	−59.38	1001.2
Mid 80s–2010	Relative change	−0.8	0.1	−9.7	−0.82	40.05
Mid 80s–2000	Changed area	−222	305.06	−278.85	53.15	142.61
Mid 80s–2000	Relative change	−0.52	0.31	−3.9	0.74	5.7
2000–2010	Changed area	−122.53	−209.38	−414.12	−112.53	858.59
2000–2010	Relative change	−0.29	−0.21	−6.03	−1.55	32.49

For the entire study period from the middle of the 1980s to 2010, 2260.64 km^2 of farmland ecosystem in the Poyang Lake Basin was transferred to other ecosystems of which 1088.48 km^2 (2.53%), 87.03 km^2 (0.2%), 310.32 km^2 (0.72%) and 774.81 km^2 (1.8%) were converted to forests, grasslands, wetlands and urban ecosystems, respectively (Table 13.9). In contrast, during the same period, a total of 1916.11 km^2 area was converted from other ecosystems types to farmland which included 1242.95 km^2 (2.91% of total farmland in 2010) of forests, 168.82 km^2 (0.4%) of grassland, 377.93 km^2 (0.88%) of wetlands and 126.41 km^2 (0.3%), of urban ecosystems (Table 13.10).

Table 13.9 Absolute and relative change of land classification types in Poyang Lake Basin. The area is given in km^2 and the rate change in percentage

Year	Type	Farmland	Forest	Grassland	Wetlands	Urban
1980s–2000	Farmland	42496.35	282.72	11.86	145.68	124.19
1980s–2000	Forest	164.85	97175.1	108.17	26.65	37.63
1980s–2000	Grassland	39.93	349.2	6738.67	12.4	4.75
1980s–2000	Wetland	123.83	7.07	7.15	7081.81	5.09
1980s–2000	Urban	13.82	3.4	0.27	11.57	2470.95
1980s–2000	Gain	342.43	642.39	127.45	196.3	171.66
1980s–2000	Loss	564.45	337.3	406.28	143.14	29.06
1980s–2000	Net change	−222.02	305.09	−278.83	53.16	142.6
2000–2010	Farmland	40938	932.32	82.75	227.72	657.32
2000–2010	Forest	1206.95	96090	165.55	63.99	282.29
2000–2010	Grassland	126.04	497.71	6189.41	18.9	32.96
2000–2010	Wetland	323.43	60.42	10.59	6845.37	38.21
2000–2010	Town	121.17	18.92	2.59	9.51	2490.38
2000–2010	Gain	1777.59	1509.37	261.48	320.12	1010.78
2000–2010	Loss	1900.11	1718.78	675.61	432.65	152.19
2000–2010	Net change	−122.52	−209.41	−414.13	−112.53	858.59
Mid 80s–2010	Farmland	40799.47	1088.48	87.03	310.32	774.81
Mid 80s–2010	Forest	1242.95	95613.2	248.85	85.09	313.62
Mid 80s–2010	Grassland	168.82	815.86	6096.73	23.78	38.68
Mid 80s–2010	Wetland	377.93	63.68	15.81	6726.78	40.67
Mid 80s–2010	Urban	126.41	18.17	2.48	19.52	2333.38
Mid 80s–2010	Gain	1916.11	1986.19	354.17	438.71	1167.78
Mid 80s–2010	Loss	2260.64	1890.51	1047.14	498.09	166.58
Mid 80s–2010	Net change	−344.53	95.68	−692.97	−59.38	1001.2

The total area of forest ecosystems loss to other ecosystems is 1890.51 km^2, of which 1242.95 km^2 (1.27% of the total forest area at the beginning of the period), 248.85 km^2 (0.26%), 85.09 km^2 (0.09%) and 313.62 km^2 (0.32%) are transferred to farmland, grassland, wetland and urban ecosystem, respectively. The area of forest ecosystems gains from other ecosystems was 1986.19 km^2 between the 1980s and 2010. 1088.48 km^2 (1.12% of the total forest area in 2010) were former farmland, 815.86 km^2 (0.84%) grassland, 63.68 km^2 (0.07%) wetland and 18.17 km^2 (0.02%)

Table 13.10 Transition matrix of percentage gain between ecosystem types in Poyang Lake Basin (%)

Year	Type	Farmland	Forest	Grassland	Wetlands	Urban
mid 80s–2000	Farmland	99.2	0.29	0.17	2	4.7
mid 80s–2000	Forest	0.38	99.34	1.58	0.37	1.42
mid 80s–2000	Grassland	0.09	0.36	98.14	0.17	0.18
mid 80s–2000	Wetland	0.29	0.01	0.1	97.3	0.19
mid 80s–2000	Urban	0.03	0	0	0.16	93.5
2000–2010	Farmland	95.84	0.96	1.28	3.18	18.77
2000–2010	Forest	2.83	98.45	2.57	0.89	8.06
2000–2010	Grassland	0.3	0.51	95.95	0.26	0.94
2000–2010	Wetland	0.76	0.06	0.16	95.53	1.09
2000–2010	Urban	0.28	0.02	0.04	0.13	71.13
Mid 80s–2010	Farmland	95.51	1.12	1.35	4.33	22.13
Mid 80s–2010	Forest	2.91	97.96	3.86	1.19	8.96
Mid 80s–2010	Grassland	0.4	0.84	94.51	0.33	1.1
Mid 80s–2010	Wetland	0.88	0.07	0.25	93.88	1.16
Mid 80s–2010	Urban	0.3	0.02	0.04	0.27	66.65

urban ecosystems, respectively. For grassland ecosystems, 1047.14 km^2 were converted to other ecosystems within the last 25 years which accounts for 14.66% of the total grassland area. In detail 168.82 km^2 (2.36%), 815.86 km^2 (11.42%), 23.78 km^2 (0.33%) and 38.68 km^2 (0.54%) of grassland ecosystem area were transferred to farmland, forest, wetland and urban ecosystem, respectively. Other ecosystems with an area of 354.17 km^2 were converted to grassland ecosystems, of which 87.03 km^2 (1.35% of the total grassland area in 2010), 248.85 km^2 (3.86%), 15.81 km^2 (0.25%) and 2.48 km^2 (0.04%) consists of former farmland, forest, wetland and urban ecosystems, respectively. The area of wetland ecosystem transferred to other ecosystems was 498.09 km^2, of which 377.93 km^2 (5.23% of the initial wetland area), 63.68 km^2 (0.88%), 15.81 km^2 (0.22%) and 40.67 km^2 (0.56%) were transferred to farmland, forest, grassland and urban ecosystem, respectively. The area of other ecosystem types converted to wetland ecosystems is 438.71 km^2 between the 1980s and 2010. Considering the transformation from each ecosystem type,

310.32 km^2 (4.33% of the total wetland area in 2010), 85.09 km^2 (1.19%), 23.78 km^2 (0.33%) and 19.52 km^2 (0.27%) were former farmland, forest, grassland and urban ecosystem. Finally, the area of urban ecosystem transferred to other ecosystems is 166.58 km^2, of which 126.41 km^2 (5.06%), 18.17 km^2 (0.73%), 2.48 km^2 (0.1%) and 19.52 km^2 (0.78%) were transferred to farmland, forest, grassland and wetland ecosystem, respectively. Urban ecosystems gained an area of in total 1167.78 km^2 which consists of 774.81 km^2 (22.13% of the urban area in 2010) of former farmland, 313.62 km^2 (8.96%) of former forests, 38.68 km^2 (1.1%) of former grassland and 40.67 km^2 (1.16%) of former wetlands. Comparing these results, sorting the ecosystems by their absolute loss and gains of ecosystem area, the order is: farmland > forest > grassland > wetland > urban for the loss of area and forest > farmland > town > wetland > grassland for the gain of area. In order to better reflect the overall changes, which consist not only of the spatial but also of the temporal dynamics of each ecosystem type, a dynamic land use model is applied to calculate the so-called dynamic degree of each ecosystem type. The dynamic degree of a single land use type (K) expresses the change of the quantity of some land use types in the study area within a certain time range. The calculation method is:

$$K = \frac{\Delta U_a - \Delta U_b}{U_a} * \frac{1}{T} * 100\% \tag{13.3}$$

K is the dynamic degree of a certain land use type during the study period, U_a is the initial area of a land use type in the region, ΔU_a and ΔU_b are the absolute values of gained and lost area of a certain land use type with the time period T, which covers the study period of this investigation. The formula for calculating the integrated dynamic degree of land use (LC) in the region is:

$$LC = [\frac{\sum_{i=1}^{n} \Delta LU_{i-j}}{2\sum_{i=1}^{n} LU_i}] * \frac{1}{T} * 100\% \tag{13.4}$$

where LU_i is the area of land use type i at the beginning of the measurement. ΔLU_{i-j} is the absolute amount of area of a land use type i to a different land use type during the measuring period. T is the length of the monitoring period. Both, individual ecosystem types, the entire basin and also the large sub-watersheds of Lake Poyang basin are used as analysis units to calculate the dynamic changes of ecosystem composition during the mid-1980s–2000 and 2000–2010 period. The calculation results are shown in Table 13.11.

From the mid-1980s to 2000, the changes of ecosystem types in Poyang Lake Basin are generally slow. Except for urban and grassland ecosystems, the dynamic changes of other types of integrated ecosystems are not significant. In contrast, the dynamic degree increases in the latter period from 2000 to 2010 for the grassland, wetland and urban ecosystem type with largest changes occurring for the urban ecosystem due to urbanization of the landscape the types of ecosystem in the Poyang Lake Basin changed greatly and the largest changes were in urban ecosystems. In the mid-1980s–2000, temporal changes of the ecosystem composition in each of the

Table 13.11 Dynamic degree of integrated ecosystem type in Poyang Lake Basin

Unit	Name	Dynamic degree (%)	
		Mid 80s–2000	2000–2010
Ecosystem type	Farmland	0.03	0.02
Ecosystem type	Forest	0.02	0.01
Ecosystem type	Grassland	0.26	0.4
Ecosystem type	Wetland	0.05	0.1
Ecosystem type	Urban	0.38	2.17
Basin	Rao River	0.02	0.05
Basin	Poyang Lake	0.02	0.08
Basin	Xiu River	0.01	0.04
Basin	Xin River	0.01	0.04
Basin	Gan River	0.01	0.05
Basin	Fu River	0.02	0.07

sub-basins of Poyang Lake Basin was not significant with dynamic degrees between 0.01 and 0.02%. During 2000–2010, dynamic change rates of ecosystem composition increased in all subbasins. The annual variation rate of all ecosystem types for the entire Poyang Lake Basin was 0.08%, which was four times more than in the previous period. Additionally, the annual change rates of ecosystem types in other watersheds increased 2.5 times than in the period mid-1980s–2000.

13.2.3 Changes of the Gravity Center of Ecosystem Distribution

Calculating the position of the gravitiy center of the ecosystem types for individual time steps allows to describe the spatial and temporal evolution of ecosystems by providing an explicite spatial interpretation. Changes in spatial ecosystem distribution cause a move of the center of gravitiy of the composition. From a trajectory of changes during a period of time, the overall trend of ecosystem changes can be obtained. The center of gravitiy for the ecosystem distribution can be calculated by:

$$X_t = \sum_{i=1}^{n} (C_{ti} g X_i) / \sum_{i=1}^{n} C_{ti} \tag{13.5}$$

$$Y_t = \sum_{i=1}^{n} (C_{ti} g Y_i) / \sum_{i=1}^{n} C_{ti} \tag{13.6}$$

Table 13.12 Shift of the center of gravity of ecosystem distribution from the mid-1980s to 2010 in Poyang Lake Basin

	Mid 80s			2010			Migration direction	Migration distance (m)
	X(°)	Y(°)	Center of gravity location	X(°)	Y(°)	Center of gravity location		
Farmland	115.607	27.571	Le'an County	115.686	27.526	Le'an County	Eastsouth	9135.56
Forest	115.665	27.888	Fengcheng	115.684	27.603	Le'an County	Southeast	31894.53
Grassland	115.507	27.338	Yongfeng County	115.629	27.588	Le'an County	Northeast	30420.56
Wetland	115.946	28.174	Fengcheng	115.756	27.904	Fengcheng	Southwest	35459.50
Urban	115.707	27.849	Fengcheng	115.779	27.862	Fengcheng	Eastnorth	7180.81

X_t, Y_t are the latitudinal and longitudinal coordinates of the center of gravity of the ecosystem distribution in the tth year. C_{ti} is the area of the ith polygon in the tth year. X_i, Y_i are the latitudinal and longitudinal coordinates of the geometric center of the ith polygon in the ecosystem. n is the number of polygon.

The center of gravity of the ecosystem distribution in the mid-1980s and in 2010 in the Poyang Lake Basin (Table 13.12) is obtained using the center of gravity formula. The center of gravity of the distribution of farmland and urban ecosystem in the Poyang Lake Basin migrates to the east. The center of forests and wetland ecosystems migrate to the south and the center of grassland ecosystems migrates northward. Analysing the migration distance, the distribution center of wetlands shifts furthest from the mid-1980s (115.946 °E, 28.174 °N) to 2010 (115.756 °E, 27.904 °N) by 35.46 km in southwestern direction. The shift of the center of gravity distribution for urban ecosystem is smallest and the coordinate of center of gravity moved 7.18 km from the mid-1980s (115.707 °E, 27.849 °N) to 2010 (115.779 °E, 27.862 °N).

References

1. Wang, Y., Z. Yibin, and D. Han. 1999. The spatial structure of landscape ecosystems: Concept, indices and case studies (in Chinese). *Advances in Earth Science* 14 (3): 235–241.
2. Xiao, D., and R. Bu. 1997. Spatial ecology and landscape heterogeneity (in Chinese). *Acta Ecologica Sinica* 17 (5): 453–461.
3. Liu, J. 1997. Study on national resources and environment survey and dynamic monitoring using remote sensing (in Chinese). *Journal of Remote Sensing* 1 (3): 225–230.

Chapter 14
Benthic Macroinvertebrates as Indicators for River Health in Changjiang Basin

Fengzhi He, Xiaoling Sun, Xiaoyu Dong, Qinghua Cai and Sonja C. Jähnig

14.1 Introduction

Rivers have been associated with development of human society. Rivers and its associated freshwater ecosystems provide multiple ecosystems services for humans [1, 2]. For example, they supply fresh water for drinking, agriculture, domestic and industrial use. Food (e.g. fish) provided by freshwater ecosystems is important protein resource for millions of people in regions such as Mekong river basin [3]. Apart from provisioning services, regulatory (e.g. carbon sequestration, flood regulation and water purification), supporting (e.g. nutrient cycling) and cultural (e.g. recreation and tourism) services provided by freshwater ecosystems are also vital for human well-beings [1]. However, freshwaters have been seriously affected by human activities due to the rapid growth of human population and demands for energy and water [4]. Thousands of dams have been built and planned on rivers globally [5, 6] and various pollutants have been released into rivers and lakes [7]. Consequently, one-

F. He (✉) · X. Sun · X. Dong · Q. Cai (✉)
State Key Laboratory of Freshwater Ecology and Biotechnology, Institute of Hydrobiology, Chinese Academy of Sciences, Wuhan, China
e-mail: fengzhi.he@igb-berlin.de

Q. Cai
e-mail: qhcai@igb.ac.cn

F. He · S. C. Jähnig
Leibniz-Institute of Freshwater Ecology and Inland Fisheries, Berlin, Germany

F. He
Institute of Biology, Freie Universität, Berlin, Germany

X. Sun
Southern University of Science and Technology, Shenzhen, China

X. Dong
School of Environmental Science and Engineering, Shenzhen Academy of Environmental Sciences, Shenzhen, China

T. Yue et al. (eds.), *Chinese Water Systems*, Terrestrial Environmental Sciences,
https://doi.org/10.1007/978-3-319-97725-6_14

third of freshwater species have been considered as threatened by IUCN Red List of Threatened Species [8].

As intermediate consumers in river ecosystems, macroinvertebrates play vital ecological roles and serve as conduits in food webs, posing strong influences on primary productivity, decomposition, nutrient cycling, and material translocations [9]. In addition, macroinvertebrates have relatively short life cycles and wide distribution range, and can rapidly respond to environmental changes within and around rivers, which makes them ideal indicators for river health [10]. During the last century, macroinvertebrates have been widely used to assess river health [11] and have been suggested being sensitive to human disturbances such as small dams [12], pollution [13, 14], and land-use change within the watershed [15]. In China, macroinvertebrates have received increasing interest during the last decade [14, 16, 17] and have been gradually integrated into aquatic assessment systems, in addition to existing chemical indicators.

As the largest freshwater lake in China and an essential part of Yangtze river ecosystem, Poyang Lake harbors a vast amount of species [18]. Due to the rapidly growing human activities, Poyang Lake and its tributaries are subject to multiple threats including hydrological alterations owing to dam construction, habitat degradation, pollution and overexploitation [18–21]. In order to improve the status of Poyang Lake, development of integrated management strategies at basin level is urgently needed. Considering the hydrological connection within the Poyang Lake Basin, maintaining a healthy status of these tributaries is not only important for freshwater species inhabiting them and for people living around them, but also vital for the Poyang Lake downstream. Here, we investigated in-stream and surrounding habitat, and biotic variables (e.g. periphyton biomass and macroinvertebrate commmunities) in Changjiang (i.e. one of the tributaries of Poyang Lake) Basin. We tested the potential of macroinvertebrates as indicators for river health and further explored the possibility to use coarse taxonomic level (i.e. family level) of macroinvertebrates in monitoring river health in Changjiang Basin, which could be a promising approach for timely assessment and management.

14.2 Methods

14.2.1 Study Area

The study area locates in Changjiang Basin (29° 35' to 30° 08'N, 117° 12' to 117° 57'E) in southeast China. Changjiang River originates from mountains in Anhui Province and is one of the two major tributaries of Rao River that flows into the Poyang Lake in Jiangxi Province. It flows through area of various land uses, including forest, agriculture and urban areas, with a drainage area of 6260 km^2. The climate in the Changjiang Basin is subtropical and influenced by summer monsoon, with an annually average precipitation of 1725 mm, and an annually average temperature of

15.6 °C. The rainy season typically occurs from April to July, accounting for about half of the annual precipitation [22].

14.2.2 *Field Sampling and Laboratory Analysis*

Fifty-seven sites were sampled within Changjiang Basin in March 2011 (Fig. 14.1), during the drought season when the flow discharge was low. At each site, a river reach of 50–100 m was selected. To evaluate the habitat quality, each sampling site was graded using qualitative habitat evaluation index (QHEI) [23]. QHEI is a visual assessment of in-stream and surrounding physical characteristics (e.g. substrate, in-stream cover, channel morphology and riparian zone) at reach scale. It has been often used to evaluate habitat quality and human disturbance on river ecosystems [24, 25].

Water samples were collected in situ and acidified to pH < 2 with H_2SO_4 for further analysis in the lab. The concentrations of total nitrogen (TN), nitrate nitrogen (NO_3-N), ammonia nitrogen (NH_4-N), total phosphorus (TP) and phosphate phosphorus (PO_4-P) were measured with a segmented flow analyzer (Skallar San++, Netherlands). The category of water quality at each site was determined using single-factor evaluation method, followed the environmental quality standards for surface

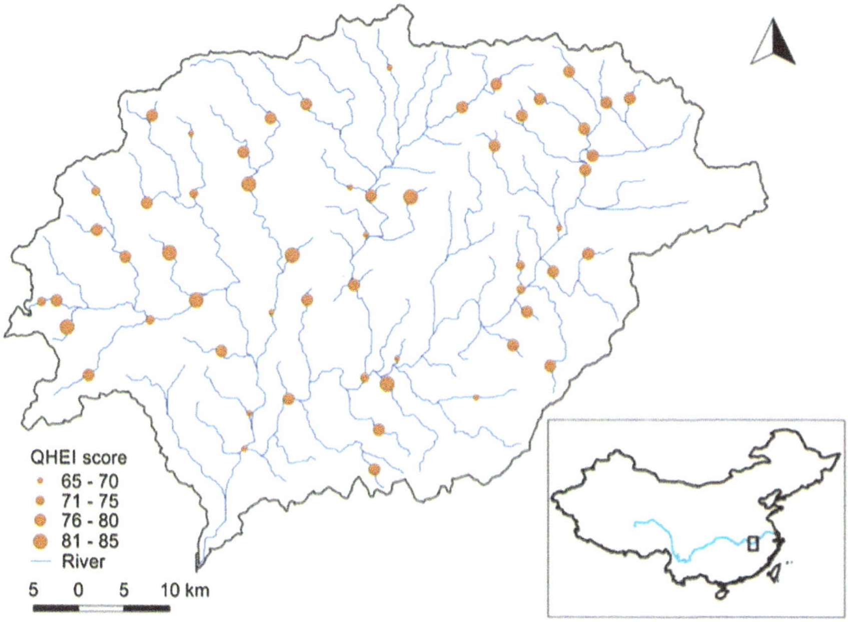

Fig. 14.1 Distribution of sampling sites in Changjiang Basin with their qualitative habitat evaluation index (QHEI) scores

water in China (GB 3838-2002), with in-stream chemical variables such as TN, NH_4-N and TP being considered.

To collect periphyton samples, 9–12 cobble-sized stones were randomly picked from the river. A 27 mm-radius corer was placed on top of each stone. Periphyton around the corer was removed and flushed away with stream water. The periphyton beneath the corer was scrubbed with a nylon brush and flushed into a 355 mL bottle with distilled water. A part of the pooled sample (80–100 mL) was filtered through a glass fiber filter (Whatman) for chlorophyll analysis and another pre-weighted glass fiber filter for ash free dry weight (AFDM) measurement. The filters with periphyton samples were stored at −20 °C for lab analysis. To determine concentration of chlorophyll, absorbance was determined at four different wavelengths (i.e. 750, 665, 645 and 630 nm) with a spectrophotometer (Shimadzu UV-1601, Japan), after 24h-extraction with 90% buffered acetone. Calculation of periphyton biomass was based on the weight loss of the glass fiber filter after incineration following Biggs and Kilroy [26].

Macroinvertebrates were collected with a Surber net (30 * 30 cm, 0.43 mm mesh size). Individuals attached and under the substrates were flushed into the net and then transferred into the sample containers, and preserved in 10% buffered formaldehyde. Five replications were conducted at each site, covering multiple habitats. In the lab, macroinvertebrates were sorted and identified to genus level (insects except for Chironomidae which were identified to sub-family level) or family level (non-insect) followed relevant references [27, 28]. The numbers of individuals were also counted.

14.2.3 Statistical Analysis

The autotrophic index (AI) was determined by the values of AFDM and chlorophyll a at each site (i.e. AI=AFDM/chlorophyll a) [29]. For benthic invertebrates, the Margalef index was calculated using the package "EcoIndR" [30] in R [31], with abundance data following the suggestion by Gamito [32]. We calculated the Modified New Walley Hawkes (MNWH) score, a modified version of Biological Monitoring Working Party (BMWP) score for each site [33]. Due to the large deviation and uncertainty within Chironomidae and Oligochaeta [33], they were not included for MNWH score calculation. Spearman's rank correlation was conducted among in-stream abiotic variables, QHEI score and biotic indices (e.g. richness (S), Margalef index (D), MNWH score) to check their congruency. All the statistical analyses were performed in R and all the maps were plotted with QGIS [34].

14.3 Results

QHEI scores of all sampling sites in Changjiang Basin were over 65, with a mean value of 76.04 (Table 14.1), indicating that the overall habitat quality in Changjiang

Basin was good (Fig. 14.1). Most headwater sites were in pristine status or slightly disturbed while downstream sites were more influenced by human activities (e.g. from villages, towns and agriculture). Among all 57 sampling sites, water quality of 24 sites were classified as level I or II according to environmental quality standards for surface water in China, while only 7 sites located close to villages and towns were classified worse than level III. The concentration of TN was the most determining factor for the classification of water quality category in Changjiang Basin. Concentration of NH_4-N showed significantly negative relationships with QHEI score ($\rho = -0.40$, $p < 0.05$, Spearman's rank correlation). However, no significant relationship between other abiotic variables (i.e. TN, NO_3-N, TP and PO_4-P) and QHEI score were detected.

Ninety-seven taxa were identified in Changjiang Basin. The composition of macroinvertebrate communities varied among sites (Table 14.2, Fig. 14.2). Within the whole basin, Orthocladiinae spp., Tanypodinae spp., *Heptagenia* sp., *Choroterpes* sp., and *Baetis* spp. were dominant taxa (i.e. proportion of abundance over 5%). Aquatic insects were the dominant groups at almost all sites, with Ephemeroptera, Plecoptera, and Trichoptera (EPT) being dominant taxa at undisturbed sites. Within human-impacted sites, Diptera and non-insect such as Oligochaeta and Gastropoda (e.g. *Bellamya* spp.) were dominant taxa.

Table 14.1 Summary of in-stream abiotic variables, autotrophic index (AI) and qualitative habitat evaluation index (QHEI) score (SD: standard deviation)

	Min	Max	Mean	SD
TN (mg/L)	0.10	1.43	0.60	0.30
NO3-N (mg/L)	0.03	1.34	0.53	0.29
NH4-N (mg/L)	0.01	0.06	0.02	0.01
TP (mg/L)	0.01	0.08	0.02	0.01
PO4-P (mg/L)	0.00	0.08	0.01	0.01
AI	139.89	1453.96	541.07	282.78
QHEI	65.00	84.50	76.04	4.37

Table 14.2 Summary of macroinvertebrate communities in Changjiang Basin (S: taxa richness; D: Margalef index; SD: standard deviation)

	Min	Max	Mean	SD
Density (ind./m^2)	291	19284	2594	2672
% EPT	0.88	88.28	45.72	24.08
% Diptera	0.69	81.99	29.06	19.94
% Non-insects	0.49	71.92	15.76	16.43
S (genus)	16	44	28.84	6.70
D (genus)	2.00	5.95	4.11	0.94
S (family)	10	32	20.86	4.60
D (family)	1.30	4.29	2.94	0.69

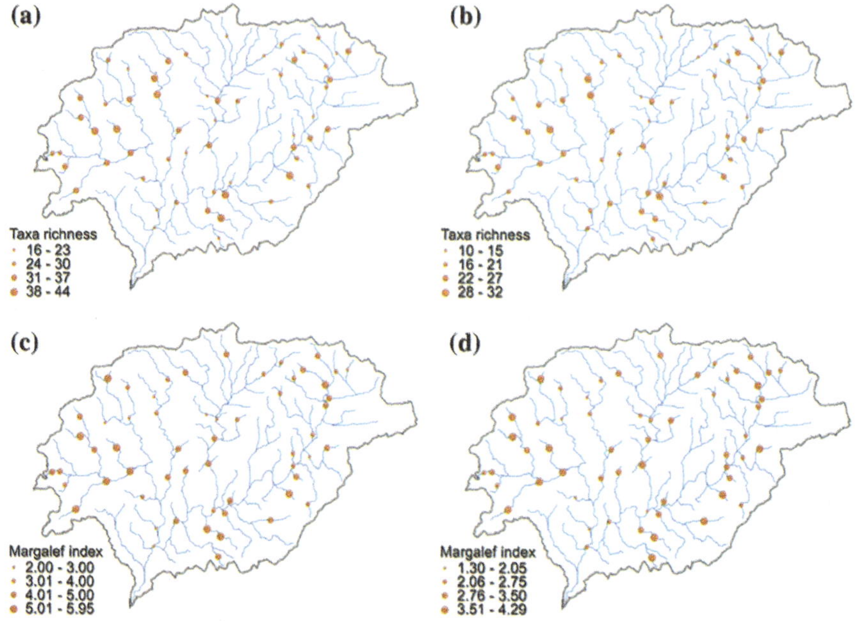

Fig. 14.2 Taxa richness and Margalef index of macroinvertebrates at sampling sites in Changjiang Basin (**a**, **c**, genus level; **b**, **d**, family level)

Table 14.3 Relationships among biotic metrics and qualitative habitat evaluation index (QHEI) score in Changjiang Basin (S: taxa richness; D: Margalef index; * $p < 0.01$, Spearman's rank correlation)

	S (genus)	D (genus)	S (family)	D (family)	MNWH
QHEI	0.59*	0.52*	0.54*	0.43*	0.49*
S (genus)		0.87*	0.91*	0.73*	0.88*
D (genus)			0.85*	0.91*	0.83*
S (family)				0.86*	0.85*
D (family)					0.73*

The biodiversity indices (i.e. taxa richness, Margalef index) of macroinvertebrates showed significantly positive correlations with QHEI score ($p < 0.01$, Spearman's rank correlation) on both genus and family level (Table 14.3). The taxa richness and Margalef index on family level had slightly lower correlation coefficients than their counterparts on genus level. In addition, MNWH score (Fig. 14.3) showed significantly positive correlations with QHEI score and biodiversity indices on both genus and family level ($p < 0.01$, Spearman's rank correlation).

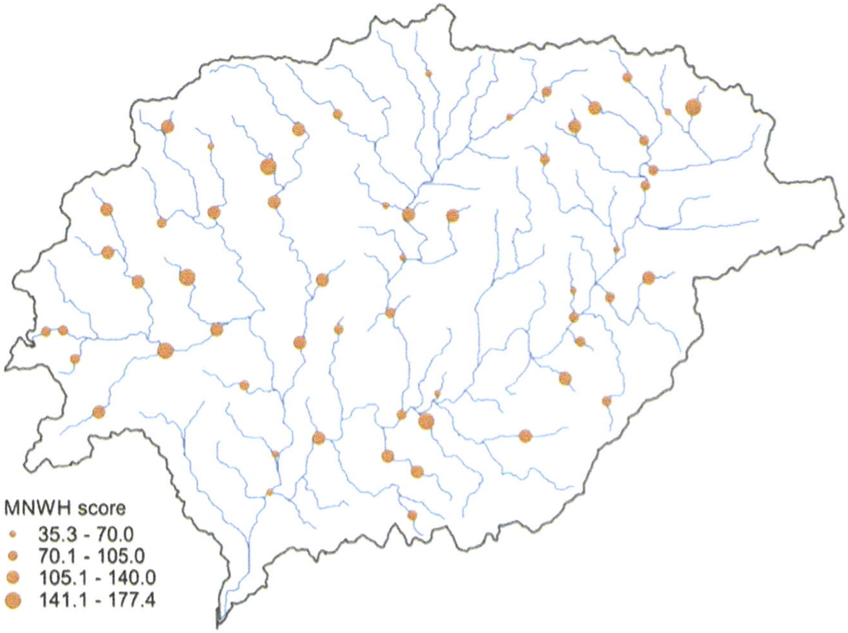

Fig. 14.3 Modified New Walley Hawkes (MNWH) scores of sampling sites in Changjiang Basin

14.4 Discussion

QHEI scores show that the overall habitat quality in Changjiang Basin is high, with some river reaches being impacted by human activities. The main disturbances on river ecosystems in the basin are domestic and agricultural activities [35]. In river reaches close to villages and towns, river morphology has been altered due to human activities (e.g. rice field, channelization). Increased concentration of NH_4-N in rivers usually indicates organic pollution from domestic or livestock waste, or influence from agricultural fertilizer [36], which imposes toxic impact on aquatic species [37]. Though TN is the main determining factor regarding water quality category, the relationships between QHEI scores and in-stream chemical variables (i.e. significantly negative correlation with NH_4-N and non-significant relationship with nutrients such as TN, TP and PO_4-P) suggest that the domestic or livestock waste has posed a major impact on river ecosystem in addition to agriculture activities. In Changjiang Basin, many villages and towns are located close to rivers, especially in downstream areas. It is not uncommon that domestic and livestock waste has been dumped or flushed directly into rivers without treatment. Such disturbances would pose impact on periphyton [38]. For example, elevated concentration of organic materials could change the composition of periphyton, leading to decline in autotrophic organisms (e.g. algae) and increase in heterotrophic microbes (e.g. fungus, bacteria) [29]. The high values of autotrophic index of periphyton at sampling sites in Changjiang Basin have confirmed this assumption.

At disturbed river reaches (i.e sampling sites located close to villages with low QHEI scores), flow was relatively slow with river bottom covered by fine-size sediments and organic matters instead of coarser mineral substrates. Taxonomic groups such as EPT are sensitive to disturbances and changes in in-stream environment [39], thereby being replaced by more tolerant groups such as Diptera and Oligochaeta. Consequently, the community structure of macroinvertebrates became more similar and simplified at disturbed reaches (e.g. dominant by Chironomidae, Oligochaeta and Bellamya spp). As expected, our results show that macroinvertebrates could be used as indicators for river health in Changjiang Basin. This stands in line with previous studies which have suggested that macronivertebrates are sensitive to various human disturbances [12, 13, 15, 40]. Due to their sensitivity to environmental change, macroinvertebrates have been integrated into river assessment protocols in many regions and countries such as North America, Europe, Australia, New Zealand and South Korea [41]. During the last decade, macroinvertebrates have received increasing interests in ecological research [42, 43] in China and have also been included as a regular monitoring group in several Chinese national water frameworks and projects (e.g. Major Science and Technology Program for Water Pollution Control and Treatment) to monitor river health.

Our results also show that it is possible to use macroinvertebrates as indicators on family level as diversity indices on that level have kept most information from diversity indices on genus level, which is consistent with previous studies on stream macroinvertebrates [44, 45]. It could help to guide a rapid monitoring framework in Changjiang Basin as identification to coarser level would require less time and staff input [44]. Such a strategy is practical, especially considering the fact that macroinvertebrate-based monitoring has been widely implemented in river assessment in China recently. Although there are some publications on classification of macroinvertebrates in China [27, 46], systematic identification keys for riverine macroinvertebrates remain lacking. It is especially problematic for identification of immature macroinvertebrates as there are usually only subtle morphological differences among related species. It is challenging for environmental agencies and related people to perform identification to species or genus level without systematic training which require extra input of both time and money. In addition, misidentification could become more frequent if high precision of taxonomic level is required, which could further lead to inaccurate evaluation. Therefore, identifying macroinvertebrates to family level for monitoring river health in Changjiang Basin could reduce the risk of misidentification without losing their ability to monitor environmental changes. Though it is feasible to use taxa richness, Margalef index and MNWH score to indicate status of river health in Changjiang Basin, detailed survey on species distribution and research on taxonomy and life history of macroinvertebrates are called to enhance our understanding on their response to environmental changes such as climate and land-use change, which have been suggested having negative impact on macroinvertebrate diversity [47, 48]. Such knowledge is required in order to develop advanced assessment approaches and management strategies, especially considering the rapid economic growth in Changjiang Basin, and subsequent expansion of urban and agricultural areas, which will pose further stress on river ecosystems.

Acknowledgements This work was funded by the German Research Foundation (DFG) and National Natural Science Foundation of China (NSFC) through the project "Integrated modelling of the response of aquatic ecosystems to land use and climate change in the Poyang Lake region, China" (JA 1827/2-1; FO 301/14-1; 40911130508) as part of the NSFC/DFG joint funding programme "Land Use and Water Resources Management under Changing Environmental Conditions". FH was supported by the SMART Joint Doctorate (Science for the MAnagement of Rivers and their Tidal systems), funded with the support of the Erasmus Mundus programme of the European Union. SCJ acknowledges funding through "GLANCE" project (Global Change Effects in River Ecosystems; 01 LN1320A; German Federal Ministry of Education and Research, BMBF).

References

1. Postel, S., and S. Carpenter, et al. 1997. Freshwater ecosystem services. In *Nature's services: Societal dependence on natural ecosystems*, 195.
2. Millennium Ecosystem Assessment. 2005. Ecosystems and human well-being: Current state and trends. In *Millennium ecosystem assessment, global assessment reports*.
3. Sabo, J.L., et al. 2017. Designing river flows to improve food security futures in the lower Mekong Basin. *Science* 358 (6368): eaao1053.
4. Vörösmarty, C.J., et al. 2010. Global threats to human water security and river biodiversity. *Nature* 467 (7315): 555–561.
5. Zarfl, C., A.E. Lumsdon, J. Berlekamp, L. Tydecks, and K. Tockner. 2015. A global boom in hydropower dam construction. *Aquatic Sciences* 77 (1): 161–170.
6. Couto, T.B, and J.D. Olden. 2018. Global proliferation of small hydropower plants-science and policy. *Frontiers in Ecology and the Environment* 16(2): 91–100.
7. Schwarzenbach, R.P., T. Egli, T.B. Hofstetter, U. Von Gunten, and B. Wehrli. 2010. Global water pollution and human health. *Annual Review of Environment and Resources* 35: 109–136.
8. Collen, B., et al. 2014. Global patterns of freshwater species diversity, threat and endemism. *Global Ecology and Biogeography* 23 (1): 40–51.
9. Wallace, J.B., and J.R. Webster. 1996. The role of macroinvertebrates in stream ecosystem function. *Annual Review of Entomology* 41 (1): 115–139.
10. Bae, Y.J., H.K. Kil, and K.S. Bae. 2005. Benthic macroinvertebrates for uses in stream biomonitoring and restoration. *KSCE Journal of Civil Engineering* 9 (1): 55–63.
11. Metcalfe, J.L. 1989. Biological water quality assessment of running waters based on macroinvertebrate communities: History and present status in Europe. *Environmental Pollution* 60 (1–2): 101–139.
12. Rehn, A.C. 2009. Benthic macroinvertebrates as indicators of biological condition below hydropower dams on west slope Sierra Nevada streams, California, USA. *River Research and Applications* 25 (2): 208–228.
13. Friberg, N., J. Skriver, S.E. Larsen, M.L. Pedersen, and A. Buffagni. 2010. Stream macroinvertebrate occurrence along gradients in organic pollution and eutrophication. *Freshwater Biology* 55 (7): 1405–1419.
14. He, F., et al. 2014. River ecosystem health assessment of the Shennongjia Forest Region. *Chinese Journal of Applied and Environmental Biology* 20 (1): 35–39.
15. Sponseller, R.A., E.F. Benfield, and H.M. Valett. 2001. Relationships between land use, spatial scale and stream macroinvertebrate communities. *Freshwater Biology* 46 (10): 1409–1424.
16. Li, F., Q. Cai, and L. Ye. 2010. Developing a benthic index of biological integrity and some relationships to environmental factors in the subtropical Xiangxi River, China. *International Review of Hydrobiology* 95 (2): 171–189.

17. Jiang, X., Z. Song, J. Xiong, and Z. Xie. 2014. Can excluding non-insect taxa from stream macroinvertebrate surveys enhance the sensitivity of taxonomic distinctness indices to human disturbance? *Ecological Indicators* 41: 175–182.
18. Fang, J., et al. 2006. Biodiversity changes in the lakes of the Central Yangtze. *Frontiers in Ecology and the Environment* 4 (7): 369–377.
19. Huang, L.L., Z.Q. Wu, and J.H. Li. 2013. Fish fauna, biogeography and conservation of freshwater fish in Poyang Lake Basin, China. *Environmental Biology of Fishes* 96 (10–11): 1229–1243.
20. Wang, Y., et al. 2013. Optimising hydrological conditions to sustain wintering waterbird populations in Poyang Lake National Natural Reserve: Implications for dam operations. *Freshwater Biology* 58 (11): 2366–2379.
21. Duan, W., et al. 2016. Water quality assessment and pollution source identification of the eastern Poyang Lake Basin using multivariate statistical methods. *Sustainability* 8 (2): 133.
22. Sun, X. 2012. Community structure of benthic macroinvertebrate in the Changjiang River and its influencing factors. PhD thesis. Institute of Hydrobiology, Chinese Academy of Sciences, Wuhan.
23. Rankin, E.T. 1995. Habitat indices in water resource quality assessments. *Biological assessment and criteria: Tools for water resource planning and decision making*, 181–208.
24. Hughes, R.M., A.T. Herlihy, and P.R. Kaufmann. 2010. An evaluation of qualitative indexes of physical habitat applied to agricultural streams in ten US states. *JAWRA Journal of the American Water Resources Association* 46 (4): 792–806.
25. Wu, W., Z. Xu, X. Yin, and D. Zuo. 2014. Assessment of ecosystem health based on fish assemblages in the Wei River basin, China. *Environmental Monitoring and Assessment* 186 (6): 3701–3716.
26. Biggs, B.F.J., and C. Kilroy. 2000. In *Stream periphyton monitoring manual*. Wellington.
27. Morse, J.C., L. Yang, and L. Tian. 1994. *Aquatic insects of China useful for monitoring water quality*. Nanjing, People's Republic of China: Hohai University Press.
28. Merritt, R.W., and K.W. Cummins. 1996. *An introduction to the aquatic insects of North America*. Kendall Hunt.
29. Camargo, J.A., A. Alonso, and M. la Puente. 2005. Eutrophication downstream from small reservoirs in mountain rivers of Central Spain. *Water Research* 39 (14): 3376–3384.
30. Guisande, C., et al. 2017. DER: An algorithm for comparing species diversity between assemblages. *Ecological Indicators* 81: 41–46.
31. R Core Team. 2016. *R: A language and environment for statistical computing*.
32. Gamito, S. 2010. Caution is needed when applying Margalef diversity index. *Ecological Indicators* 10 (2): 550–551.
33. Paisley, M.F., D.J. Trigg, and W.J. Walley. 2014. Revision of the biological monitoring working party (BMWP) score system: Derivation of presentonly and abundance-related scores from field data. *River Research and Applications* 30 (7): 887–904.
34. Quantum GIS Development Team. 2014. *Quantum GIS geographic information system. Open source geospatial foundation project*.
35. Sun, X., Q. Cai, F. Li, S. Yang, and L. Tan. 2011. Modification of river health comprehensive assessment index and its application in Changjiang River. *Journal of Ecology and Rural Environment* 27: 98–103.
36. Huang, X., et al. 2012. A high-resolution ammonia emission inventory in China. *Global Biogeochemical Cycles* 26 (1).
37. Camargo, J.A., and Á. Alonso. 2006. Ecological and toxicological effects of inorganic nitrogen pollution in aquatic ecosystems: A global assessment. *Environment International* 32 (6): 831–849.
38. He, F., X. Dong, M. Sun, and Q. Cai. 2015. Altitudinal pattern of stream periphyton biomass in tributaries of the Lancang-Mekong River: An indicator of anthropogenic impact? *Quaternary International* 380: 282–287.
39. Sánchez-Montoya, M.M., M.R. Vidal-Abarca, and M.L. Suárez. 2010. Comparing the sensitivity of diverse macroinvertebrate metrics to a multiple stressor gradient in Mediterranean streams and its influence on the assessment of ecological status. *Ecological Indicators* 10 (4): 896–904.

40. He, F., W. Jiang, T. Tang, and Q. Cai. 2015. Assessing impact of acid mine drainage on benthic macroinvertebrates: Can functional diversity metrics be used as indicators? *Journal of Freshwater Ecology* 30 (4): 513–524.
41. Buss, D.F., et al. 2015. Stream biomonitoring using macroinvertebrates around the globe: A comparison of large-scale programs. *Environmental Monitoring and Assessment* 187 (1): 4132.
42. Li, F., et al. 2012. Characterizing macroinvertebrate communities across China: Large-scale implementation of a self-organizing map. *Ecological Indicators* 23: 394–401.
43. Tonkin, J.D., et al. 2015. Climatic and catchment-scale predictors of Chinese stream insect richness differ between taxonomic groups. *PloS One* 10 (4): e0123250.
44. Marshall, J.C., A.L. Steward, and B.D. Harch. 2006. Taxonomic resolution and quantification of freshwater macroinvertebrate samples from an Australian dryland river: The benefits and costs of using species abundance data. *Hydrobiologia* 572 (1): 171–194.
45. Heino, J., and J. Soininen. 2007. Are higher taxa adequate surrogates for species-level assemblage patterns and species richness in stream organisms? *Biological Conservation* 137 (1): 78–89.
46. Zhou, C., H. Gui, and K. Zhou. 2003. Larval key to families of Ephemeroptera from China (Insecta) (in Chinese). *Journal of Nanjing Normal University* 26 (2): 65–68.
47. Kuemmerlen, M., et al. 2015. An attack on two fronts: Predicting how changes in land use and climate affect the distribution of stream macroinvertebrates. *Freshwater Biology* 60 (7): 1443–1458.
48. Schmalz, B., et al. 2015. Impacts of land use changes on hydrological components and macroinvertebrate distributions in the Poyang lake area. *Ecohydrology* 8 (6): 1119–1136.

Part V
Environmental Modelling and Information Systems

Chapter 15
Forest Type Classification in Poyang Lake Basin Based on Multi-source Data Fusion

Lu Ming

15.1 Background

Forests are among the most biologically-diverse and largest terrestrial ecosystems on Earth [1]. They play an important role in global carbon and hydrological cycles and provide a wide range of valuable ecosystem goods and services, such as food, timber and climate moderation [2, 3]. High-accuracy forest mapping including the types, spatial distribution, canopy structure, tree species composition and temporal changes and is of great importance to forest management, conservation biology and ecological restoration. Forests can be classified in different ways and to different degrees of specificity. Forest types are defined as a group of forest ecosystems with a generally similar composition that can be differentiated from other groups by their species composition, productivity or crown closure [4]. The distinction is whether the forests are composed predominantly of broad-leaved trees, coniferous trees or mixed. Identification of forest types at fine resolution is crucial to provide useful information for forest managers, as well as ecological modelers [5].

Forest inventories are regarded as the most frequent way to obtain forest properties' information with the highest accuracy. However, the traditional field survey approach is time consuming and labor intensive. Participatory sensing/citizen science has become a new cost-effective way to collect in situ forest data [6, 7], but participatory sensing approaches cannot solve all the problems, especially in isolated areas where few people live. The use of remote sensing data is still the best way to obtain accurate and timely forest information over large spatial scales and long-term temporal coverages. Currently, remote sensing of forests has developed into a new discipline. On the one hand, it improves our understanding of how and why remotely-sensed data and methods are important in forestry and forest science. On the other

L. Ming (✉)
Institute of Geographic Sciences and Natural Resources Research (CAS), Beijing, China
e-mail: lum@lreis.ac.cn

T. Yue et al. (eds.), *Chinese Water Systems*, Terrestrial Environmental Sciences,
https://doi.org/10.1007/978-3-319-97725-6_15

hand, it strengthens our awareness that a better understanding of forest ecosystems may be essential for harmonized coexistence between humans and nature [8].

A number of previous studies has presented different methods to differentiate forest types using various remote sensing data. Ren et al. [9] used a hierarchical classification method to distinguish different forest types in complex mountainous areas by incorporating high spatial resolution remote sensing images and multi-source auxiliary data. Torresan et al. [10] exploited metrics extracted from an airborne LIDAR (Light Detection and Ranging) raw point cloud to predict different forest structure types by means of classification trees. Gorgens et al. [11] utilized Airborne Laser Scanning (ALS) to discriminate different Brazilian forest types based on canopy height profiles, which revealed that it was possible to differentiate forest types using canopy height profiles derived from ALS data. Chen et al. [12] proposed a spatial feature extraction method that used the Vegetation Local Difference Index (VLDI) derived from the Normalized Difference Vegetation Index (NDVI) to increase the accuracy of forest type classification. The results showed that combining the spatial information extracted from medium-resolution images and spectral information improved both classification accuracy and visual quality. Castilla et al. [13] harmonized four independent land cover datasets and different satellite images (SPOT, Landsat and MODIS) to produce a common and simple forest map consisting of three classes of forest (needle-leaf, broad-leaved and mixed) and non-forest. Connette et al. [14] used multi-spectral Landsat OLI imagery for delineating main forest types in Myanmar's Tanintharyi Region and estimated the extent of degraded forest for each unique forest type. These studies have covered most of the commonly-used remote sensing data sources including active LIDAR, SAR, multispectral, hyperspectral, thermal systems, etc. In addition, they also have covered most of the widely-used classification methods, such as K-means, ISODATA, maximum likelihood, the spectral angle mapper, Bayesian, Support Vector Machine (SVM), neural network, random forest, etc. Despite many advances having been achieved by these studies to perform accurate forest type classification, most of the existing fine-resolution forest-type classification methods are determined by the availability of very high-resolution images and the incorporation of complex physical models associated with specific forest types. However, these datasets and methods seem to present difficulty for widespread use, particularly for larger scales.

Although there has been a growing number of satellites launched over the past few decades, the trade-off among spectral resolutions, spatial coverages and repeat frequencies still cannot be properly solved. So far, no single satellite sensor can generate images of fine spatial, temporal and spectral resolutions [15]. However, the spatial, temporal and spectral resolutions represent the ability of presenting details of the Earth's surface. Both, repeated observation and spectral detection are vital indicators to identify different forest types. Fortunately, multi-source data fusion breaks through the constraints of a single sensor and effectively integrates the advantages of multiplatform complementary observations, thus providing opportunities to achieve more accurate and comprehensive forest classification and monitoring [16]. Many data fusion methods have been proposed over the past few decades. Shen [17] proposed the integrated fusion method to obtain the complementary information from multiple temporal-spatial-spectral images. However, the temporal, spatial and spectral characteristics of objects were not completely considered. Zhang and He [18] proposed a method called Ratio Image-Based Spectral Resampling (RIBSR), which is used to accomplish data resampling in the spectral domain to conduct spatial-spectral fusion [19], but it has two disadvantages. The first one neglects the influence of sensor noises, and the other one is that it fails to utilize the correlation between hyperspectral bands. Vivone et al. [20] provided critical descriptions and extensive comparisons of some of the main state of the art pansharpening methods. Chen et al. [15] compared the advantages and disadvantages of several spatial-temporal fusion models and then proposed a Hierarchical Spatiotemporal Adaptive Fusion Model (HSTAFM) [21] to generate a fine spatial-temporal resolution image, which produces consistently lower biases and performs better than previous models.

Addressing this challenge, in this paper, we developed a novel spatial-spectral fusion model called the Segmented Difference Value method (SEGDV) to generate fine the spatial-spectral resolution image and adopted HSTAFM to conduct spatial and temporal fusion. As both spatial-spectral fusion and spatial-temporal fusion have an identical property of high spatial resolution, we may get the pixel-based information with fine spatial, temporal and spectral resolutions. Here, we present a novel spatial-temporal-spectral fusion framework through spatial-temporal fusion and spatial-spectral fusion and then use the fused information for accurate forest type classification.

15.2 Materials and Methods

Figure 15.1 presents a flowchart outlining the methods used in this mapping project, including the data collection and pre-processing, spatial-spectral fusion, spatial-temporal fusion, classification scheme and sampling design, training and validation samples' collection and classification accuracy assessment.

15.2.1 Study Area

The study area is part of the Gan River nature reserve (Fig. 15.2), which is located in the north of Wuyi mountain, between 25°56′30″ and 26°07′42″N, 116°15′01″ and 116°29′06″E in the east of Jiangxi province, China.

The study area covers an area of 1610.01 km^2. The climate is characterized by a subtropical humid monsoon pattern, high temperature and rain in the summer, warm and humid in the winter. The annual mean temperature is 17.5 °C, and the annual mean precipitation is 2100 mm. Forested area account for 95% of the whole area. It mainly includes coniferous, broad-leaved, mixed coniferous and broad-leaved forests and bamboo with high species diversity. Coniferous forests mainly consists of *Pinus massoniana* and China fir. Msin species in broad-leaved forests include *Liquidambar*

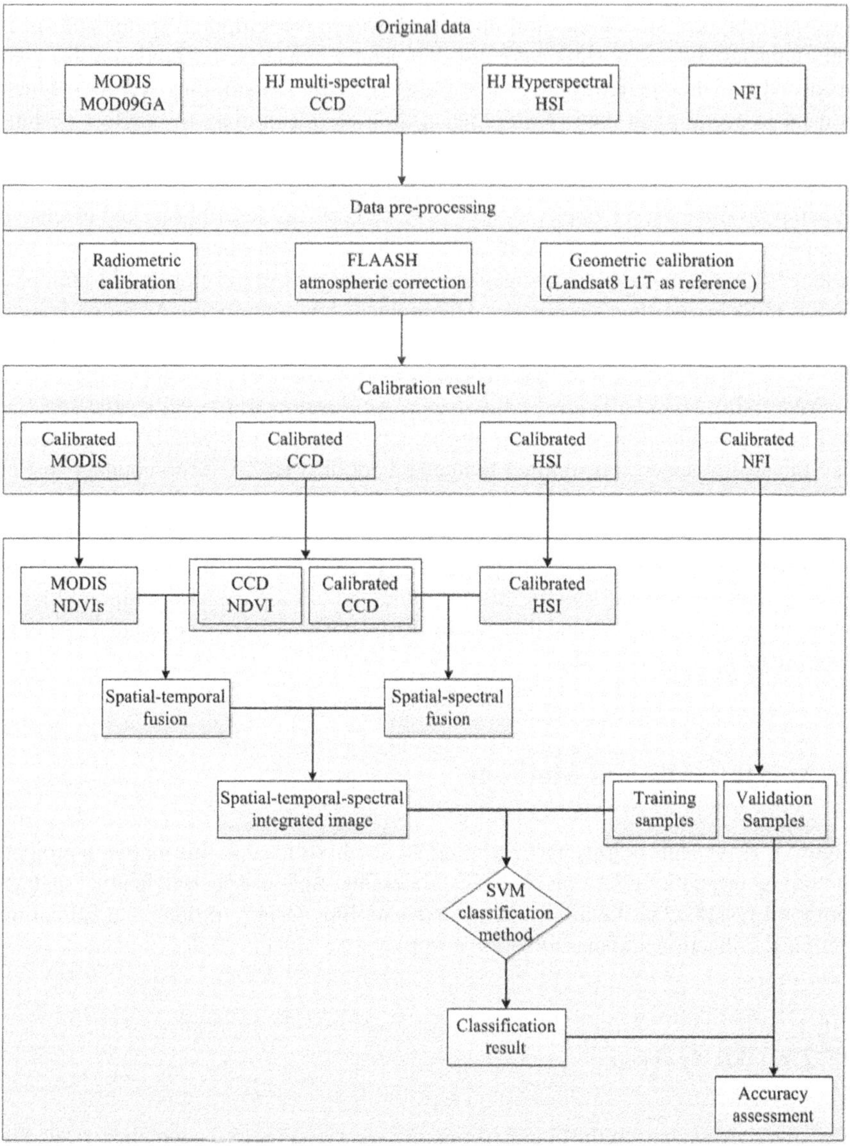

Fig. 15.1 The flowchart of the research methodology

formosana, *Castanopsis sclerophylla*, *Cinnamomum camphora*, etc. Mixed coniferous and broad-leaved forest are regarded as a succession stage of subtropical pioneer community Masson pine (*Pinus massoniana*) forest being converted into evergreen broad-leaved forest communities [22]. Bamboos include *Phyllostachys heterocycla*, *Bambusa rigida*, etc. China fir, mixed and bamboo forests are the three main forest

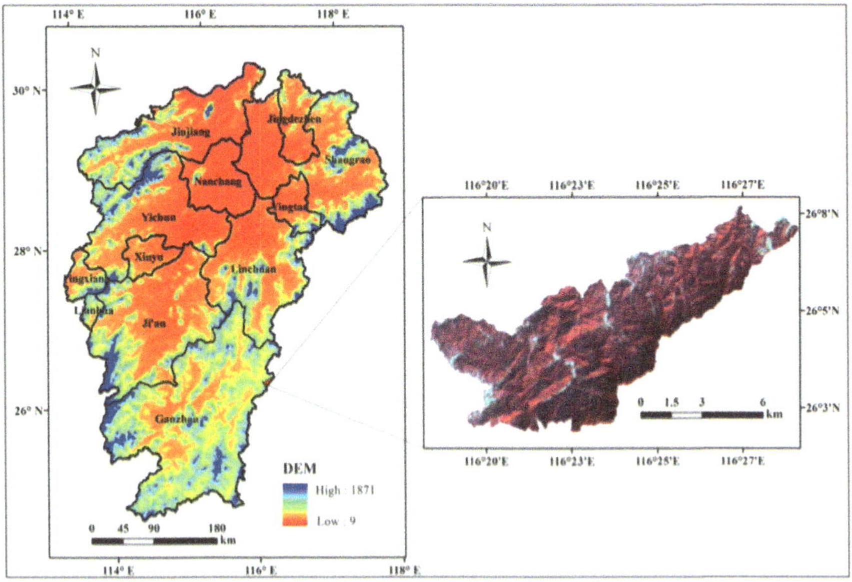

Fig. 15.2 The location of the study area in Jiangxi Province. Left: DEM of Jiangxi province. Right: the CCD image of the study area

types in the study area. Bamboos are mainly located in the south of the study area. Most of the China fir forests are located in the east and a few are located in the west of the study area. Mixed forests mainly are located in the middle-west of the study area. *Pinus massoniana*, broad-leaved forest, shrub and farm are distributed over less areas. *Pinus massoniana* mainly is located in the middle-west and east of the study area. Broad-leaved forest mainly is located in the southeast, middle and northeast of the study area and fragmentary shrub is located in the southwest and northeast of the study area. The landscape of the study area is mountainous with elevation ranging from 250 to 1389 m.

All abbreviations can be found in the list of abbreviations at the end of the manuscript (Table 15.1).

15.2.2 Remote Sensing Data

Multi-spectral Charge Coupled Device (CCD) and Hyperspectral Imager (HSI) sensors carried on the China environment (HJ) series satellite, which was launched in September 2008, were used to obtain accurate forest information with fine spatial and spectral resolutions. The CCD contains 4 spectral bands including 0.43–0.52, 0.52–0.60, 0.63–0.69 and 0.76–0.90 m, with 30 m spatial resolution per pixel. HSI contains 115 spectral bands ranging from 0.45 to 0.95 m, and the band interval is

Table 15.1 The remote sensing data used in the study. Legende: SR: spatial resolution, RP: revisiting period, BN: band numbers

	SR (m)	RP (Day)	BN (Bands)	Usefulness of data
MOD09GA	250	1	7	Capture phenological information and conduct spatial-temporal fusion
HJ-1A CCD	30	4	4	Capture spatial information and conduct spatial-spectral fusion and spatial-temporal fusion
HJ-1A HSI	100	4	115	Capture spectral information and conduct spatial-spectral fusion
Landsat 8 OLI	30	16	9	Experimental validation with data fusion

less than 1 nm, with 100-m spatial resolution per pixel. CCD and HSI data were both acquired on 20 October 2012. The data can be downloaded from the website of China center for resources satellite data and application.[1] Firstly, spectral radiometric calibration, accurate geometric correction and atmospheric correction were conducted by using the ENVI-HJ1A1B-tools. Geometric correction and resampling were conducted using the ENVI 5.1 software and the Landsat 8 L1T images on 5 October 2013 as the space reference basis, so that the CCD and HSI could be consistently matched within the spatial domain. Fast Line-of-sight Atmospheric Analysis of Spectral Hypercubes (FLAASH) was used to conduct atmospheric correction by the ENVI 5.1 software.

Time series profiles of vegetation indices, such as the Normalized Difference Vegetation Index (NDVI), can describe some important phenological information to monitor the vegetation growth status, determine whether the targeted forest is evergreen or deciduous and estimate the approximate date when the leaves green-up and fall-off. Therefore, the temporal phenological information will be an important factor to differentiate forest types, in addition to the spectral information. In order to obtain a time series dataset of the study area, all the MOD09GA reflectance images with no cloud contaminations in the year 2012 were selected, because only a few

[1]http://www.cresda.com/CN/.

CCD images are available due to contaminations by cloud and haze. MOD09GA data were downloaded from the NASA website.[2] All the data were projected to a Universal Transverse Mercator Projection (UTM) Zone 50 N coordinate system using the MODIS reprojection tool (MRT) and then resampled to 30-m spatial resolution via the cubic convolution method in the ENVI 5.1 software. A total of 21 cloud-free images was selected in 2012 (Day of Year 51, 87, 88, 94, 270, 271, 278, 284, 286, 287, 293, 294, 295, 307, 311, 318, 341, 344, 348, 360 and 366). Geometric correction and co-registration were conducted by the ENVI 5.1 software using the Landsat images as the base reference, so that all the images could be well matched in the spatial extent. The NDVI was calculated using the MODIS red and near-infrared bands according to Eq. (15.1). Median values in the two nearest days were used to replace the outliers, which are easy to separate, because they present abrupt high or low, compared to ordinary, values. After the median values process, the NDVIs in a time series could be used to better represent forest growth information.

$$NDVI = \frac{NIR - R}{NIR + R} \tag{15.1}$$

Landsat has become the longest-running civilian Earth-observing program and the world's largest collection of Earth imagery, since the first satellite was launched in 1972 [23]. It has been used to meet a wide range of information needs due to its 30-m spatial resolution and 16-day revisiting period. Hansen et al. [24] analyzed global 21st-Century forest covers' change based on Landsat data. Lehmann et al. [25] used time series Landsat data for forest cover trends' information of the Australian continent. Zhu and Woodcock [26] developed a new algorithm for Continuous Change Detection and Classification (CCDC) of land cover using all available Landsat data. Meanwhile, it is also widely used in many other applications, such as the estimation of biophysical variables and phenology information [23]. In this study, Landsat-8 OLI was used as the reference image to verify the effectiveness of the proposed method. All the remote sensing data used in this study and their detailed information can be found in Table 15.1, and their spectral distributions are provided in Fig. 15.3.

15.2.3 Chinese NFI Data

In order to obtain the area, composition and distribution of forest resources, the State Forestry Bureau of China has already organized eight forest inventories every five years since 1975.[3] The province-based unit is called the first level forest resource investigation. In each province, the local forestry bureau will carry out the detailed investigation based on the county unit, which is called the second level forest resource investigation. An important component of the investigation is to find out the

[2]https://ladsweb.modaps.eosdis.nasa.gov/.

[3]http://211.167.243.162:8085/8/chengguobaogao/showpageinit?lm=xxxz.

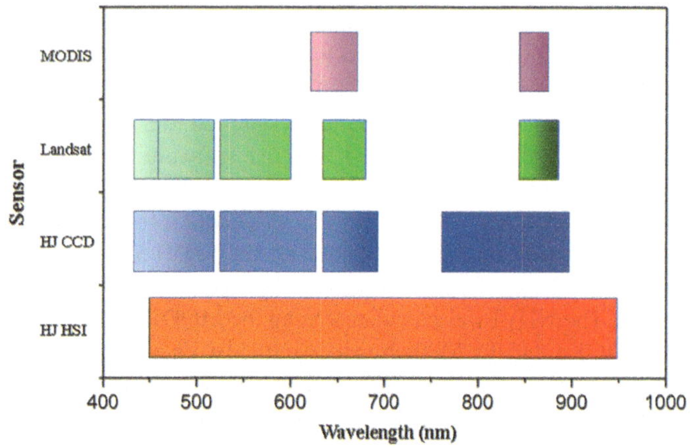

Fig. 15.3 The spectral distributions of HSI (115 bands) versus CCD, Landsat 8 and MODIS in the 400–1000-nm range. Each color means one type of sensor, and each block means one band of the corresponding sensor

categories, areas and quality of each forest type in every county, so that the investigation result can objectively reflect the relationship between the forest situation and the factors of local nature, economy and management. Many useful suggestions can be proposed as guides to effectively protect and use forest resources. In the process of investigation, the survey samples were systematically allocated based on statistical theory. Applying unified technical standards of the continuous inventory method, investigators revisited the survey sample sites periodically, processed the data and obtained regional/national forest information using statistical software. According to unified accuracy requirements, the forest stock in each county achieves greater than 80% accuracy at the 95% confidence interval. The tree species and their corresponding proportion of public forests in each county are also 80% accurate at the 95% confidence interval. The positions' accuracy is less than 0.5 mm if there are obvious objects on the ground that can be interpreted as the reference boundary. If not, the positions' accuracy is extended to less than 1 mm.

The data we used in this study was from the second level forest investigation in Jiangxi province. It is a digital forest map providing edge to edge coverage in Jiangxi province, and we set the part of the Ganjiang source nature reserve area as the reference data subset. It contains many detailed field investigation variables such as the serial number of each land parcel, village name, average elevation, terrain slope and orientation, soil properties, tree species and their percent, average tree height, diameter, age, stock volume information, and so on.

15.2.4 Spatial-Spectral Fusion Method SEGDV

In order to overcome the limitation of the CCD spectral resolution and HSI spatial resolution, spatial-spectral fusion was considered to get fine-resolution images. Generally, the spatial-spectral fusion method includes among others, the fusion of panchromatic and multi-spectral, panchromatic and hyperspectral, multi-spectral and hyperspectral images, morphological information and hyperspectral data, technique. The fusion of panchromatic and multi-spectral is the most mature. It could be classified into 4 categories: substitution based on Principal Component Analysis (PCA) [27] or Intensity Hue Saturation (IHS) [28]; fusion based on the analysis of multi-resolution [29]; fusion based on model optimization [30]; and fusion based on sparse representation [31]. The fusion methods based on PCA or IHS always cause spectral distortion. The fusion method based on sparse representation has achieved greater success than the component substitution method, but it is commonly very complex, and its efficiency is very low. Fauvel et al. [32] proposed a data fusion scheme for the classification of urban land based on the fusion of the morphological information and hyperspectral data, which succeeded in taking advantage of the spatial and the spectral information simultaneously. The fusion method based on model optimization built the relationship between panchromatic and multi-spectral images and obtained higher fusion accuracy. Hence, this model optimization method was assumed to be the best choice in this article.

The hyperspectral image commonly has high correlation between its bands. Let us denote the digital number of a pixel as $dn_{i,n}$ for band i and pixel n, $I = 1, \ldots, I$; $n = 1, \ldots, N$. I and N are the total number of bands and total number of pixels in an image, respectively [36]. The vector representation for pixel n is $dn_n = (dn_{1,n}, \ldots, dn_{I,n})^T$. The band mean vector is written as = $(1, \ldots, i, \ldots, I)^T$, where i is the mean digital number of band i. The total covariance of the image is represented by:

$$\sum_T = \frac{1}{N}\sum_{n=1}^{N}(dn_n - u)(dn_n - u)^{\mathrm{T}} \tag{15.2}$$

Figure 15.4 shows the band correlation matrix of the hyperspectral data. The diagonal line indicates the highest correlation 1, which is represented in white. The darker the tone, the lower is the absolute value of the correlation. We can see that all the hyperspectral bands are highly correlated except the former noise bands (the dark black part) [33]. Because the main component of these bands is noise and they have little relation with the signal, the value of the correlation between these bands and other bands is very low and appears dark. The contiguous bands along the diagonal line appear "in blocks" showing high correlation among them. The hyperspectral bands were separated into 4 groups according to their correlation with the 4 multi-spectral bands.

Then, the fusion between multi-spectral and hyperspectral images can be transformed to the fusion between panchromatic and multi-spectral images in each group. The commonly-used method for multi-spectral and hyperspectral fusion is RIBSR

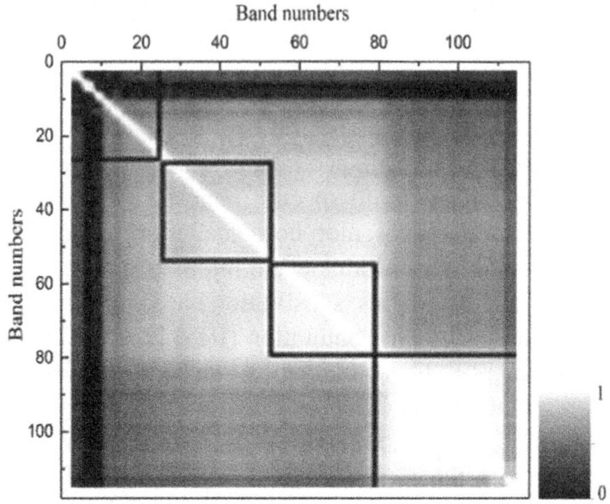

Fig. 15.4 The correlation between hyperspectral bands (the darker the tone, the lower is the absolute value of the correlation between bands) and their separation into 4 groups according to their correlation with 4 multi-spectral bands

[18, 34], but it did not consider the system noise of the sensor. SEGDV was proposed to conduct spatial and spectral fusion to generate the simulated hyperspectral image. For a fixed pixel, the spectral curve profiles of CCD and HSI were discrepant, but they must present a similar change trend, because the reflectance of an object must remain consistent, independently from the used sensor type. Under these assumption for the same spectrum range, the reflectance value difference was caused by the systematic errors between the CCD and HSI sensors. We also suppose that the systematic error is independent of wavelength, which means the error is constant for all the bands. For two random points a, b in spectral curve, we can express this as follows:

$$CCD_a = A + \alpha \tag{15.3}$$

$$CCD_b = B + \alpha \tag{15.4}$$

$$HSI_a = A + \beta \tag{15.5}$$

$$HSI_b = B + \beta \tag{15.6}$$

CCD_a, CCD_b HSI_a and HSI_b represent the reflectance value for spectrums a and b recorded by CCD and HSI, respectively. A, B, a and β represent the true value and the systematic error of CCD and HSI for spectrums a and b, respectively. After some operations, we could get:

$$CCD_a - CCD_b = HSI_a - HSI_b \tag{15.7}$$

However, the spectrum range of CCD is larger than relative to that of HSI. There are no such bands in the actual multispectral and hyperspectral images that can cover the same spectral wavelength [17]. In this paper, all the HSI bands in the spectrum wavelength range of the CCD were averaged to match the CCD. For example, HSI

Bands 26–53 (0.519–0.602 m) were used to match CCD Band 2 (0.52–0.603 m); HSI Bands 61–75 (0.632–0.692 m) were used to match CCD Band 3 (0.63–0.693 m); HSI Bands 88–110 (0.759–0.909 m) were used to match CCD Band 4 (0.76–0.903 m). Because the noise of HSI is very heavy in the first 25 bands (0.46–0.516 m) and there are some outliers after atmospheric correction in these bands, these bands corresponding to CCD Band 1 (0.43–0.52 m) were not computed. In this way, the means of each HSI group could be matched with the corresponding CCD band value. Most HSI bands have been matched with CCD, but some HSI band spectra still are not covered by CCD, such as HSI Bands 54–60 and 76–87. The hyperspectral images always have a high band correlation.

The correlation between CCD bands and HSI bands was calculated. It shows that HSI Bands 26–53 have the highest correlation coefficient with CCD Band 2; HSI Bands 54–81 have the highest correlation coefficient with CCD Band 3; HSI Bands 82–115 have the highest correlation coefficient with CCD Band 4. According to above mentioned knowledge, all the HSI bands were separated into 3 groups. The CCD value and average of HSI in the corresponding group were seen as the basis to generate simulated images with fine spatial and spectral resolution using Eq. (15.7).

15.2.5 Spatial and Temporal Fusion Model HSTAFM

Spatial-temporal fusion techniques have become an interesting tool within the remote sensing community, because they can blend multi-spectral and temporal characteristics to generate synthetic data with fine resolutions [21] [21, 37–46]. Many advances have been made in the spatiotemporal fusion models, which can be classified into four major categories: (i) transformation-based, (ii) reconstruction-based, (iii) unmixing-based and (iv) learning-based models [15]. Transformation-based models include wavelet and tasseled cap transformations [35, 36].

They mainly focus on the integration of spatial and spectral information for image enhancement, instead of constructing a distinct fusion scheme between spatial and temporal information. In the reconstruction-based models, the fusions are generated by a weighted sum of the spectrally-similar neighboring information from fine spatial, but coarse temporal resolution, and fine temporal, but coarse spatial resolution data [37–39]. The unmixing-based models rely on the pixel unmixing techniques, which downscale the coarse resolution images to generate fine-resolution synthetic images while preserving the spectral richness and fidelity [40–42]. In the learning-based models, sparse representation and dictionary learning techniques have generated wide interest [43]. One of the greatest strengths of the learning-based models is that they can predict both phenology and type changes. However, they only use the statistical relationship between the fine and coarse resolution image pair instead of taking any physical properties of remote sensing signals and combining the physical temporal change in the fusion procedure. Although many progress have been achieved, several shortcomings still remain in existing methods. In order to address the limitations of existing spatiotemporal fusion models and detail time series phenology features of forest to improve classification accuracy, the Hierarchical Spatiotemporal Adaptive Fusion Model (HSTAFM) was proposed [21]. It was used in this study to blend HJ-1A CCD and MODIS images to generate time series fusions with fine spatial and temporal resolutions.

Compared with other spatiotemporal fusion models, the HSTAFM has the following highlights: (i) it combines sparse representation techniques into the physical fusion procedure; (ii) it can predict arbitrary temporal changes including both seasonal phenology change and type change using only one image pair; (iii) it introduces a prior detection of temporal change and a two-level selection strategy of similar pixels, which ensures the accurate capturing of temporal changes.

The implementation of HSTAFM includes two major steps: (i) super-resolution of the coarse-resolution image based on sparse representation; (ii) prediction of the synthetic data by combining the fine-resolution image derived from Step (i). In the first stage, super-resolution of MODIS data was first performed to enhance their spatial resolution by CCD-MODIS image pair dictionary learning. As the transitive-resolution image derived from the first stage is much closer to the actual CCD image in spatial detail, it can be assumed that the pixel purity between the transitive-resolution image and CCD image is approximate. Therefore, the conversion coefficients from the prior/posterior to predicted time between transitive- and fine-resolution (i.e., CCD) images can be assumed to be equal.

$$V_f(x, y, b) = V_t(x, y, b) = \frac{T_2(x, y, b)}{T_1(x, y, b)} \tag{15.8}$$

where V_f and V_t represent the conversion coefficient of the fine- and transitive-resolution images. T_1 and T_2 denote the transitive-resolution images at prior/posterior time (t_1) and predicted time (t_2). (x, y) is the location of a given pixel, and b denotes the b-th band. After the conversion coefficients have been calculated according to Eq. (15.8), the initially predicted fine-resolution image at t_2 can be obtained through:

$$F_2(x, y, b) = F_1(x, y, b) \cdot V_f(x, y, b) \tag{15.9}$$

where F_1 denotes the actual fine-resolution image (i.e., CCD) at prior/posterior time (t_1) and F_2 denotes the initially predicted fine-resolution image at the predicted time (t_2). The reflectance difference is computed between the initially predicted fine-resolution image at t_2 and the actual fine-resolution image at t_1. The difference is employed to explain how much temporal change occured instead of directly identifying the specific change. Here, all possible temporal changes are categorized into two classes: significant change (including land cover change and phenology disturbance) and non-significant change (seasonal phenology change). Each class will be tackled with different strategies to select similar pixels.

After a two-level selection of similar pixels and weight calculation, the final predicted fine-resolution image at the predicted date can be computed through Eq. 15.10. Each prediction of the central pixel's reflectance will be incorporated with spatial and spectral information from its corresponding sets of similar pixels P_{ij}.

$$F(x_{\omega/2}, y_{\omega/2}, b) = \sum_{i=1}^{\omega}\sum_{j=1}^{\omega} W_{ij} \cdot P_{ij} \cdot F_1(x_i, y_j, b) \cdot V_f(x_i, y_j, b) \quad (15.10)$$

$$W_{ij} = \frac{1/(s_{ij} \cdot d_{ij})}{\sum_{i=1}^{\omega}\sum_{j=1}^{\omega} 1/(s_{ij} \cdot d_{ij})} \quad (15.11)$$

$$s_{ij} = \frac{P_{ij} \cdot \left|F_1(x_i, y_j) - F_1(x_{\omega/2}, y_{\omega/2})\right|}{\sum_{i=1}^{\omega}\sum_{j=1}^{\omega} P_{ij} \cdot \left|F_1(x_i, y_j) - F_1(x_{\omega/2}, y_{\omega/2})\right|} \quad (15.12)$$

$$d_{ij} = 1 + \sqrt{(x_i - x_{\omega/2})^2 + (y_j - y_{\omega/2})^2}/(1 + \omega) \quad (15.13)$$

where F denotes the final predicted fine-resolution image. ω denotes the moving window size. (x_i, y_j) and $(x_{\omega/2}, y_{\omega/2})$ denote the locations of candidate similar pixels and central pixels, respectively. P_{ij} is a binary matrix denoting the set of similar pixels, and V_f is the conversion coefficient matrix. W_{ij} is a combined weight determined by the spectral and distance differences [21] according to Eqs. (15.11–15.13).

HSTAFM was tested using both the simulated and observed dataset, comparing with the other three state-of-the-art algorithms including the Spatial and Temporal Adaptive Reflectance Fusion model (STARFM), the Flexible Spatiotemporal Data Fusion Model (FSDAF) and the Dictionary Learning-Based Spatiotemporal Fusion Model using only One base Landsat-MODIS image pair (SP-One) [20]. The tests revealed that the HSTAFM approach achieved the highest accuracy. Therefore, this model was used to conduct the spatial and temporal fusion in this study. The HSTAFM algorithm was realized using MATLAB 2016a software. The NDVIs of CCD and MODIS taken on 29 October 2012 were used as the basis images to generate the NDVIs of other 20 days for which MODIS images were required. Through spatial and temporal fusion, we could get the vegetation phenology information with high spatial and temporal resolution.

15.2.6 SVM Classification

After spatial-spectral fusion and spatial-temporal fusion, both fused images have an identical spatial reference and resolution, which allowed combining them to form a new dataset. After feature combination, a total of 118 variables including 95 spectral variables (Bands 26–110, covering 518–910 nm) and 23 temporal variables are derived, and all the variables are used for classification. We do not perform additional feature reduction priou to the classification since some experiments have already demonstrated that band number reduction or feature extraction (such as PCA and wavelet) of hyperspectral data cannot significantly improve the accuracy compared to just using multispectral data in the classification procedure [33]. In addition, the study area is not very large, so forest type classification can be efficiently performed using all useful variables derived from the fusion.

An SVM classifier was used to map various forest types, because SVM has been proven to be an effective way to perform hyperspectral classification [44]. In this study, classification and probability estimation were performed using an SVM classifier with a radial basis function kernel. A brief description of SVM is presented in the following. Assume there are l observations from two classes:

$$(x_1, y_1)(x_2, y_2) \ldots (x_l, y_l)|x_i \in R^N, y_i \in \{-1, 1\} \quad (15.14)$$

x_i denotes the samples, y_i is a collection of labels that represent the category of x_i, i is the i-th sample. Let us assume that two classes are linearly separable. This means that it is possible to find at least one hyperplane (linear surface) defined by a vector $w \in R^N$ and a bias $b \in R$ that can separate the two classes without errors. Finding the optimal hyperplane involves solving a constrained optimization problem using a quadratic equation. The optimization criterion is the width of the margin between the classes. The discrimination hyperplane is defined as follows:

$$f(x) = \sum_{i=1}^{l} y_i a_i k(x, x_i) + b \quad (15.15)$$

where $k(x, x_i)$ is a kernel function and the sign of $f(x)$ denotes the membership of x Constructing the optimal hyperplane is equivalent to find all nonzero a_i values, which are called Lagrange multipliers. Any data point x_i corresponding to a nonzero a_i is a support vector of the optimal hyperplane. A desirable feature of SVMs is that the number of training points that are retained as support vectors is usually quite small, thus rendering them compact classifiers.

Forest, farm and shrub are identified as the three main land covers types in the study area. The forest can be separated into 5 categories according to the dominant tree species: China fir, *Pinus massoniana*, broad-leaved, mixed and bamboo forest. Because too many broad-leaved tree species co-existed in their distributed area, it is impossible to delimit their boundaries such that all the broad-leaved tree species are incorporated into one category. A total of seven object types, China fir, *Pinus massoniana*, broad-leaved, mixed, bamboo forest, shrub and farm, were identified in the classification. All the Regions Of Interest (ROIs) for training samples and validating accuracy were processed using the ENVI 5.4 software. We used Chinese NFI data and the Statues of Forest Resources Report[4] to determine the ROIs of forest types at the local level. A total of 387 ROIs were selected, and all the reference ROIs were divided into two groups: 281 for training samples and 106 for evaluating the classification accuracy. SVM was conducted by the ENVI 5.4 software. The Radial

[4] http://www.forestry.gov.cn/.

Basis Function was set as the kernel function of SVM. The Gamma of the kernel function was set to the system default value of 0.034, and the penalty parameter was set to 100.

15.3 Results

15.3.1 Results of Spatial and Spectral Fusion

In order to acquire images with high spatial and hyperspectral resolution, the SEGDV model was used to conduct spatial-spectral fusion between multispectral CCD and hyperspectral HSI images. The algorithm was realized using the IDL 8.6 software. Because the CCD and HSI sensors were carried on the same platform and the images were both taken on December 20 2012, they have identical spatial reference and similar spectral information and could be matched together in the spatial and spectral domain. Figure 15.5 displays the standard false color composite image (NIR-R-G) of the original CCD, HSI and SEGDV fusion image. We found that the SEGDV fused image could retain the detailed spatial resolution from the CCD data while preserving the consistent spectral information of the original HSI data. We further

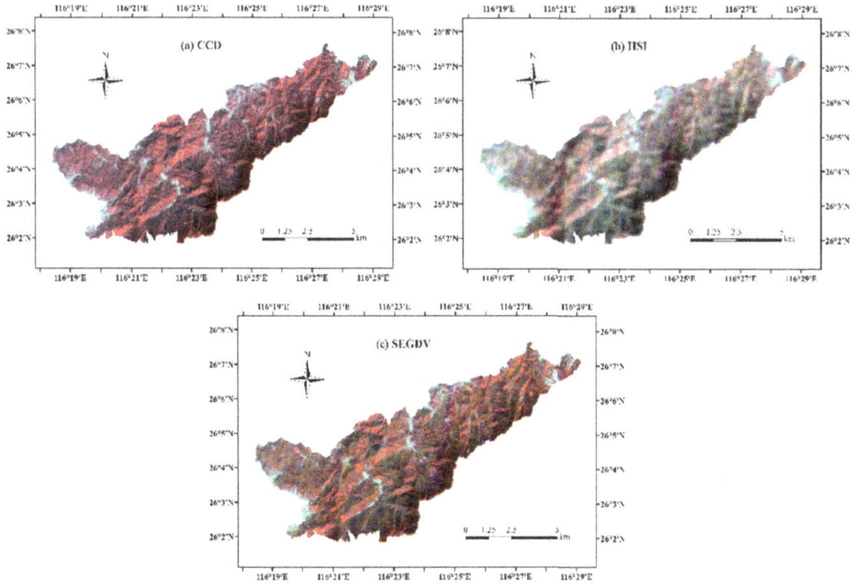

Fig. 15.5 The visual effect of the original CCD, HSI and fused image. Upper-left figure: Original CCD (Bands 4, 3, 2 composite). Upper-right figure: original HSI (Bands 100, 66, 47 composite). Lower figure: fused image (Bands 100, 66, 47 composite)

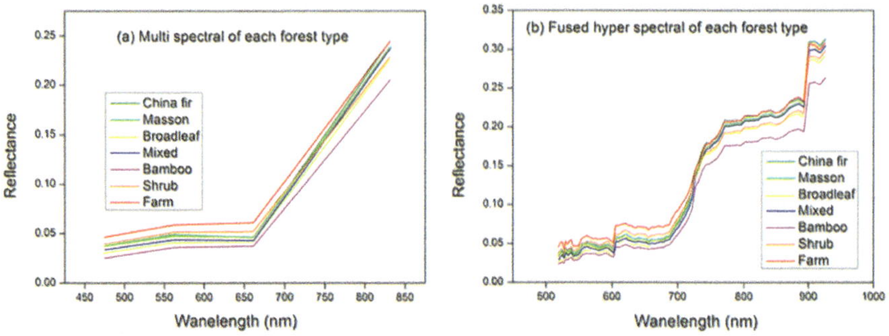

Fig. 15.6 The multiple and fused hyperspectral curves of different forest types. The multi spectral curves (left) and the fused hyperspectral curves (right) of different forest types

randomly selected one sample pixel in the homogeneous area of each category. Figure 15.6 shows the multi-spectral profiles and fused hyperspectral profiles of the selected sample. It could be found that the fused image greatly enhanced the spectral resolution compared with the multispectral CCD image, and the spectral profiles of CCD. In addition, the corresponding profiles of HSI have kept the same walking trend, which demonstrated the effectiveness of SEGDV in blending spatial-spectral information.

15.3.2 Results of Spatial and Temporal Fusion

In order to obtain detailed information about forest phenology, which is an important variable to describe and distinguish different forest types, the HSTAFM model was used to conduct spatial-temporal fusion between CCD and MODIS images [21]. The CCD and MODIS images taken on 20 December 2012 and another MODIS image taken on the predicted date were used as the basis to predict the unknown CCD on the predicted date. As shown in Fig. 15.7, we could find that the fused image could enhance the spatial information significantly compared with the original MODIS image. The seven above-mentioned sample pixels were used again to describe the time series changes of various forest types. Figure 15.8 shows that most forest spectral profiles correspond well. In accordance with the actual forest phenology. The forest began to grow in spring and the NDVI increased in February–April. The NDVI reached its peak in September and began to decrease in December. Differences among the NDVI time series profiles of different forest types are very important to differentiate different forest types.

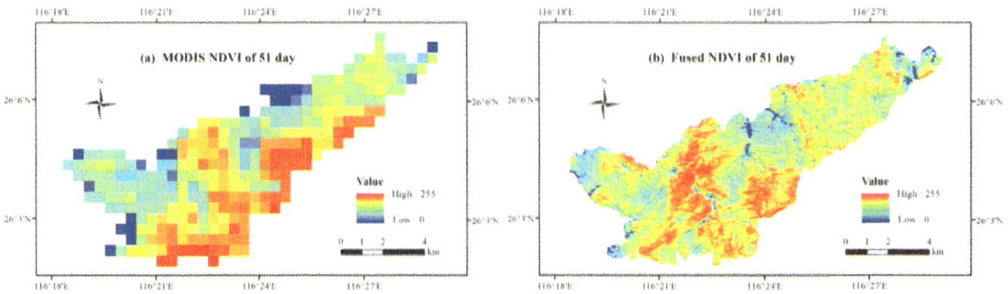

Fig. 15.7 The NDVI of MODIS (Upper-left) and (Upper-right) after spatial and temporal fusion on 20 February 2012

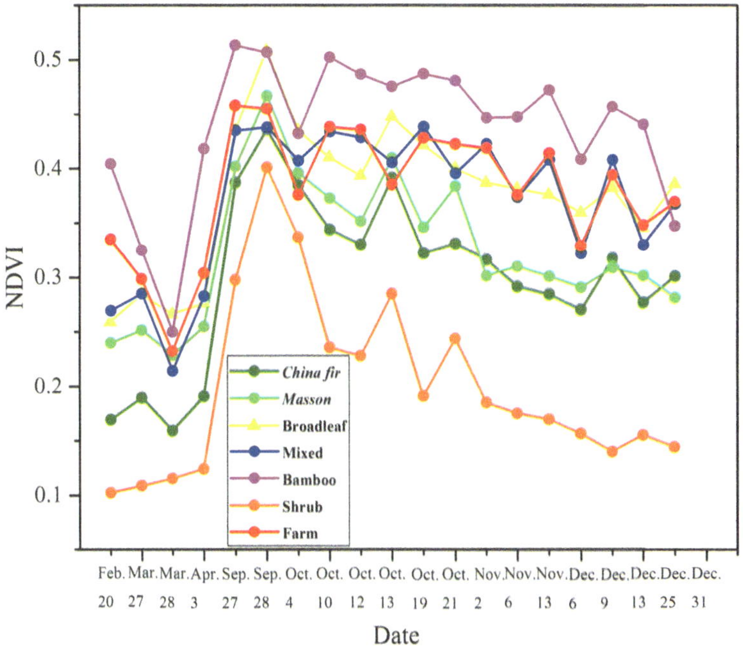

Fig. 15.8 The NDVI time series curves of China fir, *Pinus massoniana*, broad-leaved, mixed, bamboo, shrub forests and farm in 2012

15.3.3 Forest Type Classification

The classification results are evaluated to determine accuracy and reliability. The confusion matrix is used to compare the number of pixels divided into a class by the number of pixels for this class according to ground truth (Fig. 15.9). Every column denotes the ground truth and each line shows the classification results. As values increases along the diagonal, the level of accuracy increases. The overall accuracy level is the ratio of correctly classified pixel numbers to all pixel numbers. The commission error refers to pixels categorized into a class of interest that belong to

Fig. 15.9 The confusion matrix of classification. The highlighted elements represent the main diagonal of the matrix that contains the cases where the class depicted in the image classification agrees with the ground truth dataset, whereas the off-diagonal elements contain those cases where there is a disagreement in the classification

Predicted Class \ Actual Class	A	B	C	D	Σ
A	n_{AA}	n_{AB}	n_{AC}	n_{AD}	n_{A+}
B	n_{BA}	n_{BB}	n_{BC}	n_{BD}	n_{B+}
C	n_{CA}	n_{CB}	n_{CC}	n_{CD}	n_{C+}
D	n_{DA}	n_{DB}	n_{DC}	n_{DD}	n_{D+}
Σ	n_{+A}	n_{+B}	n_{+C}	n_{-D}	n

other classes. The omission error refers to pixels that belong to the real classification of the surface, but that are not correctly classified by the classifier. The producer's accuracy level is the ratio of pixels correctly classified into a class of interest to the total number of ground truth pixels of the class of interest. The user's accuracy level refers to the ratio of pixels correctly classified under the class of interest to the number of total pixels classified under the class of interest by a classifier [45].

In order to validate the effectiveness of the proposed method, comparison experiments were conducted. Classification results derived from single Landsat 8, single spatial-spectral fusions and spatial-temporal fusions were used to compare with that derived from the spatial-spectral-temporal integrated fusion image. Comparing with the NFI data (Fig. 15.10e), which served as the true distribution of various forest types, we found out that the Landsat-based classification (Fig. 15.10a) failed to separate those less distributed categories, such as broad-leaved, shrub. In addition, the China fir coverage was overestimated, and the mixed forest was underestimated. Although most forest types were correctly classified by the spatial-spectral fusion image (Fig. 15.10b), the classification map is too fragmented. As for the classification results derived from the spatial-temporal fusion image (Fig. 15.10c), the majority of forest types were correctly classified, and the fragment phenomenon improved significantly, but there still existed some obvious errors, for example the broad-leaved forest was overestimated in the middle-east, while it was underestimated in the south-east of the study area. Generally, the classification derived from the spatial-spectral-temporal fusion achieved the most plausible result (Fig. 15.10d). We could find that most forest types were correctly classified except for some local areas for which a small portion of China fir was still misclassified into broad-leaved forest and some *Pinus massoniana* was underestimated.

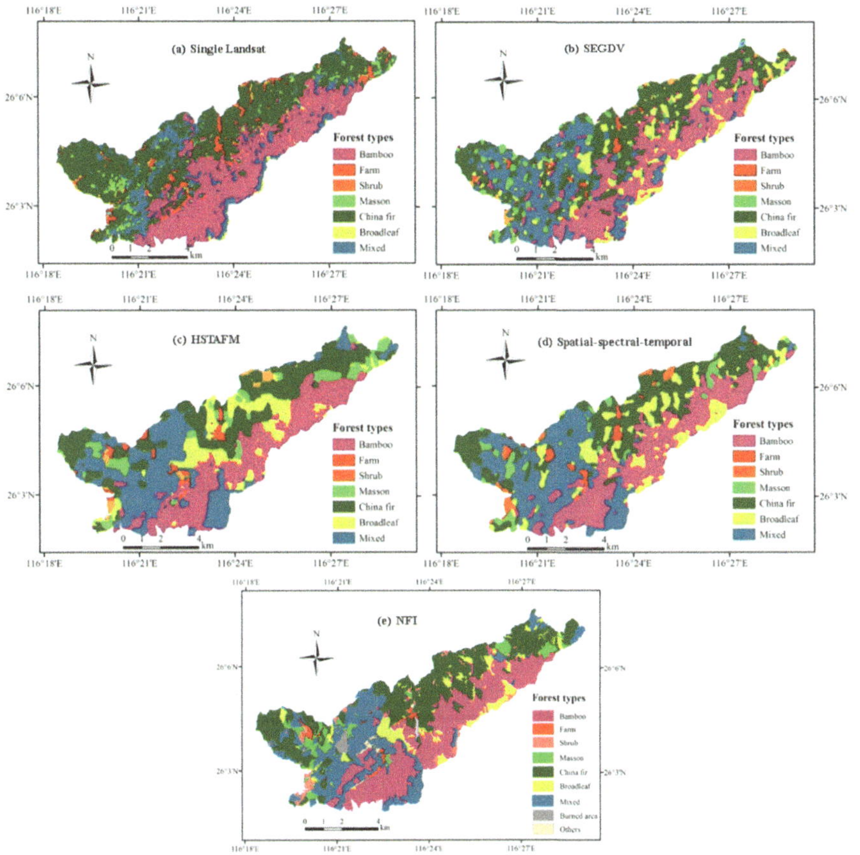

Fig. 15.10 The classification result of: upper-left figure: single landsat. Upper-right figure: spatial and spectral fusion, SEGDV. Middle-left figure: spatial and temporal fusion, HSTAFM. Middle-right figure: spatial-spectral-temporal integrated fusion. Lower figure: the NFI results

From Tables 15.2 and 15.3, the classification of most forest types have been achieved with very high accuracy exception *Pinus massoniana* and broad-leaved forest. It should be pointed out that all images were classified using exactly the same samples, and that the proposed method has achieved the highest accuracy and Kappa coefficient (Table 15.4). Another interesting point is that the accuracy of spatial-temporal fusion was better than spatial-spectral fusion, showing that the time series phenology information is more effective than spectra information for the classification of different forest types.

Table 15.2 The confusion matrix of spatial-spectral-temporal integrated fusion. Legende: the classification result of: (a) Single Landsat. (b) Spatial and spectral fusion, SEGDV. (c) Spatial and temporal fusion, HSTAFM. (d) Spatial-spectral-temporal integrated fusion. (e) The NFI results

Class	Bamboo	Farm	Shrub	Broad-leaved	Masson	China fir	Mixed	Total
Unclassi-fied	0	0	0	0	0	0	0	0
Bamboo	1076	0	0	38	0	0	39	1153
Farm	0	45	0	0	13	5	0	63
Shrub	0	0	53	0	2	0	0	55
Broad-leaved	36	0	10	144	1	71	5	267
Masson	15	3	3	0	63	19	11	114
China fir	82	0	3	21	17	751	54	928
Mixed	16	0	7	21	24	51	758	877
Total	1225	48	76	224	120	897	867	3457

Table 15.3 The commission error, omission error, producer accuracy and user accuracy of each forest type

	Commission (%)	Omission (%)	Prod. Acc (%)	User. Acc (%)
Bamboo	6.68	12.16	87.84	93.32
Farm	28.57	6.25	93.75	71.43
Shrub	3.64	30.26	69.74	96.36
Broad-leaved	46.07	35.71	64.29	53.93
Masson	44.74	47.5	52.5	55.26
China fir	19.07	16.28	83.72	80.93
Mixed	13.57	12.57	87.43	86.43

Table 15.4 The overall accuracy and kappa coefficient of each method

	Landsat image	Spatial-spectral fusion	Spatial-temporal fusion	Spatial-spectral-temporal integrated fusion
Overall accuracy	69.95%	70.95%	78.94%	83.60%
Kappa	0.59	0.61	0.72	0.78

15.4 Discussion

15.4.1 The Classification Result

The proposed spatial-spectral-temporal fusion has achieved the highest overall accuracy of 83.60% (Fig. 15.4). However, broad-leaved forest and *Pinus massoniana* did not achieve satisfactory accuracies, which could be caused by two potential reasons.

On the one hand, as shown in Fig. 15.6, the differences of spectral profiles between China fir and *Pinus massoniana* and between broad-leaved forest and shrub curves are very small. Moreover, temporal changes of China fir and *Pinus massoniana* also have very similar paces, it was the case for the temporal changes of mixed forest and farm (Fig. 15.6). This inevitably added difficulties to distinguishing these forest types. On the other hand, the NFI data were assumed to be the ground truth of forest types in this study, but in fact, they may still have some biases. As shown in Fig. 15.10e, the areas of *Pinus massoniana*, farm, shrub and broad-leaved forest are very small so that less samples would be picked up (Table 15.2). Furthermore, these forest types were sparsely and fragmentarily distributed, because the geographic environment of the Gan River nature reserve is very complex and had rich biodiversity. As their proportions and borders were not clearly differentiated, the existing mixed pixels would further add difficulties in separating them accurately. It is indeed a difficult task to differentiate forest types with a 30-m spatial resolution. A promising approach to solve this problem is to seek new earth observation data with higher spatial, temporal and spectral resolution to mitigate the effects of mixed pixels. For example, Ren et al. [46] conducted forest land type precise classification based on SPOT-5 and GF-1 images and achieved an overall accuracy of 92.28% with a Kappa coefficient of 0.8996. Unmanned Aerial Vehicles (UAV) might be another effective way to get higher resolution images. Currently, the spatial resolution of the images acquired from UAV observations can even reach 0.05 m, which will lend great support to improve the classification accuracy. On the other hand, the selection of optimum features, such as the vegetation index, the statistical index and making full use of spatial, spectral and temporal information to separate different forest types, are still pivotal tasks.

15.4.2 The Spatial-Spectral Fusion

In order to demonstrate the advantage of SEGDV, a control experiment was conducted. We should be aware of the fact that no actual hyperspectral image with high spatial resolution existed in our study area. An emerging problem is how to quantitatively validate the fusion result without reference data. Here, we resampled the HSI with 100-m spatial resolution to 300-m as a coarse resolution image. We resampled the multispectral CCD with 30-m spatial resolution to 100-m as a fine resolution image. By fusing the up-sample 300-m hyperspectral image and 100-m multispectral image via the proposed SEGDV method, we could obtain the fused 100-m hyperspectral image. In this way, the original un-resampled HSI could be used as the actual reference data to evaluate the fusion accuracy quantitatively. The commonly-used RIBSR method was used for the comparison experiment.

As shown in Fig. 15.11, the result of SEGDV provided more spatial detail than that of RIBSR, such as the bluish farm in the red circle. Additionally, the SEGDV model also achieved a better visual effect than RIBSR. Two typical land cover types, forest and farm, were selected from the observed and fused data to further investigate the

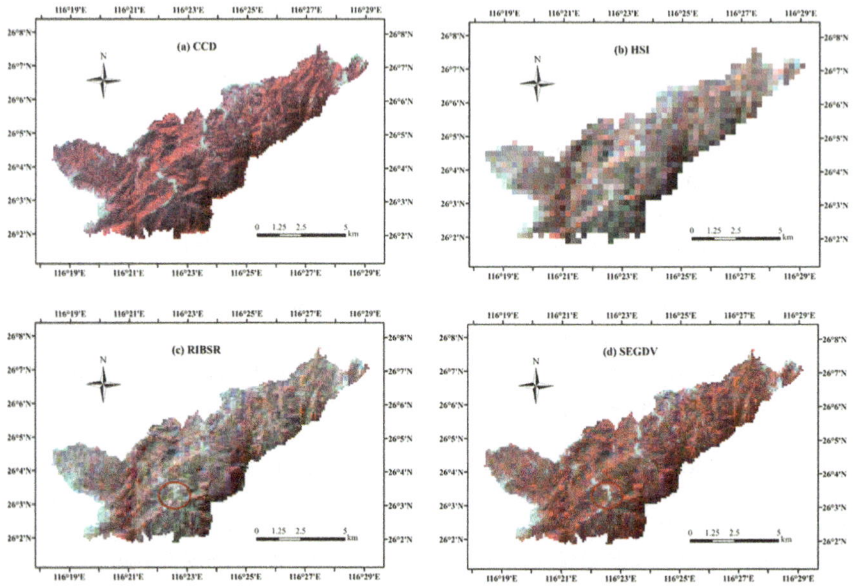

Fig. 15.11 The visual effect of **a** resampled CCD 100 m; **b** resampled HSI 300 m; **c** the RIBSR 100 m and **d** SEGDV 300 m fused images

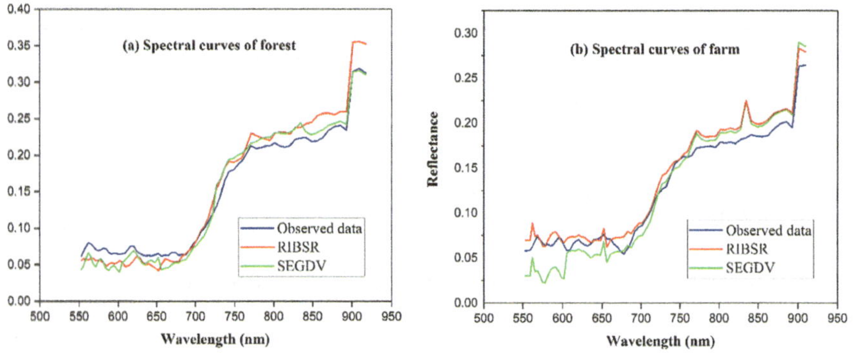

Fig. 15.12 The spectral curves of forest and farm derived from the observed HSI data, RIBSR and SEGDV fusion method

hyperspectral fidelity. As shown in Fig. 15.12, the reflectance profiles share similar trends between the observed and fused hyperspectral images for both land cover types (Fig. 15.11a, b). In addition the profile of SEGDV is closer to the real observed *I* data in most spectral ranges, especially for forest. The Root Mean Square Error (RMSE) was adopted to measure the fusion biases between the observed and fused data quantitatively. The RMSE is calculated as follow:

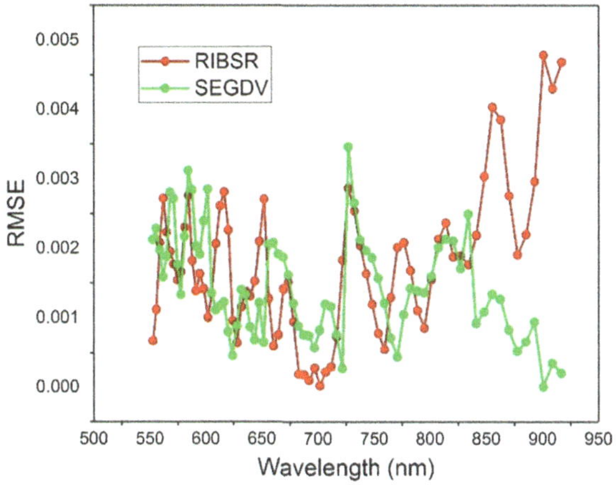

Fig. 15.13 The RMSE of the fused hyperspectral images derived for forest

$$RMSE = \sqrt{\frac{\sum_{i=1}^{n}\left(X_{fused,i} - X_{obs,i}\right)^2}{n}} \quad (15.16)$$

where n means the total number of bands, i means the i-th band, $X_{fused,i}$ means the reflectance value of the i-th band in the fused image and $X_{obs,i}$ means the reflectance value of the observed hyperspectral image. For most bands, the RMSE of SEGDV is smaller than the RIBSR, especially after 800 nm (Fig. 15.13). Above all, we conclude that SEGDV performs better than RIBSR due to two main reasons. Firstly, SEGDV considered the influence of noise. The commonly-used RIBSR model [18, 33] supposed that if I_a and I_b are two images with the same size, their ratio image I_R can be calculated as follows:

$$I_R(x, y) = I_a(x, y)/I_b(x, y) \quad x = 0, \ldots, cols - 1; \quad y = 0, \ldots, rows - 1 \quad (15.17)$$

where x,y means the location of each pixel. I_1 and I_2 are two sets of multispectral/hyperspectral images of the same scene and having the same size. $I_1[m_1]$ and $I_1[m_2]$, $I_2[n_1]$ and $I_2[n_2]$ are two different bands in I_1 and I_2, respectively. $I_1[m_1]$ and $I_2[n_1]$ have the same spectral range, so do $I_1[m_2]$ and $I_2[n_2]$. According to the fact that the ratio of reflectivity of the same kind of land cover in two given spectral ranges is almost changeless, their relationship can be depicted as follows:

$$\frac{I_1[m_1]}{I_1[m_2]} = \frac{I_2[n_1]}{I_2[n_2]} \quad (15.18)$$

However, the hyperspectral images are commonly affected by noise [47–49], and when noise exists, their relationship will not be like described above. Secondly, SEGDV considered the relationship between different bands. SEGDV separated the whole hyperspectral range into several groups. In each group, different bands highly correlate with each other, which reduces the errors in spatial and spectral fusion.

Of course, this model has remaining problems, too. For instance, it assumes that a band in a multispectral image corresponds to several bands in a hyperspectral image with a narrower wavelength range, but this is not necessarily. The relationship of the spectra range between HSI bands and CCD still needs to be considered further.

15.4.3 The Spatial-Temporal Fusion

Although the spatial resolution has been enhanced greatly after HSTAFM fusion, at the same time, some errors caused by mixed pixels remain, which eventually led to the uncertainty of the classification results. The low spatial resolution MODIS images commonly cannot capture the spatial difference in small areas due to its large spatial resolution (500 m), leading to the well-known mosaic phenomenon (Fig. 15.7a). A MODIS pixel (250 × 250 m) was split into about 64 pixels corresponding to CCD pixels (30 × 30 m), and the NDVI calculated from MODIS images became the

Table 15.5 Abbreviations

ALS	Airborne laser scanning
CCD	Charge coupled device
FLAASH	Fast line-of-sight atmospheric analysis of spectral hypercubes
FSDAF	Flexible spatiotemporal data fusion model
HSI	Hyperspectral imager
HSTAFM	Hierarchical spatiotemporal adaptive fusion model
IHS	Intensity hue saturation
LIDAR	Light detection and ranging
NDVI	Normalized difference vegetation index
NFI	National forest resources inventory
PCA	Principal component analysis
RIBSR	Ratio image-based spectral resampling
ROI	Regions of interest
SEGDV	Segmented difference value
SP-One	Dictionary learning-based spatiotemporal fusion model using only one base landsat-MODIS image pair
STARFM	Spatial and temporal adaptive reflectance fusion model
SVM	Support vector machine
VLDI	Vegetation local difference index
UTM	Universal transverse mercator projection

basis to conduct spatial-temporal fusion. The mixed pixel and mosaic problems were eventually transmitted to the fusion results as shown in Fig. 15.7b. Some pixels belong to the same category, but their value showed a big difference. Meanwhile, there are some outliers in the time series NDVI profiles (Fig. 15.8), such as the NDVI values on 6 and 9 December, which may be caused by subpixel clouds, variable illumination conditions and viewing geometries and other remnant geometric errors. Therefore, improving the quality of the data source will be an important step to produce the final classification results with high accuracy.

15.5 Conclusions

In this article, we proposed a spatial-spectral-temporal fusion framework through the SEGDV spatial-spectral fusion model and the HSTAFM spatial-temporal fusion model. The fused image with high resolution were used to classify different forest types. The entire research method could be divided into five parts. Firstly, the data preprocessing including the projection transformation, atmospheric correction and geometric correction to ensure all the images could be well matched in spatial, spectral and temporal domains. Secondly, multi-source data fusion was conducted including spatial-spectral fusion and spatial-temporal fusion. Thirdly, the fused hyperspectral and multi-temporal images were combined together to form the synthetic fusions, which contained all the spectral and temporal information. Fourth, training and validation samples were selected, and a SVM classifier was used to classify different forest types. Finally, the classification result was derived and the accuracy was estimated. Experimental results showed that compared with the classifications derived from single Landsat-8 image (69.95%), single spatial-spectral fusions (70.95%) and single spatial-temporal fusion (78.94%), the proposed method achieved the highest accuracy of 83.6%, thereby providing a new approach to sub-species classification such as the differentiation of different land cover types such as forest, grassland, crop and wetland (Table 15.5).

References

1. Pan, Y., R.A. Birdsey, O.L. Phillips, and R.B. Jackson. 2013. The structure, distribution, and biomass of the world's forests. *Annual Review of Ecology, Evolution, and Systematics* 44: 593–622.
2. Masek, J.G., D.J. Hayes, M.J. Hughes, S.P. Healey, and D.P. Turner. 2015. The role of remote sensing in process-scaling studies of managed forest ecosystems. *Forest Ecology and Management* 355: 109–123.
3. McKinley, D.C., et al. 2011. A Synthesis of current knowledge on forests and carbon storage in the United States. *Ecological Applications* 21 (6): 1902–1924.
4. Young, R.A., and R.J. Giese. 1990. *Introduction to forest science*. Wiley.
5. Qin, Y., et al. 2016. Mapping forests in monsoon Asia with ALOS PALSAR 50-m mosaic images and MODIS imagery in 2010. *Scientific Reports* 6.

6. Molinier, M., et al. 2016. Relasphone-Mobile and participative in situ forest biomass measurements supporting satellite image mapping. *Remote Sensing* 8 (10): 869.
7. Pratihast, A.K., et al. 2014. Combining satellite data and community-based observations for forest monitoring. *Forests* 5 (10): 2464–2489.
8. Wulder, M.A., and S.E. Franklin. 2012. *Remote sensing of forest environments: Concepts and case studies*. Springer.
9. Ren, C., H. Ju, H. Zhang, J. Huang, and Y. Zheng. 2016. Multi-source data for forest land type precise classification. *Scientia Silvae Sinicae* 52 (6): 54–65.
10. Torresan, C., P. Corona, G. Scrinzi, and J.V. Marsal. 2016. Using classification trees to predict forest structure types from LiDAR data. *Annals of Forest Research* 59 (2): 281–298.
11. Gaorgens, E.B., C.P.B. Soares, M.H. Nunes, and L.C.E. Rodriguez. 2016. Characterization of Brazilian forest types utilizing canopy height profiles derived from airborne laser scanning. *Applied Vegetation Science* 19 (3): 518-527.
12. Chen, C., Z. Bian, S. Li, P. Tang, and H. Wu. 2015. Improving forest type classification using the vegetation local difference index. *International Journal of Remote Sensing* 36 (14): 3701–3713.
13. Castilla, G., S. Rodrigue, R.S. Skakun, and R.J. Hall. 2016. Four national maps of broad forest type provide inconsistent answers to the question of what burns in Canada. *Remote Sensing* 8 (7): 539.
14. Connette, G., P. Oswald, M. Songer, and P. Leimgruber. 2016. Mapping distinct forest types improves overall forest identification based on multi-spectral Landsat imagery for Myanmar's Tanintharyi Region. *Remote Sensing* 8 (11): 882.
15. Chen, B., B. Huang, and B. Xu. 2015. Comparison of spatiotemporal fusion models: A review. *Remote Sensing* 7 (2): 1798–1835.
16. Zhang, L.P., and H.F. Shen. 2016. Progress and future of remote sensing data fusion (in Chinese). *Journal of Remote Sensing* 20: 1050–1061.
17. H. Shen. 2012. Integrated fusion method for multiple temporal-spatial-spectral images. In *International archives of the photogrammetry, remote sensing and spatial information sciences*, vol. 2 , p. 2.
18. Zhang, Y., and M. He. 2007. Multi-spectral and hyperspectral image fusion using 3-D wavelet transform. *Journal of Electronics (China)* 24 (2): 218–224.
19. Chen, Z., H. Pu, B. Wang, and G.-M. Jiang. 2014. Fusion of hyperspectral andmultispectral images: A novel framework based on generalization of pan-sharpening methods. *IEEE Geoscience and Remote Sensing Letters* 11 (8): 1418–1422.
20. Vivone, G., et al. 2015. A critical comparison among pansharpening algorithms. *IEEE Transactions on Geoscience and Remote Sensing* 53 (5): 2565–2586.
21. Chen, B., B. Huang, and B. Xu. 2017. A hierarchical spatiotemporal adaptive fusion model using one image pair. *International Journal of Digital Earth* 10 (6): 639–655.
22. Wang, F., et al. 2013. Community composition and spatial distribution of coniferous-broadleaved mixed forest in subtropical region, Jiangxi Province (in Chinese). *Acta Agriculturae Universitatis Jiangxiensis (Natural Sciences Edition)* 35 (4): 761–768.
23. Wulder, M.A., and N.C. Coops. 2014. Make earth observations open access: Freely available satellite imagery will improve science and environmentalmonitoring products. *Nature* 513 (7516): 30–32.
24. Hansen, M.C., et al. 2013. High-resolution global maps of 21st-century forest cover change. *Science* 342 (6160): 850–853.
25. Lehmann, E.A., J.F. Wallace, P.A. Caccetta, S.L. Furby, and K. Zdunic. 2013. Forest cover trends from time series Landsat data for the Australian continent. *International Journal of Applied Earth Observation and Geoinformation* 21: 453–462.
26. Zhu, Z., and C.E. Woodcock. 2014. Continuous change detection and classification of land cover using all available Landsat data. *Remote Sensing of Environment* 144: 152–171.
27. Chavez, P., S.C. Sides, J.A. Anderson, et al. 1991. Comparison of three different methods to merge multiresolution and multispectral data—Landsat TM and SPOT panchromatic. *Photogrammetric Engineering and Remote Sensing* 57 (3): 295–303.

28. Haydn, R., G.W. Dalke, J. Henkel, and J.E. Bare. 1982. Application of the IHS color transform to the processing of multisensor data and image enhancement. In *Proceedings of the international symposium on remote sensing of environment, first thematic conference: Remote sensing of arid and semi-arid lands*. Ann Arbor, Michigan. Center Remote Sensing Information & Analysis, Environmental Research Institute of Michigan, 1982. Cairo, Egypt.
29. Aiazzi, B., L. Alparone, S. Baronti, and A. Garzelli. 2002. Context-driven fusion of high spatial and spectral resolution images based on oversampled multiresolution analysis. *IEEE Transactions on Geoscience and Remote Sensing* 40 (10): 2300–2312.
30. Zhang, L., H. Shen, W. Gong, and H. Zhang. 2012. Adjustable model-based fusion method for multispectral and panchromatic images. *IEEE Transactions on Systems, Man, and Cybernetics, Part B (Cybernetics)* 42 (6): 1693–1704.
31. Jiang, C., H. Zhang, H. Shen, and L. Zhang. 2012. A practical compressed sensing-based pan-sharpening method. *IEEE Geoscience and Remote Sensing Letters* 9 (4): 629–633.
32. Fauvel, M., J.A. Benediktsson, J. Chanussot, and J.R. Sveinsson. 2008. Spectral and spatial classification of hyperspectral data using SVMs and morphological profiles. *IEEE Transactions on Geoscience and Remote Sensing* 46 (11): 3804–3814.
33. Xu, B., and P. Gong. 2007. Land-use/land-cover classification with multispectral and hyperspectral EO-1 data. *Photogrammetric Engineering & Remote Sensing* 73 (8): 955–965.
34. Chen, B., L. Chen, M. Lu, and B. Xu. 2017. Wetland mapping by fusing fine spatial and hyperspectral resolution images. *Ecological Modelling* 353: 95–106.
35. Acerbi-Junior, F.W., J. Clevers, and M.E. Schaepman. 2006. The assessment of multi-sensor image fusion using wavelet transforms for mapping the Brazilian Savanna. *International Journal of Applied Earth Observation and Geoinformation* 8 (4): 278–288.
36. Hilker, T., et al. 2009. A new data fusion model for high spatial-and temporalresolution mapping of forest disturbance based on Landsat and MODIS. *Remote Sensing of Environment* 113 (8): 1613–1627.
37. Gao, F., J. Masek, M. Schwaller, and F. Hall. 2006. On the blending of the Landsat and MODIS surface reflectance: Predicting daily Landsat surface reflectance. *IEEE Transactions on Geoscience and Remote Sensing* 44 (8): 2207–2218.
38. Zhu, X., J. Chen, F. Gao, X. Chen, and J.G. Masek. 2010. An enhanced spatial and temporal adaptive reflectance fusion model for complex heterogeneous regions. *Remote Sensing of Environment* 114 (11): 2610–2623.
39. Michishita, R., L. Chen, J. Chen, X. Zhu, and B. Xu. 2015. Spatiotemporal reflectance blending in a wetland environment. *International Journal of Digital Earth* 8 (5): 364–382.
40. Zhukov, B., D. Oertel, F. Lanzl, and G. Reinhackel. 1999. Unmixing-based multisensor multiresolution image fusion. *IEEE Transactions on Geoscience and Remote Sensing* 37 (3): 1212–1226.
41. Zurita-Milla, R., J.G.P.W. Clevers, and M.E. Schaepman. 2008. Unmixingbased Landsat TM and MERIS FR data fusion. *IEEE Geoscience and Remote Sensing Letters* 5 (3): 453–457.
42. Zhu, X., et al. 2016. A flexible spatiotemporal method for fusing satellite images with different resolutions. *Remote Sensing of Environment* 172: 165–177.
43. Huang, B., and H. Song. 2012. Spatiotemporal reflectance fusion via sparse representation. *IEEE Transactions on Geoscience and Remote Sensing* 50 (10): 3707–3716.
44. Mountrakis, G., J. Im, and C. Ogole. 2011. Support vector machines in remote sensing: A review. *ISPRS Journal of Photogrammetry and Remote Sensing* 66 (3): 247–259.
45. Foody, G.M. 2002. Status of land cover classification accuracy assessment. *Remote Sensing of Environment* 80 (1): 185–201.
46. Ren, C., H. Ju, H. Zhang, and J. Huang. 2016. Forest land type precise classification based on SPOT5 and GF-1 images. In *2016 IEEE international conference on geoscience and remote sensing symposium (IGARSS)*, IEEE. pp. 894–897.
47. Xu, B., and P. Gong. 2008. Noise estimation in a noise-adjusted principal component transformation and hyperspectral image restoration. *Canadian Journal of Remote Sensing* 34 (3): 271–286.

48. Acito, N., M. Diani, and G. Corsini. 2011. Subspace-based striping noise reduction in hyperspectral images. *IEEE Transactions on Geoscience and Remote Sensing* 49 (4): 1325–1342.
49. Lu, M., et al. 2017. Penalized linear discriminant analysis of hyperspectral imagery for noise removal. *IEEE Geoscience and Remote Sensing Letters* 14 (3): 359–363.

Chapter 16
Application of High Accuracy Surface Modelling to Interpolate Soil pH in Jiangxi Province

Wenjiao Shi

16.1 Study Area

The study area is located in the middle part of Jiangxi Province, China, and cover 6156.92 km^2. It includes the Ji'an municipal district, Ji'an county and Taihe county. It is a typical red soil hilly region of South China. Respectively, the precipitation in the counties is 1458, 1438 and 1381 mm per annum and the mean annual temperature is 18.1, 18.4 and 19.0 °C which are typical values for a subtropical monsoon climate. The elevation decreases from the periphery towards the center with altitude ranging from 1204.5 to 42.0 m. According to the 1/1000 000 scale soil maps reported by National Soil Census Office in 1995 (Fig. 16.1), the soils in the study area are classified into 7 groups: red soils, paddy soils, purple soils, fluvo-aquic soils, yellow soils, alluvial soils and limestone soils. The land use information was built by visual interpretation of Landsat TM 13/images for 2004/2005 with a spatial resolution of 30 by 30 m. A series of processing steps were carried out for TM 13/images, such as single band extraction and false color composition. These images were geo-referenced and ortho-rectified through field-collected ground control points and high-resolution digital elevation models achieving mean location errors of less than 1.5 pixels (i.e. 45 m). The land use types of the study area in 2004/2005 were classified into 6 groups: Woodlands, croplands, grasslands, built-up areas, water bodies and unused lands (Fig. 16.2a). According to the 1:500,000 geologic map of Jiangxi Province from the Jiangxi Provincial Bureau of Geology and Mineral Resources, underlying rock types in the study area include metamorphic rock, sedimentary rock, eruptive rock and quaternary sediment (Fig. 16.2b).

W. Shi (✉)
Institute of Geographic Sciences and Natural Resources
Research (CAS), Beijing, China
e-mail: shiwj@lreis.ac.cn

T. Yue et al. (eds.), *Chinese Water Systems*, Terrestrial Environmental Sciences,
https://doi.org/10.1007/978-3-319-97725-6_16

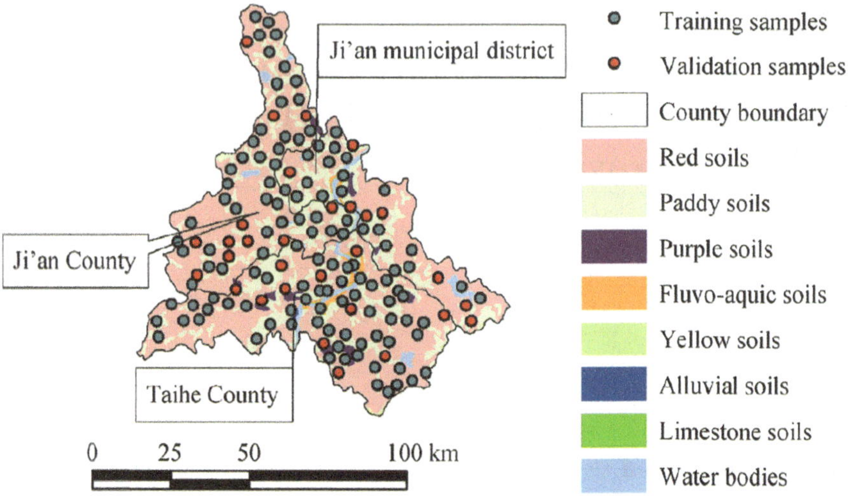

Fig. 16.1 The distribution of soil types and soil samples of the study area

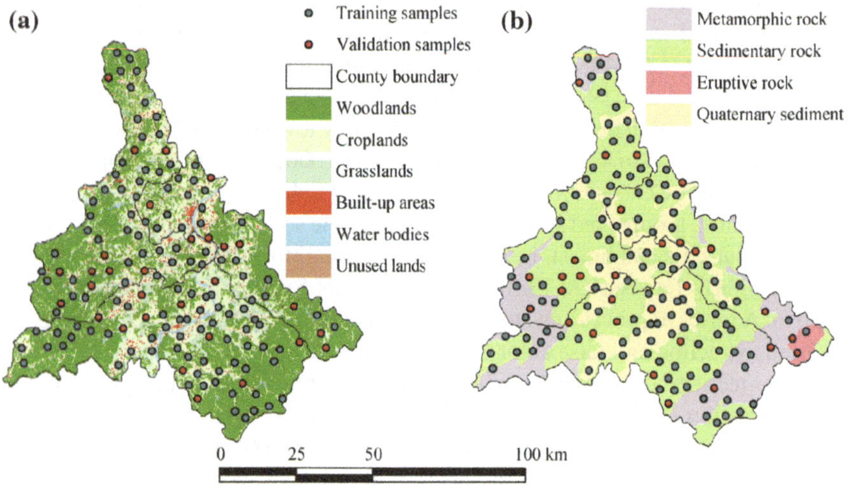

Fig. 16.2 The maps of land use types (**a**) and rock types (**b**) in the study area

16.2 Application of HASM in the Field of Soil Interpolation

Soil pH is an important soil property which has a tremendous effect on soil productivity. Acidity produces complex interactions of plant growth-limiting factors involving physical, chemical and biological properties of soil [1]. Acidic soils are sometimes considered synonymous with infertile soils because acidification causes a reduction in the availability of some nutrients but an increase of heavy metals to toxic levels.

In this case, the high rainfall in the study area might cause soils to become acid, so the applicability and importance of high accuracy surface modelling in soil pH was studied in this paper.

16.2.1 HASM

According to the fundamental theorem of surfaces, a surface is uniquely defined by the first and the second fundamental coefficients (Henderson and Taimina 1998). If a surface is a graph of a function $z = u\,(x, y)$, the basic theoretical equations of HASM could be formulated as [2–5]:

$$u_{xx} = \Gamma_{11}^{1} u_x + \Gamma_{11}^{2} u_y + \frac{L}{\sqrt{EG - F^2}} \tag{16.1}$$

$$u_{yy} = \Gamma_{22}^{1} u_x + \Gamma_{22}^{2} u_y + \frac{N}{\sqrt{EG - F^2}} \tag{16.2}$$

where:

$$E = 1 + u_x^2 \tag{16.3}$$

$$F = u_x u_y \tag{16.4}$$

$$G = 1 + u_y^2 \tag{16.5}$$

$$L = \frac{\mathrm{u}_{xx}}{\sqrt{1 + \mathrm{u}_x^2 + \mathrm{u}_y^2}} \tag{16.6}$$

$$N = \frac{\mathrm{u}_{yy}}{\sqrt{1 + \mathrm{u}_x^2 + \mathrm{u}_y^2}} \tag{16.7}$$

$$\Gamma_{11}^{1} = \frac{1}{2}\left(GE_x - 2FF_x + FE_y\right)\left(EG - F^2\right)^{-1} \tag{16.8}$$

$$\Gamma_{22}^{1} = \frac{1}{2}\left(2GF_y - GG_x - FG_y\right)\left(EG - F^2\right)^{-1} \tag{16.9}$$

$$\Gamma_{11}^{2} = \frac{1}{2}\left(2EF_x - EE_y - FE_x\right)\left(EG - F^2\right)^{-1} \tag{16.10}$$

$$\Gamma_{22}^{2} = \frac{1}{2}\left(EG_y - 2FF_y + FG_x\right)\left(EG - F^2\right)^{-1}. \tag{16.11}$$

$$E_x = \frac{E(i+1,j) - E(i-1,j)}{2h}, F_x = \frac{F(i+1,j) - F(i-1,j)}{2h}, \quad (16.12)$$
$$G_x = \frac{G(i+1,j) - G(i-1,j)}{2h}$$

$$E_y = \frac{E(i,j+1) - E(i,j-1)}{2h}, F_y = \frac{F(i,j+1) - F(i,j-1)}{2h}, \quad (16.13)$$
$$G_y = \frac{G(i,j+1) - G(i,j-1)}{2h}$$

If the maximum lengths of the computational domain in the x and y directions are respectively L_x and L_y, the computational domain can be included in the rectangular domain $[0, L_x] \times [0, L_y]$. If h represents interpolation step length and $I + 2$ and $J + 2$ represent the lattice numbers in direction x and in direction y, the central point of lattice $(0.5h + (i-1)h, 0.5h + (j-1)h)$ could be expressed as (x_i, y_j), in which $i = 0, 1, \ldots, I, I+1$ and $j = 0, 1, \ldots, J, J+1$. $u(x+h, y)$ and $u(x-h, y)$ could be formulated as the following Taylor expansion in series:

$$u(x+h, y) = u(x, y) + h\frac{\partial u(x, y)}{\partial x} + \frac{h^2}{2!}\frac{\partial^2 u(x, y)}{\partial x^2} + \frac{h^3}{3!}\frac{\partial^3 u(x, y)}{\partial x^3} + O(h^4) \quad (16.14)$$

$$u(x-h, y) = u(x, y) - h\frac{\partial u(x, y)}{\partial x} + \frac{h^2}{2!}\frac{\partial^2 u(x, y)}{\partial x^2} - \frac{h^3}{3!}\frac{\partial^3 u(x, y)}{\partial x^3} + O(h^4) \quad (16.15)$$

Subtracting Eq. (16.15) from Eq. (16.14), results:

$$u(x+h, y) - u(x-h, y) = 2h\frac{\partial u(x, y)}{\partial x} + \frac{2h^3}{3!}\frac{\partial^3 u(x, y)}{\partial x^3} + O(h^5) \quad (16.16)$$

Therefore,

$$u_x(x, y) = \frac{\partial u(x, y)}{\partial x} = \frac{u(x+h, y) - u(x-h, y)}{2h} + \frac{h^2}{3!}\frac{\partial^3 u(x, y)}{\partial x^3} + O(h^4) \quad (16.17)$$

For sufficiently small h, the finite difference approximation of $u_x(x, y)$ and $u_y(x, y)$ could be expressed as:

$$u_x(x, y) \approx \frac{u(x+h, y) - u(x-h, y)}{2h} \quad (16.18)$$

$$u_y(x, y) \approx \frac{u(x, y+h) - u(x, y-h)}{2h} \quad (16.19)$$

Equation (16.18) plus Eq. (16.19), results in:

$$u(x+h,y)+u(x-h,y)=2u(x,y)+\frac{2h^2}{2!}\frac{\partial^2 u(x,y)}{\partial x^2}+O(h^4) \tag{16.20}$$

Therefore,

$$u_{xx}(x,y)=\frac{\partial^2 u(x,y)}{\partial x^2}=\frac{u(x+h,y)-2u(x,y)+u(x-h,y)}{h^2}+O(h^2) \tag{16.21}$$

For sufficiently small h, the finite difference approximation of $u_{xx}(x,y)$ and $u_{yy}(x,y)$ could be expressed as,

$$u_{xx}(x,y)\approx\frac{u(x+h,y)-2u(x,y)+u(x-h,y)}{h^2} \tag{16.22}$$

$$u_{yy}(x,y)\approx\frac{u(x,y+h)-2u(x,y)+u(x,y-h)}{h^2} \tag{16.23}$$

If $\{\bar{u}_{i,j}\}$ are the sampled values of u at sampling points $\{(x_i,y_j)\}$, $u^n_{i,j}(n\geq 0$, $0\leq i\leq I+1$ and $0\leq j\leq J+1)$ are the nth iteration values of lattices whose centers are points of (x_i,y_j),in which $u^0_{i,j}=\tilde{u}_{i,j}$ and $\{\tilde{u}_{i,j}\}$ are the interpolated values based on the sampled values $\{\bar{u}_{i,j}\}$. In terms of numerical mathematics [6], the n + 1th iterative formulation of finite difference of basic equations of HASM given by (16.22) and (16.23) could be formulated as [7–11]:

$$\frac{u^{n+1}_{i+1,j}-2u^{n+1}_{i,j}+u^{n+1}_{i-1,j}}{h^2}=(\Gamma^1_{11})^n_{i,j}\frac{u^n_{i+1,j}-u^n_{i-1,j}}{2h}+\left(\Gamma^2_{11}\right)^n_{i,j}\frac{u^n_{i,j+1}-u^n_{i,j-1}}{2h}+\frac{L^n_{i,j}}{\sqrt{E^n_{i,j}+G^n_{i,j}-1}} \tag{16.24}$$

$$\frac{u^{n+1}_{i,j+1}-2u^{n+1}_{i,j}+u^{n+1}_{i,j-1}}{h^2}=(\Gamma^1_{22})^n_{i,j}\frac{u^n_{i+1,j}-u^n_{i-1,j}}{2h}+\left(\Gamma^2_{22}\right)^n_{i,j}\frac{u^n_{i,j+1}-u^n_{i,j-1}}{2h}+\frac{N^n_{i,j}}{\sqrt{E^n_{i,j}+G^n_{i,j}-1}} \tag{16.25}$$

where $n\geq 0; 0<i<I+1; 0<j<J+1$

$$E_{i,j}^{n} = 1 + \left(\frac{u_{i+1,j}^{n} - u_{i-1,j}^{n}}{2h} \right)^{2}; \tag{16.26}$$

$$F_{i,j}^{n} = \left(\frac{u_{i+1,j}^{n} - u_{i-1,j}^{n}}{2h} \right) \left(\frac{u_{i,j+1}^{n} - u_{i,j-1}^{n}}{2h} \right); \tag{16.27}$$

$$G_{i,j}^{n} = 1 + \left(\frac{u_{i,j+1}^{n} - u_{i,j-1}^{n}}{2h} \right)^{2}; \tag{16.28}$$

$$L_{i,j}^{n} = \frac{u_{i+1,j}^{n} - 2u_{i,j}^{n} + u_{i-1,j}^{n}}{h^{2}\sqrt{1 + \left(\frac{u_{i+1,j}^{n} - u_{i-1,j}^{n}}{2h} \right)^{2} + \left(\frac{u_{i,j+1}^{n} - u_{i,j-1}^{n}}{2h} \right)^{2}}}; \tag{16.29}$$

$$N_{i,j}^{n} = \frac{u_{i,j+1}^{n} - 2u_{i,j}^{n} + u_{i,j-1}^{n}}{h^{2}\sqrt{1 + \left(\frac{u_{i+1,j}^{n} - u_{i-1,j}^{n}}{2h} \right)^{2} + \left(\frac{u_{i,j+1}^{n} - u_{i,j-1}^{n}}{2h} \right)^{2}}}; \tag{16.30}$$

$$(\Gamma_{11}^{1})_{i,j}^{n} = \frac{G_{i,j}^{n}(E_{i+1,j}^{n} - E_{i,j}^{n}) - 2F_{i,j}^{n}(F_{i+1,j}^{n} - F_{i-1,j}^{n}) + F_{i,j}^{n}(E_{i,j+1}^{n} - E_{i,j-1}^{n})}{4(E_{i,j}^{n}G_{i,j}^{n} - (F_{i,j}^{n})^{2})h}; \tag{16.31}$$

$$(\Gamma_{22}^{1})_{i,j}^{n} = \frac{2G_{i,j}^{n}(F_{i,j+1}^{n} - F_{i,j-1}^{n}) - G_{i,j}^{n}(G_{i+1,j}^{n} - G_{i-1,j}^{n}) - F_{i,j}^{n}(G_{i,j+1}^{n} - G_{i,j-1}^{n})}{4(E_{i,j}^{n}G_{i,j}^{n} - (F_{i,j}^{n})^{2})h}; \tag{16.32}$$

$$(\Gamma_{11}^{2})_{i,j}^{n} = \frac{2E_{i,j}^{n}(F_{i+1,j}^{n} - F_{i-1,j}^{n}) - E_{i,j}^{n}(E_{i,j+1}^{n} - E_{i,j-1}^{n}) - F_{i,j}^{n}(E_{i+1,j}^{n} - E_{i-1,j}^{n})}{4(E_{i,j}^{n}G_{i,j} - (F_{i,j}^{n})^{2})h}; \tag{16.33}$$

$$(\Gamma_{22}^{2})_{i,j}^{n} = \frac{E_{i,j}^{n}(G_{i,j+1}^{n} - G_{i,j-1}^{n}) - 2F_{i,j}^{n}(F_{i,j+1}^{n} - F_{i,j-1}^{n}) + F_{i,j}^{n}(G_{i+1,j}^{n} - G_{i-1,j}^{n})}{4(E_{i,j}^{n}G_{i,j}^{n} - (F_{i,j}^{n})^{2})h}; \tag{16.34}$$

$$u_{0,j}^{n+1} = u_{0,j}^{0}(0 \le j \le J+1);\ u_{i,0}^{n+1} = u_{i,0}^{0}(0 \le i \le I+1); \tag{16.35}$$

$$u_{I+1,j}^{n+1} = u_{I+1,j}^{0}(0 < j < J+1);\ u_{i,J+1}^{n+1} = u_{i,J+1}^{0}(0 < i < I+1) \tag{16.36}$$

$u_{0,j}^{n+1}$, $u_{i,0}^{n+1}$, $u_{I+1,j}^{n+1}$ and $u_{i,J+1}^{n+1}$ are boundary conditions of HASM. The matrix formulation of equation set (16.25) and (16.26) can be expressed as,

$$A_1 U^{n+1} = b_1 \tag{16.37}$$

$$A_2 U^{n+1} = b_2 \tag{16.38}$$

where $U^{n+1} = (u_{1,1}^{n+1}, \ldots, u_{1,J}^{n+1}, u_{2,1}^{n+1}, \ldots, u_{2,J}^{n+1}, \ldots, u_{I,1}^{n+1}, \ldots, u_{I,J}^{n+1})$, A_1 and b_1^n are respectively coefficient matrix and right-hand item vector of equation (16.37); A_2 and b_2^n are respectively coefficient matrix and right-hand item vector of Eq. (16.38).

If $Z = \begin{bmatrix} A_1 \\ A_2 \end{bmatrix}, q^n = \begin{bmatrix} b_1^n \\ b_2^n \end{bmatrix}$, the following equality-constrained least squares problem can be developed to make the interpolated values equal to or approximate to the sampled values at the sampling points,

$$\min \begin{bmatrix} A \\ B \end{bmatrix} F^{n+1} - \begin{bmatrix} c \\ d \end{bmatrix} |_2$$
$$S \times F^{n+1} = t \tag{16.39}$$

where $C(k, (i-1) \cdot J + j) = 1$ and $d(k) = \overline{u}_{i,j}$, which means that the sampled value is $\bar{u}_{i,j}$ at the kth sampling point (x_i, y_j).

To solve the equation system given by Eq. 16.39 which is the least squares problem, a positive weight λ is introduced on the basis of the well known method of weights [12]. The parameter λ is the weight of the sampling points and determines the contribution of the sampling points to the simulated surface [13]. For sufficiently large λ, he algorithm (16.39) can be transferred into unconstrained least squares approximation [12]:

$$\min_f \left\| \begin{bmatrix} Z \\ \lambda C \end{bmatrix} U^{n+1} - \begin{bmatrix} q^n \\ \lambda d \end{bmatrix} \right\|_2 \tag{16.40}$$

or

$$\begin{bmatrix} Z^T & \lambda C^T \end{bmatrix} \begin{bmatrix} Z \\ \lambda C \end{bmatrix} U^{n+1} = \begin{bmatrix} Z^T & \lambda C^T \end{bmatrix} \begin{bmatrix} q^n \\ \lambda d \end{bmatrix} \tag{16.41}$$

16.2.2 Validation

Three indices were calculated from the measured and interpolated values at each validation sample site for each test data set. As there are n sites belonging to the validation data set, the mean error (ME), the mean absolute error (MAE) and the root mean square error (RMSE) are determined from the measured values $z(x_i)$ and the predicted value $z*(x_i)$. They are given by:

$$ME = \frac{1}{n} \sum_{i=1}^{n} \{z*(x_i) - z(x_i)\} \tag{16.42}$$

$$MAE = \frac{1}{n} \sum_{i=1}^{n} \{|z*(x_i) - z(x_i)|\} \tag{16.43}$$

$$RMSE = \sqrt{\frac{1}{n}\sum_{i=1}^{n}\{z*(x_i) - z(x_i)\}^2} \quad (16.44)$$

The ME is a measure of the bias of the interpolation which should be close to zero for unbiased methods, and the MAE and RMSE are accuracy measures of the interpolation which should be as small as possible for accurate interpolation. The ME, MAE and RMSE were estimated for each resampled validation sets, and respectively the averages of the MEs, MAEs and RMSEs of the 20 validation sets were used to determine the performance of each interpolator.

If outliers exist in the data sets, the noise-to-signal ratio of the data may be large which could affect the interpolation performance. A more revealing measure of performance can be obtained by computing the rank of each interpolation method for every data point, which is based on the point value of interpolation square error of a method [12, 14–16]. Thus, the rank of the ith method at the jth data point is computed by the rank of each interpolation method for each validation point j ($j = 1, \ldots$n). The mean rank of the ith method R_i is defined by:

$$R_i = \frac{1}{n}\left\{\sum_{i=1}^{l} r_{ij}\right\} \quad (16.45)$$

and the standard deviation of the ranks of ith method Si is defined as:

$$S_i = \sqrt{\frac{\sum_{i=1}^{n}\left(r_{ij} - R_i\right)^2}{n-1}} \quad (16.46)$$

The optimal method is determined by the plot of mean rank (R_i) versus the standard deviation of the ranks (S_i). An optimal result should exhibit a low mean rank in addition to a low standard deviation. A lower mean rank represent a, on average, more accurate interpolation, and a lower standard deviation of the ranks means that the interpolation accuracy has smaller dispersion.

16.2.3 Verification Results

The histograms of RMSEs of the prediction for the validation sets using the four interpolators are shown in Fig. 16.3. The histograms show that RMSEs have large dispersion around the mean value of RMSEs (Table 16.1) for all four techniques, so it is necessary to use the modified jackknifing technique for validation.

To compare the performance of the interpolators, we calculated the mean value of MEs, MAEs and RMSEs of 20 validation sets (Table 16.1). MEs of the four methods are all close to 0 for all cases, which indicates that they are all relatively unbiased in interpolating the soil pH values. HASM achieves slightly better MAE (0.16), when

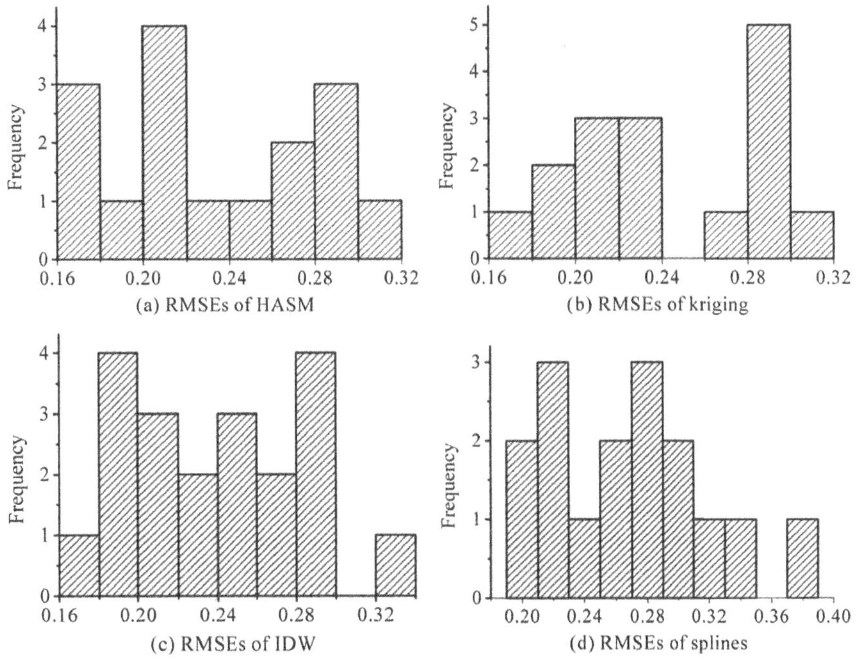

Fig. 16.3 Histograms of RMSEs for 20 validation sets using **a** HASM, **b** kriging, **c** IDW and **d** splines

Table 16.1 The means of MEs, MAEs and RMSEs of 20 validation sets for HASM, kriging, IDW and splines

	HASM	Kriging	IDW	Splines
ME	0.01	0.01	0.01	0.00
MAE	0.16	0.17	0.18	0.20
RMSE	0.22	0.23	0.24	0.26

compared with kriging (0.17), IDW (0.18) and splines (0.20). Similarly, the RMSE is smallest for HASM (0.22), followed by kriging (0.23), IDW (0.24) and splines (0.26). Consideration the performance of HASM in comparison with classical techniques indicates a considerable improvement in prediction of soil pH.

Further comparison of the ranks of every validation point for the four methods were calculated to determine if a small number of outliers affect ME, MAE and RMSE in each validation set. Mean ranks and the standard deviation of the ranks were calculated for every validation set. We present the plot of the mean value of mean ranks and the mean value of the standard deviation of the ranks for 20 validation sets (Fig. 16.4). Similar to the previous criteria, HASM showed lowest mean rank and the standard deviation of ranks, of all the four interpolation methods. Kriging is ranked

Fig. 16.4 The mean value of mean ranks versus the mean value of the standard deviations of ranks from 20 validation data sets derived by HASM, kriging, IDW and splines method

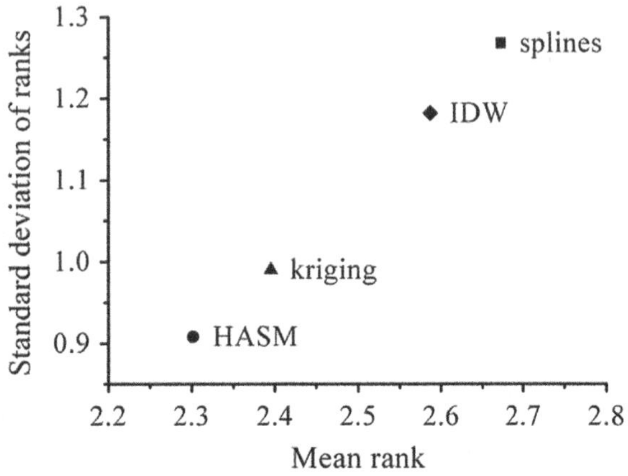

second with much lower mean rank and standard deviation than the other classical methods. IDW is the third most accurate method with the second largest standard deviation of ranks, whereas splines have the largest mean rank with the highest standard deviation of ranks. In conclusion, HASM is the most accurate technique of the four methods for interpolating soil pH values.

16.2.4 Comparison of the Interpolated Maps by the Four Techniques

The patterns of soil pH distribution produced by different interpolators based on data from 150 points are shown in Fig. 16.5. Comparing the maps from the four techniques, kriging map shows rather gradual transitions with low level of details and its smoothing effect results in the lowest maximum value and the highest minima of all four technique (Fig. 16.5b). The ranges of soil pH in the prediction map by kriging are somewhat smaller because the interpolator creates smoothed surfaces that have a smaller range of value than the observations. On the contrary, splines have a serious oscillation problem which make its minimum output (3.99) lower and the maximum output (6.75) higher than those of other technique (Fig. 16.5d). IDW interpolation produces many "Bull's eyes" on the soil pH surface (Fig. 16.5c). HASM obtains soil pH surface without narrower ranges, oscillation problem and "Bull's eyes" (Fig. 16.5a).

It is apparent that soil pH is relatively high in the centre and eastern part of the study area as compared with other parts. This could be explained by the distribution of soil types. pH values in paddy soils, which are dominant in the centre and eastern part of the catchment are significantly higher. Therefore, the lower density of soil sampling and the abrupt boundaries of soil types in this study (Fig. 16.1) may be the reasons for the prediction of soil pH without more details.

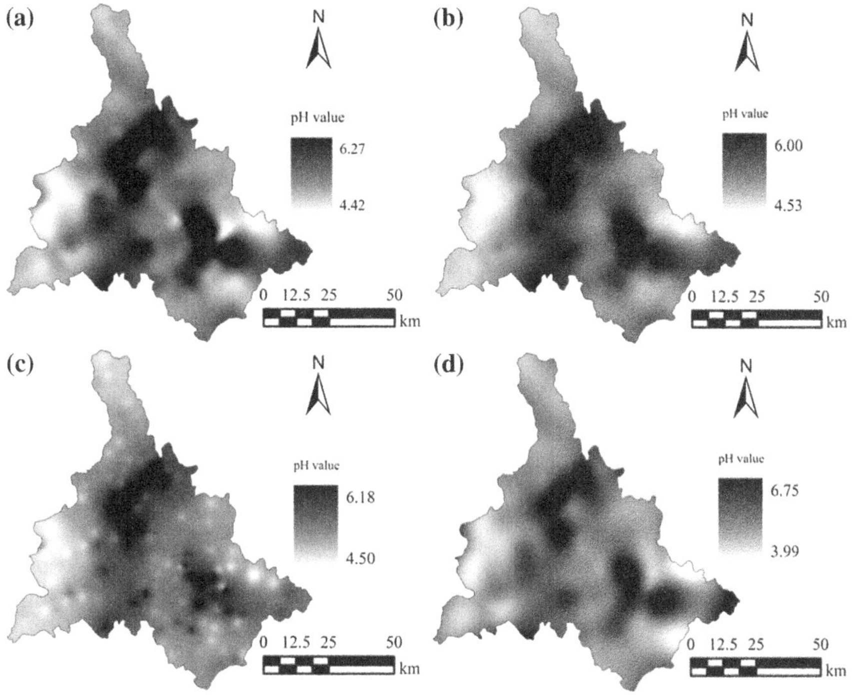

Fig. 16.5 Interpolated maps of soil pH by four different methods, **a** HASM, **b** OK, **c** IDW and **d** splines

16.2.5 Discussion

This study reports on an initial attempt to apply a new technique based on fundamental theorem of surfaces, namely high accuracy surface modelling (HASM) method to interpolate the spatial pattern of a soil property, and to evaluate the performance of the proposed method. Our case study showed that the proposed technique is more accurate for predicting the spatial patterns of soil pH than the classical methods, kriging, IDW and splines. It is also the most accurate method to avoid outliers. The evaluation of the maps obtained from HASM, kriging, IDW and splines showed clearly that the map from HASM is most realistic. The generally superior performance of HASM is due to the less error incorporated into HASM when using the development of least squares problem in Eq. (16.39) and enough iteration number in the procedure. Therefore, HASM can be considered as an alternative and more precise method for interpolating soil properties.

Distribution of soil pH is controlled by several environmental factors like acid deposition, parent rock, altitude, soil types, land use types and vegetations [17]. The soil pH data may vary significantly within a short horizontal distance, making it difficult to interpolate accurately using HASM if ignoring such environmental features. This problem is equally true with kriging and auxiliary variables can be used in several ways, such as in co-kriging, stratified kriging and regression kriging, which are well-known in soil science [15, 18–20]. Similarly, HASM can also incorporate secondary variables to improve the interpolation efficiency. The ancillary variables can be a digital elevation model (DEM) or terrain attribute, remote sensing images, cheap-to-measure auxiliary variables and some related categorical variables, such as parent rock types, soil types, land use types and vegetation types. The accuracy with which a variable is estimated could be improved by combining HASM with continuous variables or categorical information.

16.3 Application of HASM-MG in the Field of Soil Interpolation

HASM has two problems of high computational complexity and high memory when solving partial differential equations for soil interpolation. Compared with other methods, HASM is computationally expensive, which will restrict its wide application in practice. The multi-grid (MG) HASM was used to analyze the spatial distribution of soil pH. The interpolation accuracy of HASM-MG was compared with Kriging, IDW and Spline method for comparison.

16.3.1 HASM-MG

The former part of HASM algorithm is the same as HASM, so the description of HASM algorithm is omitted here.

$$A_h = \begin{bmatrix} Z^T & \lambda C^T \end{bmatrix} \begin{bmatrix} Z \\ \lambda C \end{bmatrix}, b_h^n = \begin{bmatrix} Z^T & \lambda C^T \end{bmatrix} \begin{bmatrix} q^n \\ \lambda d \end{bmatrix}, \tag{16.47}$$

h is the simulation step, the problem is solving linear algebraic equations of:

$$A_h U^{n+1} = b_h^n \tag{16.48}$$

The previous HASM model [5] needed to occupy a large amount of memory to store the matrix of the finest grid layer coefficients. In view of the distribution law of non-zero elements in the coefficient matrix, the multi-grid equations (16.48) [9] is used to store the diagonal elements of different grid layers (A_h, A_{2h} ...), thus greatly reducing the storage capacity. The general steps of the solution process are:

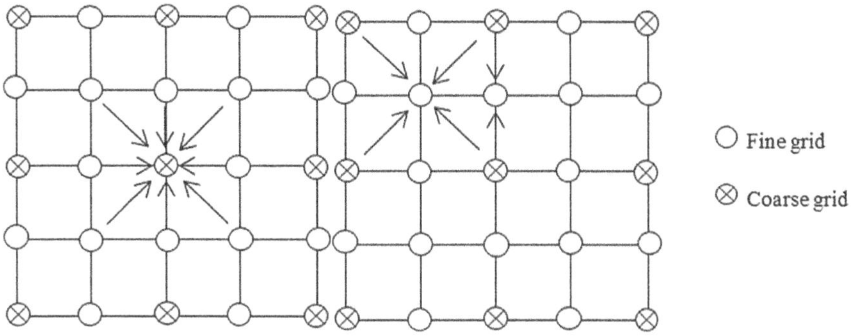

Fig. 16.6 Information fusion on multigrids: Left: Information on the fine grid ○ is transferred to the coarse grid. Right: Information on the coarse grid ⊗ is transferred to the fine grid

1. A series of coarse grid layers are established, and their pixel sizes are $2h \times 2h$, $4h \times 4h, \ldots$, respectively. Repeating the process of establishing equations (16.48) on the $h\times h$ grid layer to obtain $A_{2h}, A_{4h}, \ldots$;
2. The equation at the finest grid level is $A_h U^{n+1} = b_h^n$, and $U_h^0 = U^n$ is the initial value of iteration. Iterating using Gauss-Seidel iteration to obtain approximate solution U_h^1;
3. Calculating r_h^0, $r_h^0 = b_h - A_h U_h^1$;
4. The limit operator passes the information of r_h^0 to the next level grid (Fig. 16.6a); $b_{2h} = I_{2h}^h r_h^0$ is the right end of the term vector of the next grid layer linear equations, and I_{2h}^h represents limit operator.
5. Repeating steps 2 to 4 until the coarsest grid layer has been reached. The coarsest grid layer equation is $A_H U^{n+1} = b_H^n$, H is the coarsest grid layer step, directly solve this equation, and get U_H^1;
6. The limit operator passes the information of U_H^1 to the upper level grid (Fig. 16.6b). The grid equation is $O_X = P_H^X U_H^1$. X is the upper level grid layer step. P_H^X represents continuation operator. $U_X^0 = U_X^1 + O_X$, U_X^1 is the approximate solution during the pass down;
7. $A_X U^{n+1} = b_X^n$, U_X^0 is the initial value, Iterating the v2 step using Gauss-Seidel iteration to obtain approximate solution U_X^2;
8. Repeating steps 6 to 7 steps until it returns to the finest grid layer to get the approximate solution U_h^2 on the finest grid layer;
9. $U_h^0 = U_h^2$ is the initial value. Repeating steps 2 to 8 and keeping U_h^2 updated until you get satisfactory accuracy.

The calculation time of HASM-MG increases linearly with the total number of rasters, which solves the problem that the amount of computation of HASM grows geometrically with the total number of raster. MG greatly improves the calculated speed and solves the high computational complexity of HASM.

16.3.2 Verification

The accuracy of the interpolation method was evaluated by using four indexes, the mean error (ME), the mean absolute error (MAE), the root mean square error (RMSE) and the correlation coefficient (r) between the measured value of the verification point and the simulated value. ME, MAE, and RMSE are calculated using the same equations as shown previously (Eqs. (16.42), (16.43) and (16.39)).

16.3.3 Verification Results

Three sets of different datasets were randomly generated for simulation and verification. The verification point datasets were 20% (30, $V_{20\%}$), 40% (60, $V40_{\%}$) and 60% (90, $V_{60\%}$) respectively for the total number of 150 sampling points. The simulated datasets are 80% (120, $S_{80\%}$), 60% (90, $S_{60\%}$) and 40% (60, $S_{40\%}$) of the total number of sampling points (150). The MEs, MAEs, RMSEs, and r were used to compare the predictive performance of different methods (Table 16.2). The MEs of these four methods are all close to zero, indicating that they are unbiased. The MAEs and RMSEs of the three data sets HASM-MG are the smallest, r is the largest, indicating that HASM-MG is the most accurate method among the four methods.

Table 16.2 The statistical measures ME, MAE, RMSE and R of HASM-MG, Kriging, IDW and spline

		HASM-MG	Kriging	IDW	Spline
$V_{20\%}$	ME	−0.02	0.06	0.03	0.01
	MAE	0.14	0.19	0.16	0.22
	RMSE	0.20	0.25	0.21	0.29
	R	0.82	0.73	0.81	0.66
$V_{40\%}$	ME	0.01	0.05	0.03	−0.03
	MAE	0.19	0.24	0.21	0.23
	RMSE	0.27	0.31	0.29	0.32
	R	0.64	0.48	0.55	0.58
$V_{60\%}$	ME	−0.04	0.01	0.02	−0.05
	MAE	0.20	0.26	0.22	0.30
	RMSE	0.28	0.34	0.30	0.40
	R	0.67	0.46	0.62	0.59

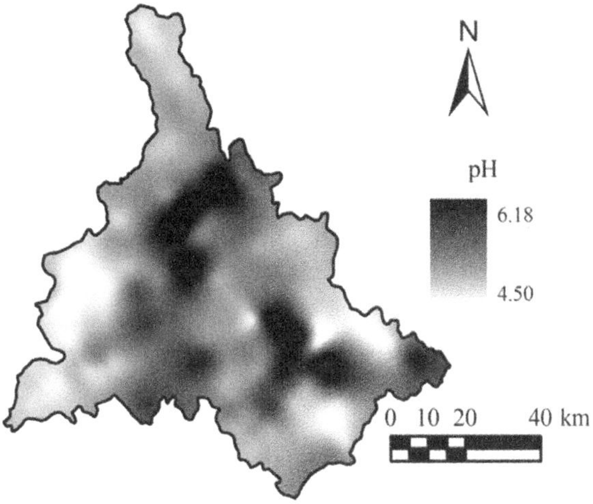

Fig. 16.7 Interpolated maps of soil pH by HASM-MG

16.3.4 The Maps of HASM-MG

As can be seen from Fig. 16.7, the soil pH in the central and southeastern part of the study area is relatively high compared to other areas, which is largely related to the soil type distribution (Fig. 16.2b). The pH of paddy soil was significantly higher than that of red soil, while most of the central and southeastern part of the study area were paddy soils, so the soil pH in central and southeast was significantly higher than that in other areas.The soil pH of Jiangxi is classified as follows: Grade 2 7.5–8.5 (slightly alkaline), Grade 3 6.5–7.5 (neutral), Grade 4 5.5–6.5 (Acidic), grade 5 4.5–5.5 (acidic), grade 6 ≤ 4.5 (strongly acidic). According to the second soil survey in the Ji'an area in the 1980s 0.28% of the soils have a soil pH refering to grade 1, 0.92% of the second grade, 2.10% of the third grade, 17.35% of the fourth grade, grade 5 accounted for 68.58%, grade 6 accounted for 10.77% (Table 16.3). It can be seen that in the 1980s, most of the soil pH in the Ji'an region ranged from slightly acidic (5.5–6.5), acidic (4.5–5.5) and strongly acidic ($\leq$4.5) pH values. Soil pH in 2007 was mainly distributed in slightly acidic and acidic range (Fig. 16.7). Slightly acidic soil (5.5–6.5) accounted for 32.07% of the total area and acidic (4.5–5.5) accounted for 67.93% of the total area. Compared with the situation in the 1980s in Ji'an area, the distribution of soil pH in acidic (4.5–5.5) was very similar. In contrast, in the 80s, the acid soil (10.77%), neutral soil (2.10%), Alkaline soils (0.92%) and strongly alkaline soils (0.28%) have almost disappeared within the study area in 2007 and were converted into acidic soils or slightly acidic soils. This shows that soils of strong acids and bases in the study area have evidently developed toward their intermediate state, acidic or slightly acidic, due to long-term anthropogenic activities such as tillage and fertilization.

Table 16.3 Soil pH distribution area statistics of Ji'an and the study area (1 mu ≈ 666.6 m^2)

Soil classification	Total area	Grade 1	Grade 2	Grade 3	Grade 4	Grade 5	Grade 6
Natural soil (10,000 mu)	2847.21	9.85	25.58	50.41	384.98	1996.84	379.55
Cultivated soil (10,000 mu)	746.73	0.15	7.46	25.13	238.65	467.82	7.52
Area summary (10,000 mu)	3593.94	10	33.04	75.54	623.63	2464.66	387.07
Ji'an area in the 80's (%)		0.28	0.92	2.10	17.35	68.58	10.77
Study area in 2007 (%)		0	0	0	32.07	67.93	0

16.3.5 Conclusions

The theoretical roots of HASM-MG's high precision and practicability demonstrated in interpolation of soil properties are:

1. When simulating soil pH surface, HASM first calculates the coefficient matrix and constant matrix of its numerical simulation equations according to the sampling data, and assures its accuracy by least square method and multiple iteration steps.
2. HASM can iterate enough times to obtain not only the first basic coefficient and the second basic coefficient, but also the simulation of surface with higher accuracy. Subsequently, common problems of traditional methods such as kriging and spline to simulate soil pH could be solven;
3. The HASM model is solved by the multi-grid method. The computation time of the HASM model increases linearly with the increase of the total number of grids. This solves the problem that previously the computation time of the HASM model increases non-linearly with the total number of grids in the past. Under the premise of speed, the simulation accuracy is also guaranteed and the practicality of interpolating high-precision surface modeling into soil properties is a big step forward.

16.4 Application of HASM for Soil Interpolation Using Auxiliary Variables

Due to the difference of soil environment, the soil properties may change significantly within small areas. If the high accuracy surface modelling is simulated using auxiliary information, the accuracy may greatly improve. Therefore, it is one of the most important research directions in the field of high accuracy surface modelling of soil attributes that auxiliary information can be combined with HASM to interpolate soil attributes.

16.4.1 High Accuracy Surface Modelling of Soil Properties Based on Land-Use Information (HASM-LU)

Land use types can be considered as a combination of soil types, vegetation cover and anthropogenic effects on land surfaces. Several studies have indicated that land use information is significantly related to the spatial pattern of soil properties. The spatial distribution pattern of soil organic carbon is approximately consistent with land use types.

Observations on soil pH, alkali-hydrolyzable N (AN), total C, N, K, Al, Ca, Mg and Zn were interpolated. To evaluate the performance of HASM-LU, it was compared with four other interpolators: HASM, ordinary kriging with land use information (OK-LU), stratified kriging (SK) and regression-kriging using a generalized linear model (RK-GLM).

16.4.1.1 Verification

Jackknife validation proposed by Deutsch and Journel (1992) [21] was used for the validation of the interpolators in this study. The procedure involves randomly splitting the data into the interpolation and validation subsets, estimating the spatial values using the interpolation subset and comparing the interpolated value at every validation point with its measured value. A total of 120 training points were randomly created as interpolation data sets, and the remaining 30 samples were used as validation data sets. We used the most common indices, the mean error (ME), the mean absolute error (MAE) and the root mean square error (RMSE) as measures of interpolation quality.

As n sites belong to the validation sample set, the mean error (ME), the mean absolute error (MAE) and the root mean square error (RMSE) are determined from the measured values $z(x_i)$ and the interpolated value $z*(x_i)$. Their formulas are the same as in Eqs. (16.42), (16.43) and (16.39).

The prediction efficiency (PE), a reduction-in-error index, referred to as goodness-of-prediction [22, 23], was also determined to assess the quality of soil property maps. It is calculated by:

$$\mathrm{PE}(\%) = \left(\mathrm{MSE}_{\mathrm{average}} - \mathrm{MSE}_{\mathrm{interpolation}}\right) \times 100/\mathrm{MSE}_{\mathrm{average}} \qquad (16.49)$$

Where $\mathrm{MSE}_{average}$ represents the mean square error (MSE) by using the average of the training data as estimate for all validation data points; $\mathrm{MSE}_{interpolation}$ is the square of the RMSE from comparing the interpolated values with measured values. Positive prediction efficiencies can be interpreted as the percent reduction in MSE compared to the field average approach [24]. This index can be used to compare the map quality for the interpolations of different soil properties from the two estimators n because it is a relative index in contrast to ME, MAE and RMSE.

16.4.1.2 ANOVA Analysis

Table 16.4 showed analysis of variances to compare the effects of different land use types on the tested soil properties. We found that the variances σ_m^2 of most soil properties (except for Zn) among the land use categories were significant at the 0.01 or 0.05 level, but that of Zn was significant at the 0.1 level. ANOVA analysis revealed that variances of tested soil properties among land use types might play an important role in their spatial prediction in the study area.

16.4.1.3 Geostatistical Analysis

The analyses of spatial correlation of residuals reflected good performance after removing the local mean within the same land use type (Fig. 16.8). Variograms of residuals for most of the soil variables tended to show a shorter range and a smaller sill, which indicated that the drift had indeed been removed [25]. However, variograms of original values and residuals of soil were similar for Zn, which indicates that to the spatial distribution of soil Zn might not be influenced significantly by the land use types in the study area.

16.4.1.4 Comparison of the Performance of HASM-LU, HASM, OK-LU, SK and RK-GLM

A comparison of HASM-LU, HASM, OK-LU, SK and RK-GLM was performed for assessing the feasibility of HASM-LU for interpolating soil properties. To evaluate the performance of the five methods, the MEs, MAEs and RMSEs were calculated from comparing the predicted values with the measured values (Table 16.5). Most of the MEs for soil properties (except for C and Zn) interpolated with land use information, such as OK-LU, HASM-LU and RK-GLM were much closer to 0 than the original HASM-Model without auxiliary data (see Sect. 16.2), which indicated that land use information could help the interpolator to obtain a more unbiased model. Although SK was also a method including land use information, most of MEs of SK (except for Al) were not closer to 0 than HASM. In general, most of the MAEs from HASM-LU were much lower than those of OK-LU (except for N and Mg) and RK-GLM (except for Mg and Zn). Also, MAEs of HASM-LU (except for pH and Zn) were much lower than HASM. Among the five methods, lowest RMSEs were also obtained by HASM-LU. For example, the RMSEs of AN with

Table 16.4 ANOVA analysis for testing the significance of land use type effects on the variances of soil properties. Concentrations are given in mg kg^{-1}

Soil property	Source of variance	df	Sum of squares	Mean square	F value	Sig.
pH	Between land use types	2	2.35	1.18	10.03	0.00**
pH	Within land use types	147	17.26	0.12		
pH	Total	149	19.61			
AN	Between land use types	2	66648.95	33324.47	14.16	0.00**
AN	Within land use types	147	346016.39	2353.85		
AN	Total	149	412665.34			
C	Between land use types	2	386.79	193.39	3.84	0.02*
C	Within land use types	147	7410.55	50.41		
C	Total	149	7797.33			
N	Between land use types	2	3.79	1.90	13.90	0.00**
N	Within land use types	147	20.06	0.14		
N	Total	149	23.85			
K	Between land use types	2	453.74	226.87	4.24	0.02*
K	Within land use types	147	7859.17	53.46		
K	Total	149	8312.91			
Al	Between land use types	2	8928.50	4464.25	20.16	0.00**
Al	Within land use types	147	32553.08	221.45		
Al	Total	149	41481.58			
Ca	Between land use types	2	466.23	233.12	64.01	0.00**
Ca	Within land use types	147	535.39	3.64		
Ca	Total	149	1001.62			
Mg	Between land use types	2	140.50	70.25	16.72	0.00**
Mg	Within land use types	147	617.75	4.20		
Mg	Total	149	758.24			
Zn	Between land use types	2	6605.64	3302.82	2.59	0.08
Zn	Within land use types	147	187322.39	1274.30		
Zn	Total	149	193928.03			

* represents the significance is <0.05;
** represents the significance is <0.01.

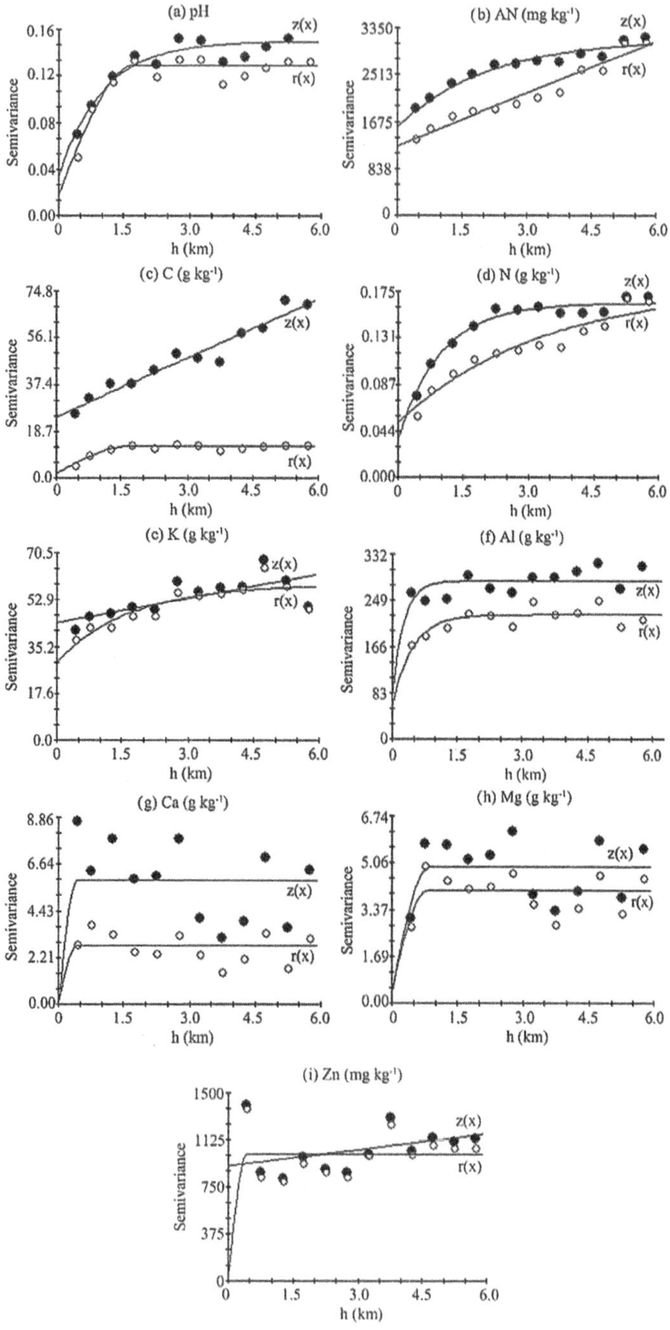

Fig. 16.8 Semivariograms of original values and the residuals for soil properties: **a** pH, **b** AN, **c** C, **d** N, **e** K, **f** Al, **g** Ca, **h** Mg and **i** Zn.

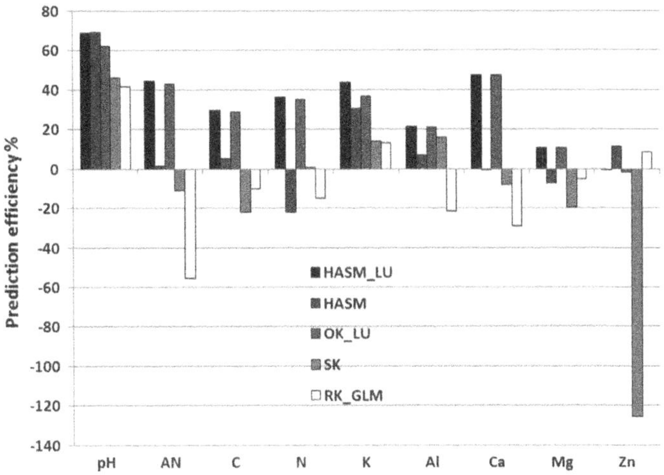

Fig. 16.9 Comparison of the PEs for interpolation of 9 soil properties by HASM-LU, HASM, OK-LU, SK and RK-GLM

HASM-LU were smaller than HASM, OK-LU, SK and RK-GLM by 33.4%, 1.6%, 41.5% and 67.59%, respectively. Rather high MAEs and RMSEs were found in most cases of soil properties which were interpolated by SK when compared with OK-LU and HASM-LU. This is because the number of available data points to estimate the variogram is not large enough to give reliable estimates [18]. HASM with land use information (HASM-LU) generally achieves the lowest RMSEs (except for Zn) when compared with the other methods. This may be due to the insignificant correlation between soil Zn with land use information (Table 16.4). All the aforementioned results have suggested that HASM combined with land use information (HASM-LU) is an effective method to interpolate soil properties.

The PEs of the five methods are shown in Fig. 16.9. The largest PEs of HASM-LU, HASM, OK-LU, SK and RK-GLM were 68.89%, 69.23%, 62.05%, 46.14% and 41.75% in the case of soil pH, respectively. This indicated that the five methods reduced 68.89, 69.23, 62.05, 46.14 and 41.75% of MSE compared to the field average approach. The smallest PE of HASM was −21.86% for soil N. However, smallest PE values of HASM-LU, OK-LU and SK were −0.48%, −1.96% and −125.64% for Zn, respectively. The smallest PE value of RK-GLM was −55.61% for AN. For AN, C, N, K, Al, Ca and Mg, the PE values of HASM were much smaller than those of the methods, which include land use information (HASM-LU). The largest differences in PE occurred when comparing HASM-LU with HASM for N (57.95%), with OK-LU (7.18%) for K, with SK (125.16%) for Zn and with RK-GLM (100.21%) for AN. Nevertheless, for soil Zn, the performance of the method with land use information (OK-LU, HASM-LU, SK and RK-GLM) did not perform better than that without land use information, which might be caused by the larger significant level (0.08) in the ANOVA analysis in comparison to other soil attributes (Table 16.4) and similar spatial variation between original data and residual data (Fig. 16.8i). This indicated that the methods considering land use information were not appropriate for Zn, and land use information might not be the major factor affecting its spatial distribution. For the four methods considering land use information, the performance of SK and RK-GLM are generally not better than OK-LU and HASM-LU. The poorer performance of SK is due to the limited sample data and in case of RK-GLM, it is due to solely using land use information as the independent variable in the generalized linear model. It is shown that superior performance of OK-LU and HASM-LU resulted from the consideration of different land use types.

16.4.1.5 Comparison of Soil Property Maps Obtained by HASM-LU, HASM, OK-LU, SK and RK-GLM

In order to compare the map quality generated by HASM-LU, HASM, OK-LU, SK and RK-GLM, soil AN maps from the five methods were taken as examples (Fig. 16.10a–e). HASM-LU (Fig. 16.10a), OK-LU (Fig. 16.10c), SK (Fig. 16.10d) and RK-GLM (Fig. 16.10e) gave many details of alkali hydrolyzable nitrogen in different land use types of the whole map. However, the method excluding land use information (Fig. 16.10b: HASM) exhibited worse interpolation quality than the methods including land use information especially in the presence of sharp boundaries among different land use types. Hence, the methods combined with land use information are able to describe local spatial variability more accurately.

To comparison the differences of the performance between the four methods-HASM, OK-LU, SK and RK-GLM, with HASM-LU to interpolate soil AN are shown in Fig. 16.11a, b, c and d, respectively. The difference between HASM and HASM-LU (Fig. 16.11a) had similar spatial pattern with land use types (Fig. 16.2a), which showed that the difference between adding and not adding land use information using HASM. OK-LU method had the most similar pattern with HASM-LU (Fig. 16.11b), but larger differences existed at some sample points (Fig. 16.1). Figure 16.11c showed that the soil AN values were estimated lower in mid-eastern woodland by SK than those of HASM-LU, while Fig. 16.11d showed that the soil AN values were estimated lower in most of cropland by RK-GLM than those of HASM-LU.

Table 16.5 and Fig. 16.9 showed that HASM-LU is an accurate method in interpolating soil properties, and Figs. 16.10 and 16.11 suggest that HASM-LU can improve the interpolation performance of soil properties and enhance the map accuracy.

16.4.1.6 Discussion

HASM is a surface modelling method with high accuracy in DEM construction and for interpolating other smoothed surfaces. However, the method may not perform well if abrupt boundaries of environmental correlates exist. HASM-LU, which invokes HASM to interpolate the residual surface and combines it with land use information to deal with abrupt boundaries, can therefore be seen as an improved technique for

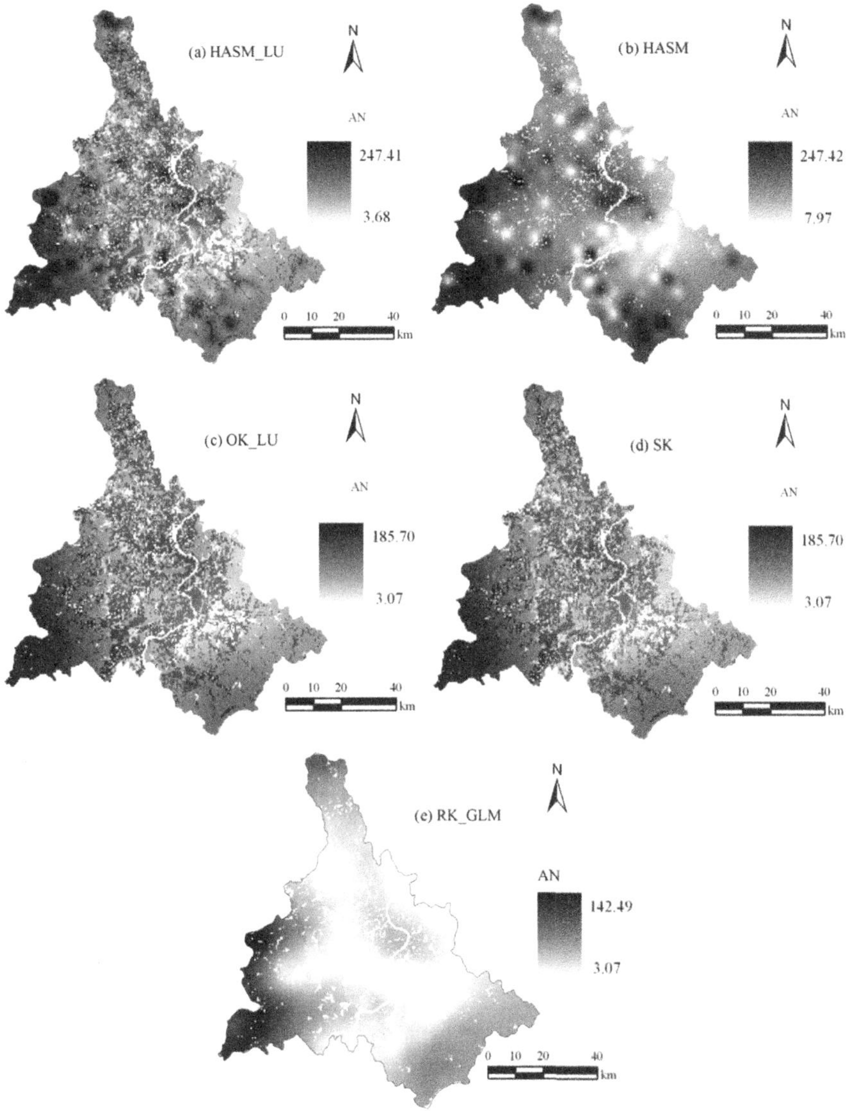

Fig. 16.10 Soil AN (mg kg^{-1}) maps obtained by **a** HASM-LU, **b** HASM, **c** OK-LU, **d** SK and **e** RK-GLM

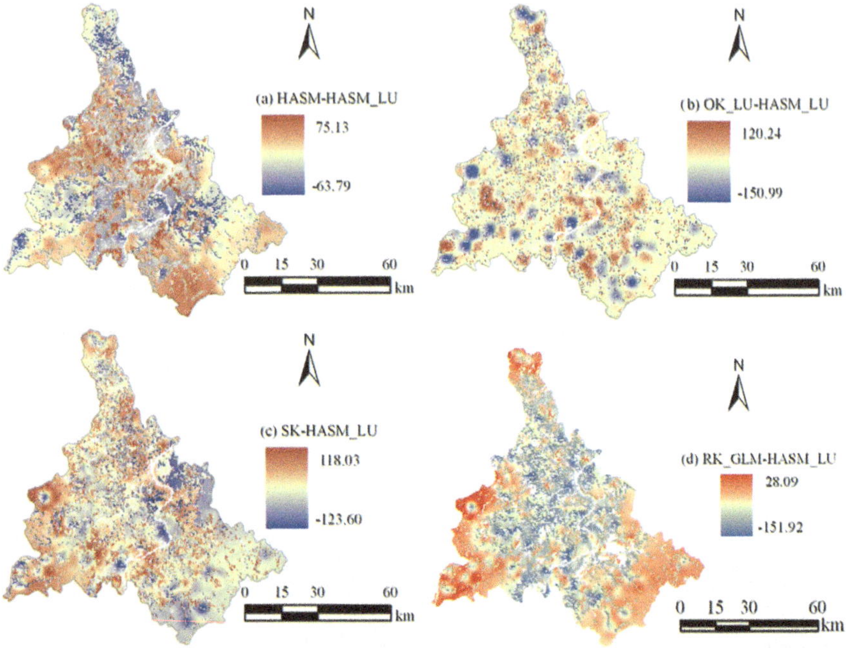

Fig. 16.11 Performance differences of **a** HASM, **b** OK-LU, **c** SK, and **d** RK-GLM in comparison to HASM-LU for interpolating soil AN

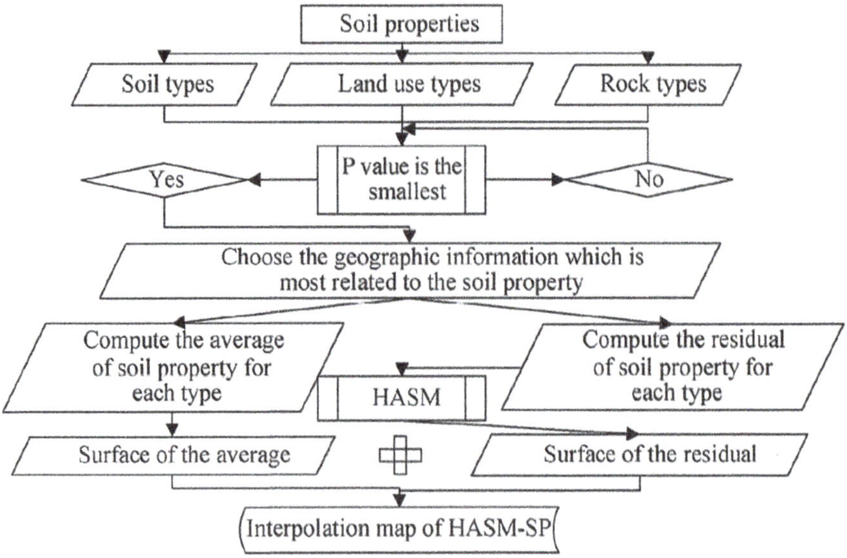

Fig. 16.12 The framework of HASM-SP method

Table 16.5 The MEs, MAEs and RMSEs of HASM-LU, HASM, OK-LU, SK and RK-GLM

Method	Index	pH	AN (mg kg^{-1})	C (g kg^{-1})	N (g kg^{-1})	K (g kg^{-1})	Al (g kg^{-1})	Ca (g kg^{-1})	Mg (g kg^{-1})	Zn (mg kg^{-1})
HASM-LU	ME	−0.01	0.65	−0.2	0	−1.45	−5.38	−0.09	−0.58	−12.58
HASM		−0.01	4.73	0.1	0.01	−2.15	−6.1	−0.54	−0.87	−12.53
OK-LU		0.01	1.47	−0.15	0	−1.35	−5.32	−0.09	−0.59	−12.31
SK		−0.01	−4.72	−0.92	−0.04	−2.3	−3.49	−0.45	−0.69	51.96
RK-GLM		−0.01	1.68	2.23	−0.01	−1.53	−4.51	−0.11	−0.52	−13.31
HASM-LU	MAE	0.14	29.84	4.13	0.22	4.68	10.72	1	1.49	27.98
HASM		0.14	41.58	4.83	0.34	5.5	11.75	1.29	1.56	26.63
OK-LU		0.15	30.37	4.15	0.22	4.99	10.79	1.01	1.49	28.69
SK		0.19	40.7	5.28	0.29	6.23	11.59	1.34	1.72	53.36
RK-GLM		0.21	52.43	5.29	0.35	6.05	14.24	1.75	1.48	25.99
HASM-LU	RMSE	0.19	36.69	5.07	0.31	5.94	13.71	2.6	2.23	36.87
HASM		0.19	48.95	5.88	0.43	6.61	14.93	3.6	2.44	34.62
OK-LU		0.2	37.28	5.1	0.32	6.31	13.77	2.61	2.23	37.14
SK		0.25	51.92	6.67	0.39	7.35	14.19	3.73	2.58	55.25
RK-GLM		0.26	61.49	6.34	0.42	7.40	17.06	4.08	2.42	35.15

surface modelling of soil properties. Secondly, HASM-LU takes into account land use information which is a determinant of soil properties. The improvement results from capitalizing on the significant variation of soil properties among different land use types and the introduction of abrupt land use boundaries into the interpolation of soil property maps. Summarizing, HASM assisted by land use information (HASM-LU) can be considered as a satisfactory method for surface modelling of soil properties.

16.4.2 High Accuracy Surface Modelling of Soil Properties

High accuracy surface modeling (HASM) is a method which can be applied to soil property interpolation. In this study, we present a method of HASM combined with geographic information for soil property interpolation (HASM-SP) to improve the accuracy of the devived surfaces.

16.4.2.1 HASM-SP

Each observation $z\left(x_{i,l,k}, y_{j,l,k}\right)$ of one specific soil property at grid *(i, j)* with the lth type of kth geographic information can be expressed by:

$$z(x_{i,l,k}, y_{j,l,k}) = m(G_{l,k}) + r(x_{i,l,k}, y_{j,l,k}) \tag{16.50}$$

where $m(G_{l,k})$ is the mean value of $z\left(x_{i,l,k}, y_{j,l,k}\right)$ in the kth type of the lth kind of geographic information, and $r(x_{i,l,k}, y_{j,l,k})$ is the residual computed by subtracting the mean value of the kth type of the relative lth geographic information from the original value of soil property. We assume that $m(G_{l,k})$ and $r(x_{i,l,k}, y_{j,l,k})$ are independent mutually and the variation of $r(x_{i,l,k}, y_{j,l,k})$ is homogeneous over the whole study area.

The residuals are used to interpolate the surface of residuals in the whole study area by HASM. The interpolated values of residuals are finally summed to soil property averages of the relevant types of geographic information to obtain the interpolated values of the HASM-SP interpolation method. The framework of HASM-SP is shown in Fig. 16.12 and the processes is described as follows:

1. Based on ANOVA analysis or linear mixed model, we analyzed the geographic information which affected the spatial distribution of soil properties most significantly, and chose the geographic information (l) which was most related to each type of soil property.
2. According to the soil property values, we calculated the averages and residuals of soil property values for each type (k) of geographic information (l).
3. According to the spatial distribution of the soil property averages for each kind of most related geographic information (l), we mapped the average distribution of soil property $m(G_{l,k})$ for each type (k).

Table 16.6 Comparisons of the accuracy among HASM-SP, OK, OK-Geo and SK spatial interpolation schemes

Method	Index	AP ($mg\ kg^{-1}$)	Li ($mg\ kg^{-1}$)	pH (/)	AN ($mg\ kg^{-1}$)	TK ($g\ kg^{-1}$)	Cr ($mg\ kg^{-1}$)
HASM-SP	ME	3.01	−5.21	−0.01	0.65	−0.19	−4.08
OK		3.08	−3.49	0.01	5.20	−0.20	−4.31
OK-Geo		2.88	−4.22	0.01	1.47	−0.17	−3.95
SK		3.50	−5.99	−0.01	−4.72	0.43	15.69
HASM-SP	MAE	6.78	15.32	0.14	29.84	0.54	13.46
OK		7.13	19.87	0.15	41.88	0.58	14.01
OK-Geo		7.53	17.29	0.15	30.37	0.56	13.71
SK		7.21	19.91	0.19	40.70	0.84	22.95
HASM-SP	RMSE	8.67	22.39	0.19	36.69	0.67	17.28
OK		9.25	26.88	0.20	49.42	0.73	18.45
OK-Geo		9.03	24.12	0.20	37.28	0.71	17.55
SK		8.69	27.54	0.25	51.92	1.03	26.68

4. According to the related residuals, HASM was used to simulate the residual surface of soil property$r(x_{i,l,k}, y_{j,l,k})$.
5. Finally, we added up average surface and residual surface as HASM-SP simulation results.

16.4.2.2 Verification Results

In order to assess the accuracy of HASM-SP for interpolating soil properties, we compared the performance of HASM-SP, ordinary kriging (OK), ordinary kriging combined with geographic information (OK-Geo) and stratified kriging (SK) [18]. We calculated the MEs, MAEs and RMSEs as measures of interpolation quality by comparing the predicted values with the measured values (Table 16.6). We found that most interpolation schemes combined with geographic information, especially including HASM-SP and OK-Geo, showed less bias than traditional methods. HASM-SP and OK-Geo had less MAEs than OK for most of soil properties except AP. SK also had larger MAE than OK except for AN. Similarly, HASM-SP had the lowest RMSEs, followed by OK-Geo and OK. SK also had the largest RMSEs except for AP. The higher accuracy of HASM-SP is due to the consideration of the abrupt boundary of geographic information, describing the spatial variation characters of soil properties in different geographic types accurately. Though SK is an interpolation scheme, which combines geographic information, the distributions of soil types, land use types and rock types of the study area is complicated and consists of many tiny

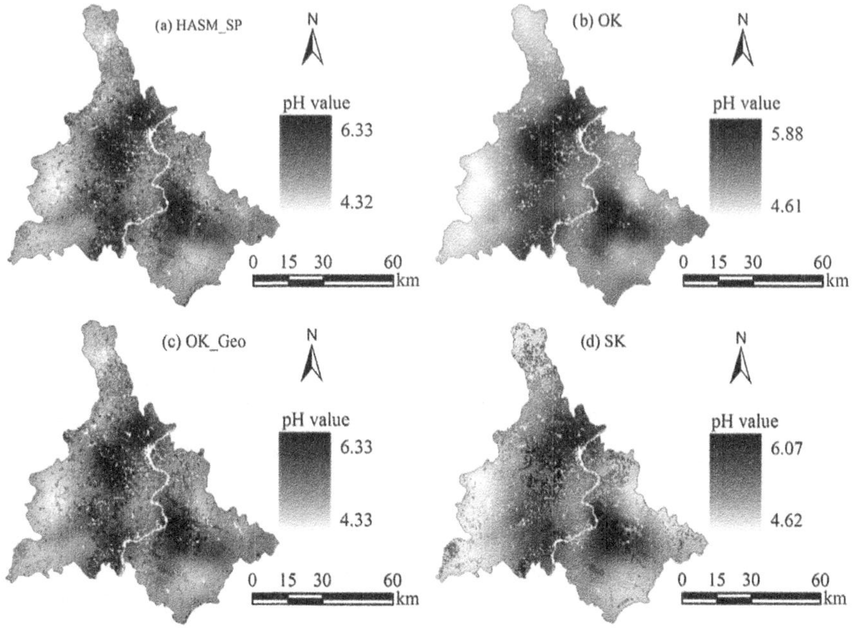

Fig. 16.13 Comparisons of the soil pH maps interpolated by HASM-SP (**a**), OK (**b**), OK-Geo (**c**) and SK (**d**)

patches. Consequently the numbers of data points available for each type to estimate the variogram were not large enough to provide reliable estimates for the SK method [18, 26].

16.4.2.3 The Maps of HASM-SP

We took soil pH as an example, and compared the spatial interpolation maps of HASM-SP, OK, OK-Geo and SK (Fig. 16.13).

The interpolation surface of OK was most smoothing, and the range of OK interpolation map was the narrowest (4.61–5.88). This was OK due to the smoothing effect, of which reduces surface variation in comparison to the measured values. In contrast, HASM-SP (Fig. 16.13a), OK-Geo (Fig. 16.13c) and SK (Fig. 16.13d), all combining geographic information eliminated the smoothing effect of OK (Fig. 16.13b), and had significant variation at the different types of abrupt boundary. For example, HASM-SP, OK-Geo and SK gave more details of soil pH distribution to different land use types, especially at abrupt boundaries. The values of cropland and grassland from the HASM-SP map were significantly higher than those of woodland, which was in accordance with the real condition of the study area (Table 16.6). In contrast, soil pH values of OK interpolation map especially in the grassland and a part of cropland did not contain this discrete information. Therefore, the interpolation methods, which use the geographic information describe the local variation more accurately. These results showed that combining appropriate geographic information as secondary variable could significantly improve the interpolation performance of local variations.

References

1. Fageria, N., and V. Baligar. 2008. Ameliorating soil acidity of tropical oxisols by liming for sustainable crop production. *Advances in Agronomy* 99: 345–399.
2. Yue, T., Z. Du, and J. Liu. 2004. High accuracy surface modeling and error analysis (in Chinese). *Progress in Natural Science* 14(2): 83–89.
3. Yue, T., and Z. Du. 2005. High accuracy surface modelling: the core module of CAD and GIS (in Chinese). *Progress in Natural Science* 15(4): 423–432.
4. Yue, T., and Z. Du. 2006. Comparison and analysis of errors between high accuracy surface modelling and classical models (in Chinese). *Process in Natural Science* 16(8): 986–991.
5. Yue, T., Z. Du, D. Song, and Y. Gong. 2007. A new method of surface modelling and its application to DEM construction. *Geomorphology* 91(1–2): 161–172.
6. Quarteroni, A., R. Sacco, and F. Saleri. 2010. *Numerical mathematics*, vol. 37. Springer Science & Business Media.
7. Yue, T., and Y.J. Song. 2008. The YUE-HASM method. In: *Proceeding of the 8th international symposium on spatial accuracy assessment in natural resources and environmental sciences*, Shanghai, 2008, pp. 148–153.
8. Yue, T., Z. Du, and D. Song. 2007. High accuracy surface modelling: HASM4 (in Chinese). *Journal of Image and Graphics* 12(2): 343–348.
9. Yue, T., Z. Du, and Y. Song. 2008. Spatial models and geographic information systems, pp. 3315–3325.
10. Yue, T., and S. Wang. 2010. Adjustment computation of HASM: A high-accuracy and high-speed method. *International Journal of Geographical Information Science* 24(11): 1725–1743.
11. Yue, T., et al. 2013. A multi-grid method of high accuracy surface modelling and its validation. *Transactions in GIS* 17(6): 943–952.
12. Van Loan, C. 1985. On the method of weighting for equality-constrained least-squares problems. *SIAM Journal on Numerical Analysis* 22(5): 851–864.
13. Shi, W., et al. 2009. Surface modelling of soil pH. *Geoderma* 150(1–2): 113–119.
14. Laslett, G.M., A.B. McBratney, P. Pahl, and M.F. Hutchinson. 1987. Comparison of several spatial prediction methods for soil pH. *European Journal of Soil Science* 38(2): 325–341.
15. Odeh, I.O.A., A.B. McBratney, and D.J. Chittleborough. 1995. Further results on prediction of soil properties from terrain attributes: Heterotopic cokriging and regression-kriging. *Geoderma* 67(3–4): 215–226.
16. Triantafilis, J., A.I. Huckel, and I.O.A. Odeh. 2001. Comparison of statistical prediction methods for estimating field-scale clay content using different combinations of ancillary variables. *Soil Science* 166(6): 415–427.
17. Boruvka, L., L. Mladkova, V. Penizek, O. Drabek, and R. Vasat. 2007. Forest soil acidification assessment using principal component analysis and geostatistics. *Geoderma* 140(4): 374–382.
18. Stein, A., M. Hoogerwerf, and J. Bouma. 1988. Use of soil-map delineations to improve (co-)kriging of point data on moisture deficits. *Geoderma* 43(2–3): 163–177.
19. Stein, A., and L.C.A. Corsten. 1991. Universal kriging and cokriging as a regression procedure. *Biometrics* 575–587.
20. Knotters, M., D.J. Brus, and J.H.O. Voshaar. 1995. A comparison of kriging, co-kriging and kriging combined with regression for spatial interpolation of horizon depth with censored observations. *Geoderma* 67(3–4): 227–246.

21. Deutsch, C.V., and A.G. Journel. 1992. Geostatistical software library and user's guide, p. 384.
22. Agterberg, F.P. 1984. Trend surface analysis. In: *Spatial statistics and models*. Springer, pp. 147–171.
23. Gotway, C.A., R.B. Ferguson, G.W. Hergert, and T.A. Peterson. 1996. Comparison of kriging and inverse-distance methods for mapping soil parameters. *Soil Science Society of America Journal* 60(4): 1237–1247.
24. Mueller, T.G., and F.J. Pierce. 2003. Soil carbon maps. *Soil Science Society of America Journal* 67(1): 258–267.
25. Hengl, T., G.B.M. Heuvelink, and A. Stein. 2004. A generic framework for spatial prediction of soil variables based on regression-kriging. *Geoderma* 120(1–2): 75–93.
26. Shi, W., J. Liu, Z. Du, and T. Yue. 2011. High accuracy surface modeling of soil properties based on geographic information (in Chinese). *Journal of Geographical Sciences* 66(11): 1574–1581.

Chapter 17
Simulation Analysis Platform for the Poyang Lake Basin Ecosystems

Yapeng Zhao

17.1 Introduction

17.1.1 Simulation Analysis Platform Overview

The ecological health of Poyang Lake basin maintains the ecological security in the entire basin and is an important guarantee for the sustainable economic, social and ecological development of the entire region. The comprehensive simulation and analysis of the Poyang Lake Basin environment is of particular significance for the assessment, management and management of the Poyang Lake eco-environment.

The comprehensive simulation and analysis of Poyang Lake requires comprehensive and accurate observation data and efficient and accurate simulation analysis of the data. With the continuous improvement of earth observation technologies such as satellite remote sensing, the precision and frequency of observation have been continuously improved, and large amounts of data have been acquired. These data have been exponentially growing and become "big spatial data". These data provide sufficient and accurate data sources for the analysis and simulation of the Poyang Lake geographical environment and lay a solid foundation for the comprehensive analysis and simulation of the environment. As a new model of surface modeling with high performance, HASM [1] has been developed into a comprehensive surface simulation with remote sensing data as the driving field and high-precision ground-based data as the control conditions. The analysis model can provide good simulation and analysis support for the comprehensive simulation of Poyang Lake Basin. However, the simulation analysis of Poyang Lake Basin needs to deal with large and complex spatial datasets which, on the one hand, pose challenges to the processing power of current personal computers. Subsequently, traditional processing methods

Y. Zhao (✉)
Institute of Geographic Sciences and Natural Resources
Research (CAS), Beijing, China
e-mail: zhaoyp@lreis.ac.cn

T. Yue et al. (eds.), *Chinese Water Systems*, Terrestrial Environmental Sciences,
https://doi.org/10.1007/978-3-319-97725-6_17

are not sufficient to deal with these data. On the other hand, HASM as a calculation and memory-intensive model fail to cope with the modeling of big spatial data, too. To comprehensively analyze and analyze the Poyang Lake Basin, it is necessary to establish a high-performance Poyang Lake Basin analysis and simulation platform capable of managing big spatial datasets.

17.1.2 Existing Research Basis

Two research institutes provided the prerequisites for this study. On the one hand, the Institute of Geographic Sciences and Natural Resources Research of the Chinese Academy of Sciences (IGSNRR-CAS) has in-depth experience with research on ecological and environmental protection on a domestic and international level. It has a high-performance cluster and conducted a number of important simulation experiments dealing with research questions in Jiangxi Provice. In addition, the CAS Institute of Computing Center owns the most advanced high-performance computing devices and can provide strong support for high-performance hardware devices. Hence, a successful long-term cooperation between both institutes is existing and provides beneficial conditions for to conduct this research project in terms of data foundation, cooperation experience, scientific research capability and scientific research equipment.

The project team members have experience with the HASM software since many years including scientific publications on numerous treatises on HASM and obtaining a number of invention patents. Recently, parallel algorithms for HASM and GPU high-performance algorithms have been developed and the results were published in international peer-reviewed journals [2, 3]. At the same time, the scientific visualisation techniques for HASM have been improved by scientific work published in doctoral dissertations and published papers.

The dynamic geographical representation of environmental elements is an important part of the simulation analysis platform. Based on the comprehensive analyzing and analysis of the theoretical constrains of classical surface modeling method and the development of a modern mathematical theory that overcomes these limitations, the research group of Professor Tianxiang Yue (IGSNRR-CAS) is able to transfer the geographical expression of ecological environment elements into a mathematical "surface". Based on the principle of differential geometry, a driving force field is established based on global approximation data (including remote sensing data and global model coarse-resolution simulation data) and high-precision local data (including monitoring network data and survey sampling data) as the optimal control conditions. These approach of Accuracy Surface Modeling (HASM) [4] effectively solved the long-standing problems of modeling errors in distorted surfaces and formed a rather complete theoretical system of simulation models of the Earth's surface ecological environment. Their comprehensive book "Surface Modeling: High Accuracy and High Speed Methods" was published by the Taylor & Francis Publishing Group in 2011 [5].

17.1.3 Research Content

The project aims to establish a high-performance platform for comprehensive simulation and analysis of Poyang Lake Basin in light of the large spatial data generated in the Poyang Lake Basin. A detailed and dynamic comprehensive simulation analysis of the Poyang Lake Basin is conducted to support government decision-making, researchers and environmental assessment. Therefore, based on the Poyang Lake Basin environmental data, this study uses the HASM model and to set up an efficient system simulation and analysis platform for Poyang Lake Basin. Summarizing, the research can be divided into the following four aspects:

1. The comprehensive integration of Poyang Lake Basin resources and environmental data: Environmental simulations, especially dynamic environmental simulations in high resolutions, require an as accurate, complete, real-time, multi-type data support as possible using different data sources. The large variety of large amounts of data, its processing, management and integration to the simulation system and analysis is crucial but technically challenging.
2. Construction of high performance distributed HASM model for Poyang Lake Basin for comprehensive simulation and analysis: The dynamic and advanced simulations of Poyang Lake Basin will inevitably involve a large variety of large-scale data processing and analysis steps. However, the traditional simulation analysis methods of simulation and analysis models have limited capabilities to deal with this tasks regarding large-scale data. Therefore, it is necessary to develop a high-performance distributed model. This section intends to high-precision simulation of surface Poyang Lake basin surfaces to study the establishment of a high-performance distributed integrated simulation and analysis platform.
3. Simulation analysis platform visualization: Interacting with simulation-related data and simulation results, visual analysis is an important part of the simulation and analysis platform. Visualization allows users to have a good intuitive visual experience, making it easy to understand the data and results of simulation and analysis quickly as well as to acquire the relevant knowledge. This section intends to study efficient interactive visualization capabilities and visualization based on efficient query analysis to observe and analyze simulation results.
4. Release and promotion of data products: A simulation analysis platform is created and set-up integrating database, population distribution simulation and forecasting module, land cover change monitoring module, climate change module, forest carbon storage [6] simulation module and change driving force module as well as interfaces to achieve data upload, data preprocessing, data correction, data integration, data modeling and data visualization and other functions, which will help the majority of users to use the platform and to promote its data products.

17.2 Platform Construction Program and System Design

17.2.1 System Platform to Build the Overall Framework

The goal of the "Poyang Lake Basin Ecosystem Simulation and Analysis Platform" is to establish the prototype system of ecosystem simulation analysis at the national level by using the high accuracy surface modeling method HASM, computer technology and network technology. The system platform is established based on a full integration of resources and environment data of Poyang Lake Basin. Combining with the remote sensing information model, the optimal control factors such as topography, landform and precipitation are introduced to simulate population distribution, climate change, land use and ecosystem status, trends and future scenarios. A comprehensive dynamic simulation of the Poyang Lake Basin and the regional ecological environment is conducted effectively supporting government decision-making, research and the comprehensive assessment of the environment.

The overall framework of the simulation platform as shown (Fig. 17.1). In the process of implementation, the platform uses ArcGIS, Matlab, Python, Visual Stidio and other tools to better integrate their respective advantages and finally to achieve a high-performance, interactive and easily portable simulation analysis platform (Fig. 17.2).

17.2.1.1 Function of System Modules

The Poyang Lake Basin Ecosystem Simulation and Analysis Platform integrates different types of resources and environment elements and allows a comprehensive simulation analysis of climate [7], vegetation ecosystem [8], forest carbon storage [6] and population spatial distribution [9–11].It provides the query, display, statistics, cartographic function and dynamic update of factor elements of regional environmental factors under the guidance of the principle and the concept of sustainable

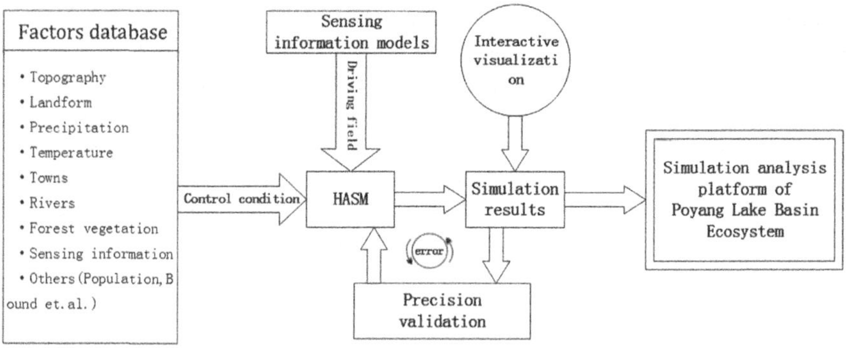

Fig. 17.1 The overall framework of the simulation platform

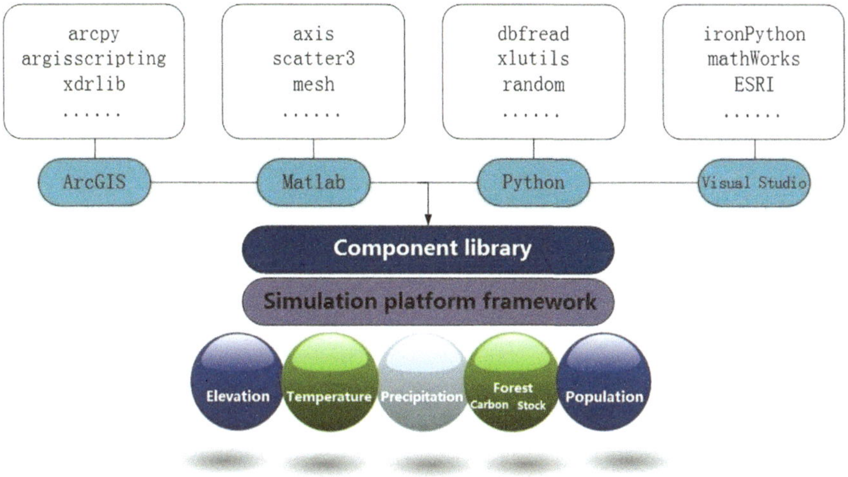

Fig. 17.2 Technical framework of the simulation platform

development. Furthermore it is a scientific basis for future ecological planning, urban and industrial development of Poyang Lake Basin under different climate change scenarios and provides data and technical support for the construction of ecological civilization in Jiangxi Province.

The Poyang Lake Basin ecosystem simulation analysis platform contains the following five functional modules:

1. system management module: features include project management, data management, thematic mapping and file browsing.
2. System management module: Features include project management, data management, thematic mapping and file browsing.
3. Climate change module: To achieve elevation, temperature, precipitation simulation analysis process. Features include technical route description, parameter settings in detail, the results of simulation analysis, data list, accuracy verification, production topics.
4. Forest carbon storage simulation module: Realize the simulation process of forest carbon storage. The functions include description of technical routes, detailed parameter setting, vegetation type simulation and production special topic.
5. Population spatial distribution simulation module: To achieve the spatial distribution of population simulation process. Features include technical route description, parameter settings in detail, the results of simulation analysis, production of topics.

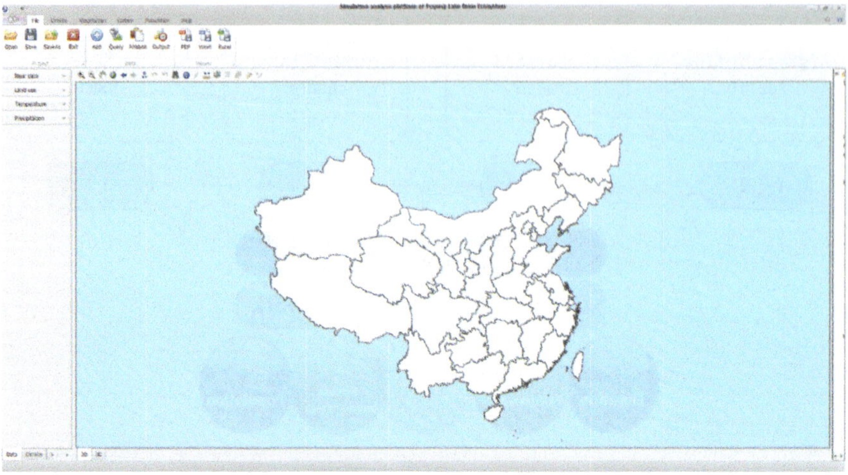

Fig. 17.3 Interface of the system management module

17.3 System Function Module Design

17.3.1 System Management Module

The features of the database management module include the control elements project file open, save, save as, exit, add data, PDF browsing, Word browsing and Excel browsing. The database management module mainly manages the thematic data of the system basic data and other modules. Thematic data and thematic maps generated by the other module functions can be displayed on the module. At the same time, the corresponding query and query and the inquiry of formulating functions can be conducted. The query results can be retrieved in form of a table to facilitate data management (Fig. 17.3).

17.3.1.1 Project Management

ArcMap project files, vector raster datasets, GPS navigation site data items and other data file structures can be opened for data display, query and other processing steps. The provided control elements are shown in Fig. 17.4.

17.3.1.2 Data Management

The data management functions include data import, editing, and deletion of the system space database (Fig. 17.5). There is a large amount of spatial information data

Fig. 17.4 Project management menu

Fig. 17.5 Data management menu

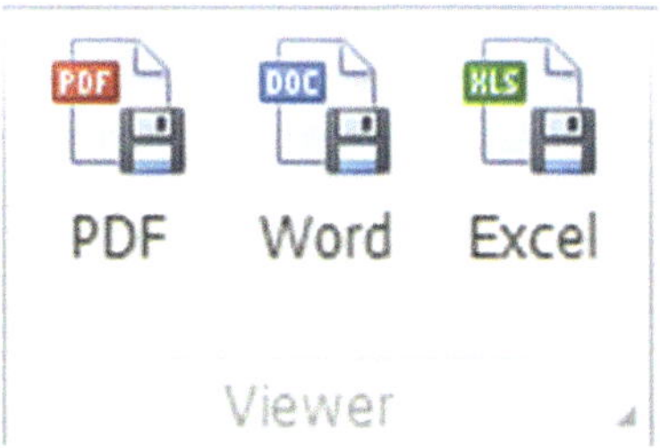

Fig. 17.6 File browse menu

in the system database, including the basic geographic data which can be management with this module. If a certain data is updated and needs to be replaced by the latest data, the user can import this data by specifiyng the data folder path, select the update data item. At the same time in the thematic data management, you can edit the datasets including data name and data attributes. The data management module allows to delete unnecessary data, too. If you delete data, the system will warn you to confirm that data deletion is irreversible. To ensure the normal operation of the system, all data management functions only provide functions being in the administrator interace.

17.3.1.3 File Browsing

The file browsing function module (Fig. 17.6) supports the pdf, word, excel file types in a user-friendly way to facilitate the understanding of the corresponding technical documents. After the user selects the appropriate browsing menu, the selected file information can be displayed on the layer.

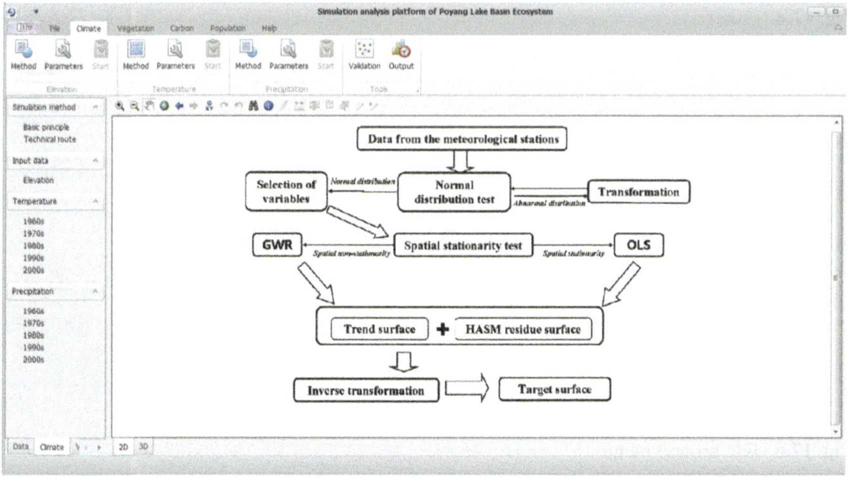

Fig. 17.7 Climate change module interface

17.3.2 Climate Change Module

Climate change module employs data on elevation, temperature and precipitation simulation withtin the analysis process. Features of the module include: Technical route description, parameter settings in detail, the results of simulation analysis, accuracy verification, production topics. The detailed operating parameters of the HASM model with high accuracy are set based on the altitude of the Poyang Lake basin, the temperature data of the stations and precipitation data to simulate and analyze the spatial distribution of temperature and precipitation in the area and finally show the original data and simulation results in the map display.

The main inferface of the climate change module main interface is shown in Fig. 17.7. The climate change module is divided into elevation, temperature, precipitation, tools and other four sub-modules. The left panel provides technical descriptions of the available datasets.

17.3.2.1 Elevation Simulation Analysis

The elevation simulation analysis module consists of the three control elements "method", "parameters setting" and "start" to configure the elevation interpolation scheme (Fig. 17.8). Firstly, click on "Parameters", enter the detailed parameters required for HASM operation and soon after pressing "Start" the final simulation results will appear (Fig. 17.9).

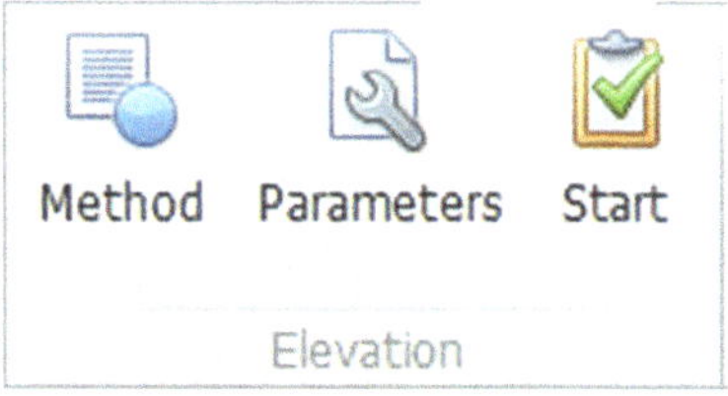

Fig. 17.8 DEM simulation analysis menu

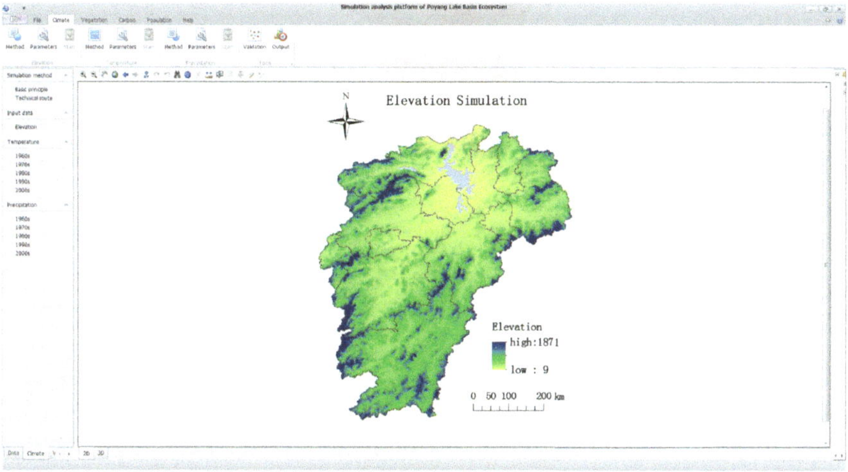

Fig. 17.9 Visualisation of the elevation simulation

Fig. 17.10 Temperature simulation analysis menu

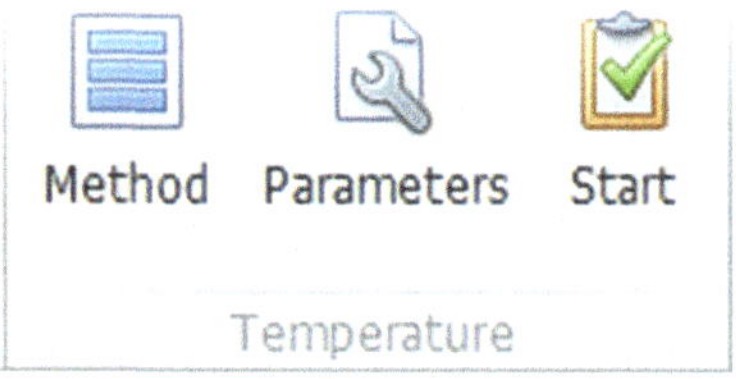

17.3.2.2 Temperature Simulation Analysis

The temperature simulation analysis module includes three control elements (Fig. 17.10) to choose the simulation method as well as the corresponding parameter settings to simulate temperatures (Fig. 17.11). Click the "Parameters", enter the detailed parameters required for HASM operation, click on the "Start" will appear after the final simulation results.

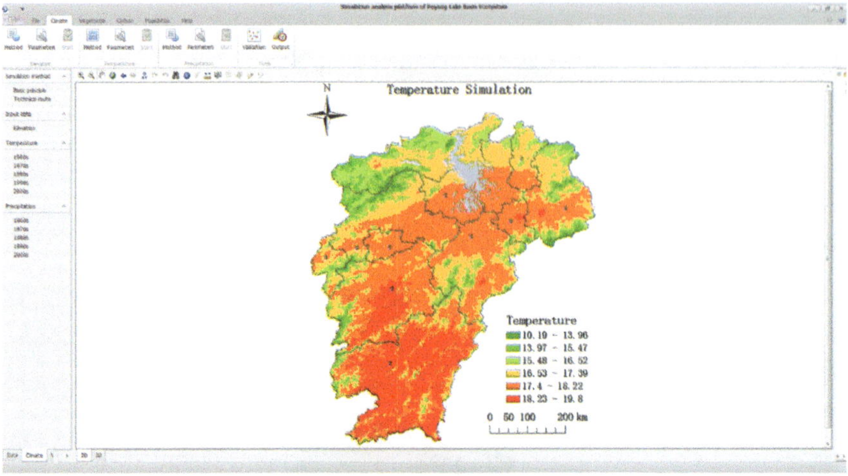

Fig. 17.11 Visualisation of the temperature simulation

Fig. 17.12 Precipitation simulation analysis menu

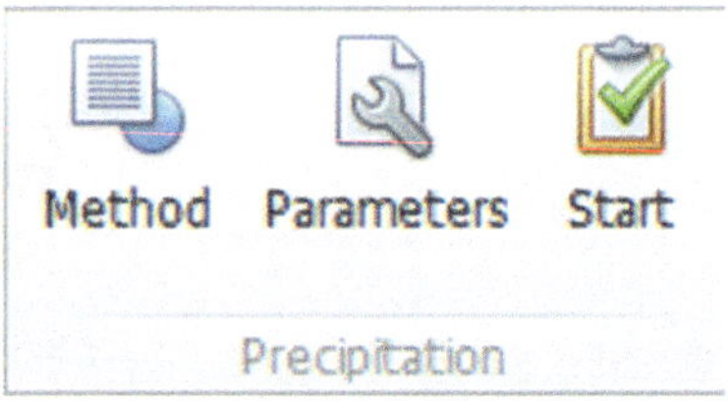

17.3.2.3 Precipitation Simulation Analysis

The precipitation simulation analysis module includes several control elements (Fig. 17.12) to choose the simulation method as well as the corresponding parameter settings for simulating spatial-temporal precipitation patterns (Fig. 17.13). Click the "Parameters", enter the detailed parameters required for HASM operation, click "Start" will appear after the final simulation results.

17.3.3 Vegetation Ecosystem Module

The vegetation ecosystem module mainly implements the vegetation type simulation process. The functions include technical route description, detailed parameter setting and vegetation type simulation. The control elements of the vegetation ecosystem modules include options for vegetation type simulation methods, parameter settings and output analysis (Fig. 17.14).

Based on the temperature data and precipitation data in Poyang Lake Basin, the detailed operating parameters of HASM model with high precision are set to

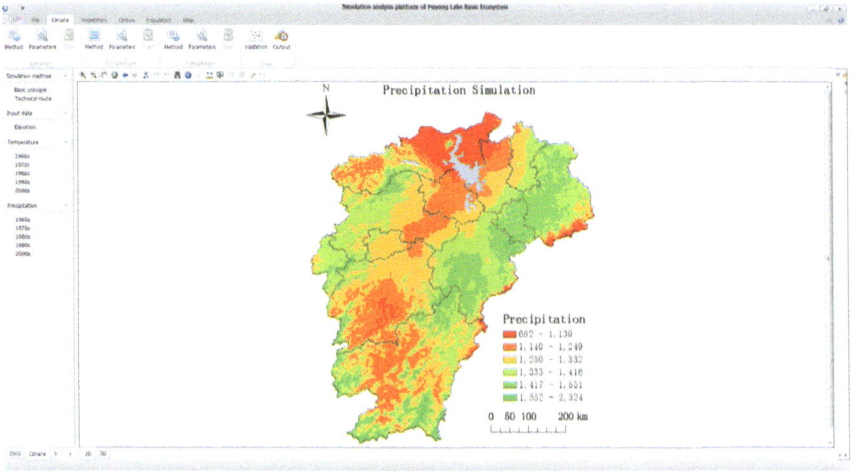

Fig. 17.13 Simulation of precipitation—precipitation simulation

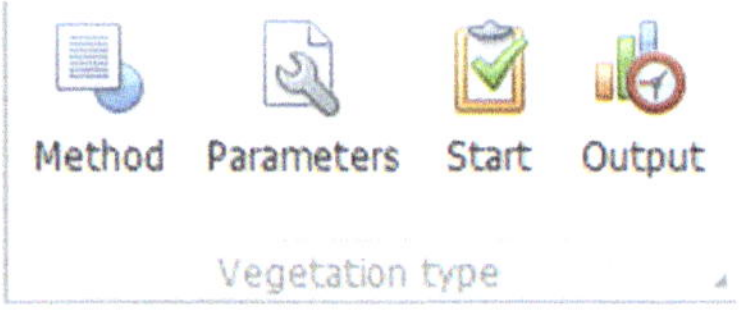

Fig. 17.14 Vegetation ecosystem menu

simulate and analyze the distribution of vegetation types in the region and display the simulation result data in the map display area (Fig. 17.15). Click the "Parameters", enter the detailed parameters required for HASM operation, click "Start" will appear after the final simulation results.

17.3.4 Forest Carbon Stock Simulation Module

The forest carbon stock module mainly implements the forest carbon stock simulation process. The functions of the forest carbon stock module include technical route description, detailed parameter setting and vegetation type simulation. The control elements allow to choose the carbon stock simulation method and parameter settings, start the carbon stock simulation and process the output (Fig. 17.16). The detailed operating parameters of high-precision HASM model are set up based on forest inventory data and remote sensing information model data of Poyang Lake Basin to simulate and analyze the spatial distribution of forest carbon stock in the region and display the simulation results data in the map display (Fig. 17.17).

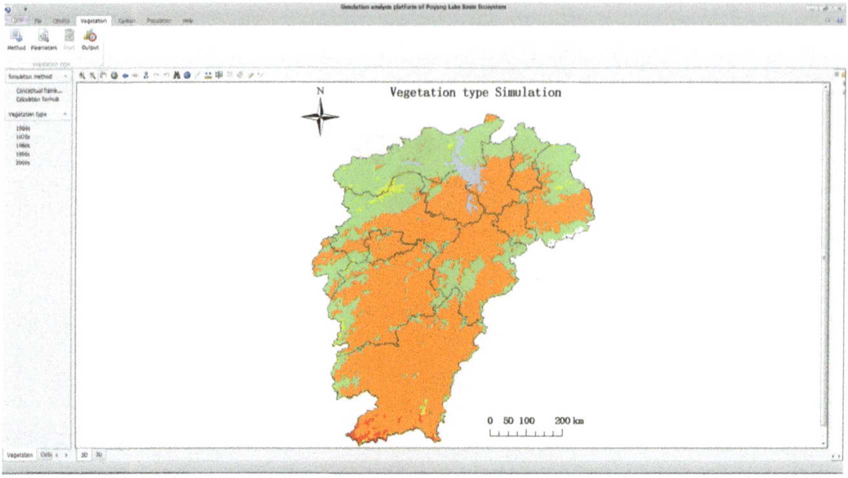

Fig. 17.15 Visualisation of the vegetation ecosystem simulation

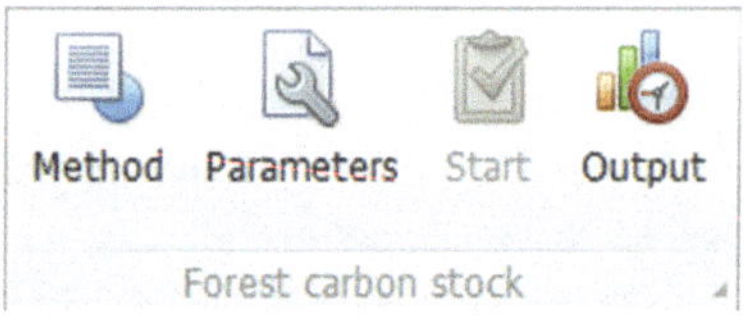

Fig. 17.16 Forest Carbon stock simulation menu

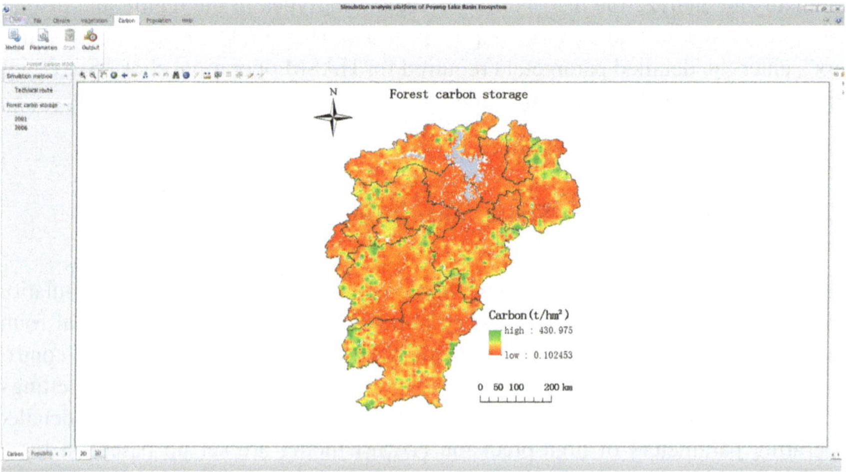

Fig. 17.17 Simulation of forest carbon stock—carbon stock simulation

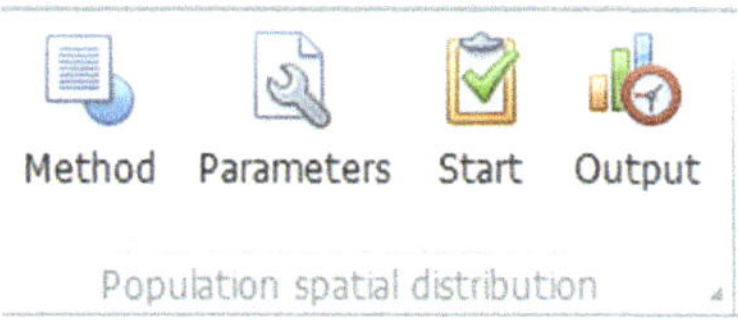

Fig. 17.18 Population spatial distribution simulation menu

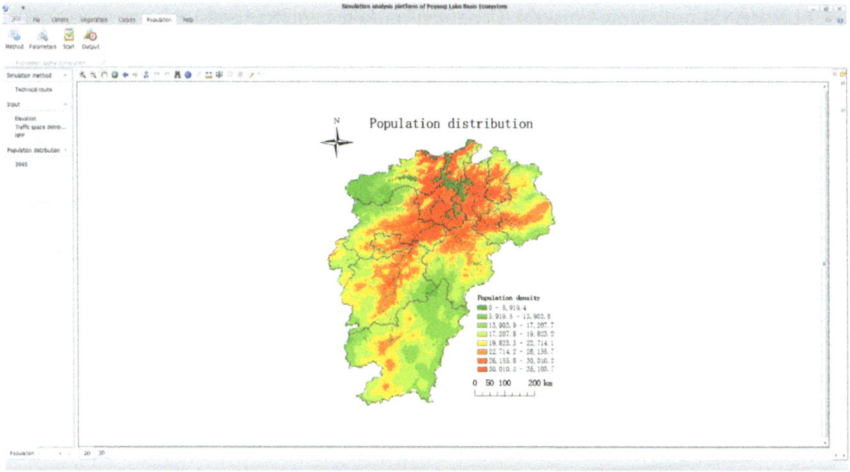

Fig. 17.19 Visualisation of the population spatial distribution simulation

17.3.5 Population Spatial Distribution Simulation Module

The module of population spatial distribution simulation mainly realizes the simulation process of population spatial distribution. Its functions include a description of technical routes, detailed setting of parameters, simulation and analysis of results as well as production of maps (Fig. 17.18). Based on the elevation of Poyang Lake basin, NPP data of NPP, traffic space distribution data, urban distribution data and water distribution data, the detailed operating parameters of HASM model with high precision are set up to simulate and analyze the spatial distribution of population, Finally, the simulation result data is displayed as maps in the simulation platform system (Fig. 17.19).

References

1. Yue, T., and Z. Du. 2006. Comparison and analysis of errors between high accuracy surface modelling and classical models (in Chinese). *Process in Natural Science* 16 (8): 986–991.
2. Zhao, N. 2014. *Robustness of high accuracy surface modelling (HASM) (in Chinese).*

3. Yan, C. 2012. *Surface simulation and application based on high accuracy surface modelling (HASM) (in Chinese)*.
4. Yue, T., et al. 2015. A review of recent developments in HASM (in Chinese). *Environmental Earth Sciences* 74 (8): 6541–6549.
5. Yue, T., Z. Fan, C. Chen, X. Sun, and B. Li. 2011. Surface modelling of global terrestrial ecosystems under three climate change scenarios. *Ecological Modelling* 222 (14): 2342–2361.
6. Wang, Y. 2016. *High accuracy surface modelling for forest carbon stock simulation in China (in Chinese)*. PhD thesis., University of Chinese Academy of Sciences.
7. Fan, Z., T. Yue, and Y. Song. 2009. Hanging trends of temperature and precipitation based on YUE-HASM method (in Chinese). *Geographical Research* 28 (3): 643–652.
8. Du, Z., Z. Fan, and T. Yue. 2013. Simulation and analysis of provincial scale climate and vegetation ecosystem based on HASM (in Chinese). *Geographical Research* 32 (9): 1591–1601.
9. Yue, T., Y. Wang, Q. Zhang, N. Liu, and W. Tang. 2012. The future scenario simulation analysis of population spatial distribution in Beijing (in Chinese). *Geo-Information Science* 10 (4): 479–488.
10. Wang, Y. 2006. *Surface modeling method for digital simulation of Chinese population spatial distribution (in Chinese)*.
11. Liu, J., et al. 2003. Digital simulation of population density in China (in Chinese). *Acta Ecologica Sinica* 28 (1): 17–24.

Chapter 18
Virtual Geographical Environment-Based Environmental Information System for Poyang Lake Basin

Changqing Yan, Karsten Rink, Lars Bilke, Erik Nixdorf, Tianxiang Yue and Olaf Kolditz

18.1 Introduction

With the rapid development of remote sensing, and earth observation technology, increasingly huge amount of geospatial data became available. These data enables the monitoring and evaluation of the environment and provides data source for numerous environment models, which in turn generate huge simulation results data. Utilization of both, remote sensing data and simulation results will be of great value for environmental related researchers, the public and local governments to support relevant research and environmental related decision- making processes. With these data, the government could establish the scientific rules for the exploitation and utilization of resources, to maintain the sustainable development of local economy and society. However, these data are heterogeneous and complex, exist in a variety of formats and is difficult to understand and analyze. Therefore, it is essential to build a system, which can integrate these data in a unified context, combine the relevant data sets, and present these data in an intuitive way.

In the field of hydrology, there is a great demand for those systems, since water pollution has been increased severely with the rapid development of Chinese economy in recent years, which threatens the health and wellbeing of the local people. In this context, an environmental information system (EIS) will support the government

C. Yan (✉)
Shandong University of Science and Technology, Taian, China
e-mail: yancq@lreis.ac.cn

C. Yan · K. Rink · L. Bilke · E. Nixdorf · O. Kolditz
Helmholtz Centre for Environmental Research, Leipzig, Germany

T. Yue
Institute of Geographic Sciences and Natural Resources Research (CAS), Beijing, China

O. Kolditz
Technical University Dresden, Dresden, Germany

T. Yue et al. (eds.), *Chinese Water Systems*, Terrestrial Environmental Sciences,
https://doi.org/10.1007/978-3-319-97725-6_18

to make scientific decisions on the environmental management to protect the water from pollution on the sound base.

Just as aforementioned in previous chapters, Poyang Lake, the largest fresh water lake of China, has been increasingly polluted and the ecosystem in the area of Poyang Lake decreased obviously in recent years. Aside of a population of approximately 39.66 million people living in the Poyang Lake Basin, it is further the home to more than a half million migratory birds and is a favorite destination for birding. It plays a significant role for the ecological security in China. Therefore, it is rather crucial to ensure that the environmental management of Poyang Lake is based on scientific results and to carry out effective research itself.

However, building an environmental information system is never trivial, which involves challenges from both data processing and information presentations, and is cross-discipline. Utilizing a virtual geographic environment (VGE) provides an opportunity to build such a system, which can visualize geospatial data or geographic process in a vivid manner in a realistic 3D virtual environment [1].

Therefore, in this chapter, a VGE based environmental information system named the Poyang Lake Basin environmental information system (PyEis) is developed for Poyang Lake Basin and Poyang Lake.

This system does not only support the decision-making processes in environmental management, but also provides a rich-information platform for relevant sciences. It can integrate heterogeneous data and simulation results, such as lot lake level time series data changes and combine them in a unified context to present all the information in an intuitive manner.

18.2 Framework

Building such an EIS is challenging, as not only technical challenges such as the file format conversion involved in heterogeneous data sets processing need to be addressed, but also conceptual challenges about choosing a proper way to visualize a certain data set needs to be solved so that the users have an intuitive feeling on approaching the system. Undoubtedly, realistic 3D visualization allows the users an immersive view on data, which helps to provide the concerned information in an understandable way. Hence, a virtual geographic environment presenting geographic environmental process or phenomena is an effective tool to achieve this. Nevertheless, it still needs numerous efforts to build a virtual geographic environment based environmental information system and integrate the heterogeneous datasets.

To add heterogeneous data sets into a real-world scene and build such an environmental information system, we developed an overall framework for data integration and an outline for EIS building, which is illustrated in Figs. 18.1 and 18.2. This system building and data integration framework combine techniques from the software packages ArcGIS, OpenGeoSys Data Explorer, Paraview, MiddleVR and Unity.

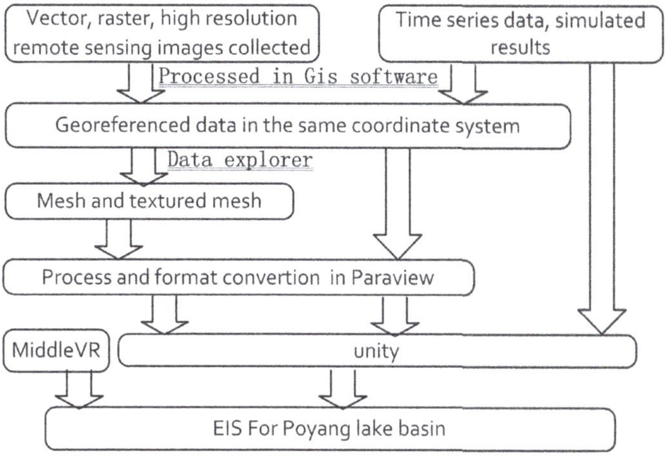

Fig. 18.1 Framework for Poyang Lake Basin EIS

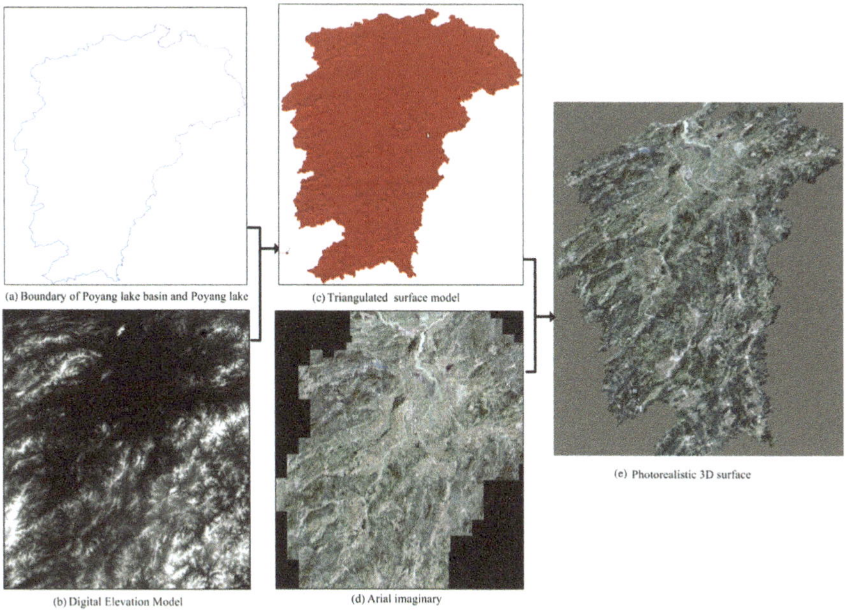

Fig. 18.2 The outline for data sets integration

18.2.1 Relevant Software

The OpenGeoSys Data Explorer is used for the detailed visualization of geoscientific data in a 3D space [2], which acts as the graphical interface of the interdisciplinary software project OpenGeoSys (OGS) originally developed for the simulation of THMC processes within fractured-porous media [3]. It supports a variety of formats for geoscientific data sets, such as vector data, raster textures (e.g. DEM data) and satellite imagery. In terms of data integration, the OpenGeoSys Data Explorer was applied to export the relevant datasets as vtk data, which is supported as import file format for Paraview.

Paraview is an open-source software for data analysis and visualization, which allows quick creations of visualizations and visualization parameters modification in the VR environment. Combined with the custom data conversion tools [4], it can create a satisfying scenery and export it in an adequate format.

Unity is a cross-platform game engine developed by Unity Technologies, which can be used to create a broad range of 2D and 3D virtual environments for game development and other 3D applications that are deployable to the web, game consoles, mobile devices and personal computers. However, the basic scope of operation of Unity does not include any interaction functionality [4], although it can be implemented by scripting. Combining unity and developed presentation and interaction techniques, a highly immersive geographic environment based environmental information system can be built with a high efficiency.

MiddleVR is a generic virtual reality middleware designed to work with different 3D applications, which provides a graphical configuration tool to set up VR systems independently of specific software. In addition, MiddleVR implements interaction devices, stereoscopic rendering and clustering even if the used software does not support this functionality. Integrated in Unity as a plugin, MiddleVR enables all these features in unity, which facilitates the development of an interactive 3D application.

It is worth noting that geographic information system (GIS) software is indispensible in the system building. All the relevant geographic operations should be accomplished by using GIS software, such as ArcGIS. Applying all these technologies, we can integrate the heterogeneous data sets in an understandable and immersive way as shown exemplarily in Fig. 18.2.

18.2.2 Workflow for EIS Building and Data Integration

Prior to the begin of system building, the required heterogeneous data sets should be collected and prepared, which means that a variety of formats can't be used directly. Most of these data are geospatial data, which may have different coordinate systems. Hence, geoprocessing such as format conversion, coordinate system conversion, projection and interpolation are needed which can be carried out using a geographic information system (GIS) software, such as ArcGIS and QGIS. Time series data generated by the model may need to be convert into numerous images which are labelled by the time step each image refers to.

After finishing the geoprocessing, the georeferenced geospatial data sets are imported into the aforementioned software OGS data explorer. We create a triangular mesh, map the mesh with the data of the digital elevation model, texture the mesh and export the file as vtk, which is supported by Paraview (Fig. 18.2). The next step is to import the generated vtk file into Paraview, load the texture, and check if it is correctly textured. If it is correct, export it in a format named fbx by employing a plugin to export the scene to fbx format, which was developed by UFZ [5]. The final step is to import the fbx file into unity, make some necessary visual establishment, establish the parameters for time series data, and define the viewpoints and camera paths. Combined with MiddleVR, menu options can be set-up to make the system more user-friendly. If everything works as expected, the executable system can be generated. This workflow is open and scalable to allow an ongoing integration of prepared data sets. The workflow can be illustrated by the following example (Fig. 18.1).

Regarding the data sets integration, the process described in Fig. 18.2 can be explained as follows. After preparing the georeferenced Poyang Lake Basin boundary by ArcGIS software, it was imported into the OpenGeoSys Data Explorer for mesh generation. Then the mesh was mapped by the DEM covering the whole Poyang Lake Basin area, textured by the high resolution imaginary from Google Earth, and finally imported for realistic presentation in Unity. With these two outlines, we can build our EIS for Poyang Lake Basin and integrate the available data sets.

18.3 Implementation and Results of the EIS

The current system can be divided into 6 parts, including the digital elevation model textured by Google Earth images for the whole Poyang Lake Basin, the river network of Poyang Lake Basin, the seasonal lake water level dynamics, the Poyang Lake level changes at daily frequency, climate change data from 1961 to 2010 including precipitation and temperature as well as data from the ecological system of the basin.

18.3.1 Digital Elevation Model Textured by Google Earth Images

Poyang Lake Basin spans from the bank of the middle reaches of Yangtze River in the north to mountainous areas in the south and east. Mountains surround Poyang Lake Basin on three sides, with the Mufu Mountains, Jiuling Mountains, and Luoxiao Mountains on the west; Huaiyu Mountains and Wuyi Mountains on the east; and the

Fig. 18.3 Digital elevation model of Poyang Lake Basin

Jiulian Mountains and Dayu Mountains in the south. The mountains and the lake form the rather complex topography of Poyang Lake Basin.

The dataset provided by the SRTM satellite mission was used to extract the digital elevation model of Poyang Lake Basin with a resolution of 90 m (Fig. 18.3). According to the aforementioned outlines, the boundary data was used to generate the mesh, then the mesh was mapped by the DEM, finally the mapped DEM was textured by high-resolution images from Google Earth.

The finite element mesh for Poyang Lake Basin is composed of 2,076,772 elements of which 2,068,153 are triangles. The mesh of Poyang Lake itself comprises 179,718 elements (172,008 triangles). Both meshes were overlaid to verify whether the georeferencing was conducted correctly (Fig. 18.4). With respect to the high resolution satellite image provided by Google Earth, we had to download it block by block, then mosaic the images into one and georeference the final mosaick image. This georeferenced image acts as the texture of the Poyang Lake Basin mesh.

After successfully accomplishing the aforementioned steps using the OGS data explorer, the datasets were further processed in ParaView before importing them to the Unity environment. Figure 18.5 shows the visualization of the textured DEM

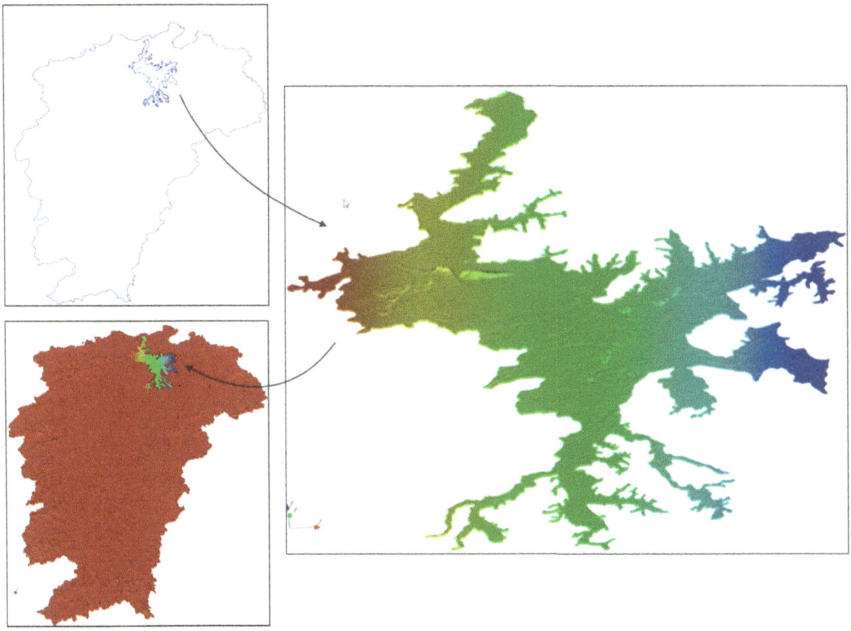

Fig. 18.4 Mesh of Poyang Lake Basin and Poyang Lake

Fig. 18.5 Topography of Poyang Lake Basin visualized by Unity

and the DEM with no texture dataset in Unity respectively (see also Fig. 18.2c, d). The visualized mapped DEM demonstrates that the middle area of the catchment is relatively flat and has a low altitude whereas it is more hilly and mountainous at the other sides.

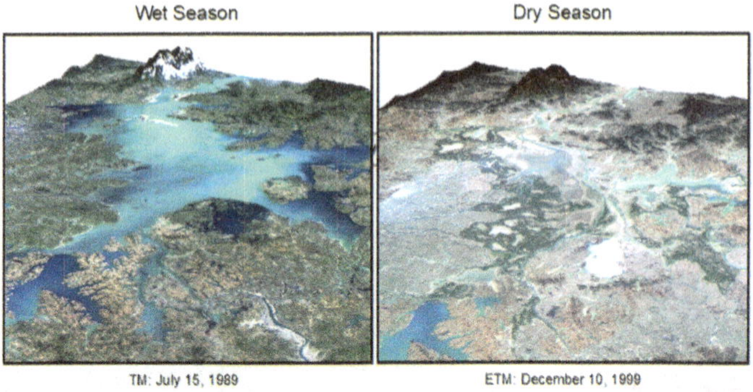

Fig. 18.6 Water level during wet (left) and dry season (right) in 1999

18.3.2 Water Level Dynamics of Poyang Water Level

The area of Poyang Lake changes significantly in every year between the wet and dry season.[1] In recent years the average size of the lake decreased significantly (Fig. 18.6). In a normal year the area of the lake reaches 3,500 km^2. In early 2012, due to drought, sand quarrying, and the practice of storing water at the Three Gorges Dam, the area of the lake surface reached a minimun area of only about 200 km^2.

18.3.2.1 Annual Water Level Changes

To investigate annual dynamics of the water level surface extension of Poyang Lake, wetland land cover types including the open water surface were extracted from a 15 years record of remote sensing data between 2000 and 2014 (Fig. 18.7), which, in general, was a period with comparatively dry climate in the basin [6].

To integrate these data, the coordinate system needs to be converted to be similar with the generated mesh. Thereafter, these data was used to texture the mesh and the textured mesh was exported as vtk files. After processed in Paraview and Unity using relevant plugins for conversion, the time series data of the land cover can be visualized as animation in our EIS (Figs. 18.8 and 18.9).

Figure 18.8 are the animation screenshots of time series annual change data sets at Poyang Lake. From these pictures, the spatiotemporal changes of Poyang Lake can be observed easily. The distribution of land cover in Fig. 18.8 clearly shows that the blue water area decreased notably during the dry season, while the share of other areas increased. Figure 18.9 is the consecutive enlarged views for annual changes of land cover in the Poyang Lake area. The enlargement of view in unity supports a better visualisation of the changes in Poyang Lake. Figure 18.9 exhibites a clear land use distribution change in a near sight. It can be observed that the water area in blue

[1] http://threegorgesdamfromspace.blogspot.de/p/poyang-lake.html.

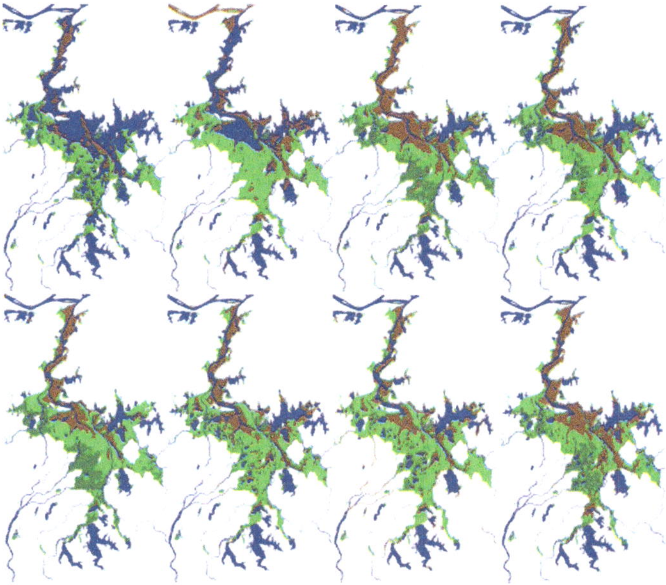

Fig. 18.7 Dynamic changes of Poyang Lake from year 2000 to 2014, showing even years only

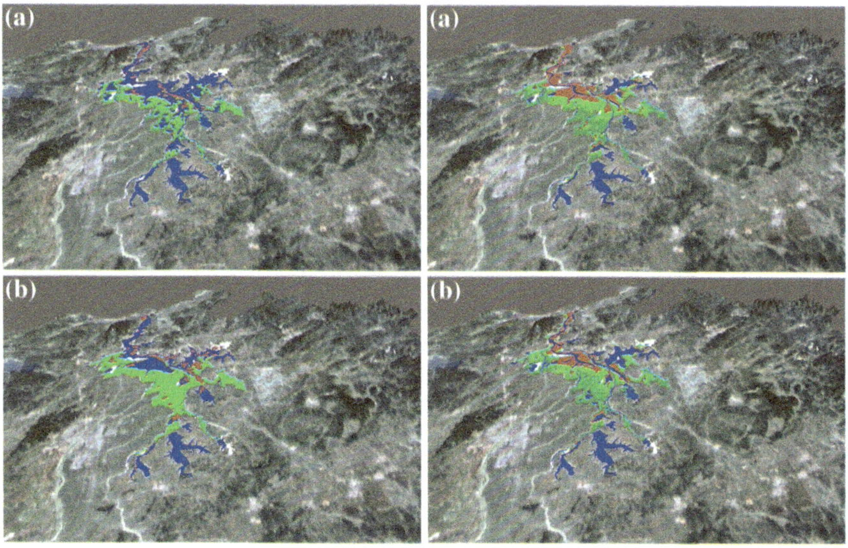

Fig. 18.8 Left: Animation No.1 of dynamic water level changes in Poyang. **a** Is the 0th frame of the animation, and **b** for the 5th frame (for further explanations see running text). Right: Poyang Lake dynamic changes animation 2, **a** is the view of the 10th frame and **b** is the 14th frame

Fig. 18.9 Poyang Lake dynamic changes animation

decreased and the mud flat area in yellow increased. This trend helps the authorities to make decisions on a scientitic base.

All the scenes above are presented with lighting. To observe the role of light on the scene, the lighting of the scene was switched off to obtain the Figs. 18.10 and 18.11.

It can be observed that realistic scenes can still be generated without lighting. Although the style scenes differ with the level of lighting, the results remain comparable. For instance, Fig. 18.11 shows a realistic and complex topography by presenting scenes on the lake and the mountainous areas of the Poyang Lake Basin. These scenes also exhibit highly realistic views as those with lighting on.

18.3.2.2 Daily Water Level Dynamics

A high resolution time series data set with temporal resolution of two days and a spatial resolution of 2000 m was obtained from a numerical hydrological model [7]. This simulation period was from June 13, 2006 to June 28, 2007. The simulation results from the model were provided as a large ASCII file with a size of over 4 GB, which was split into 184 files for each time step. After processing the files using GIS, they were integrated into the EIS. The Figs. 18.12 and 18.13 show the

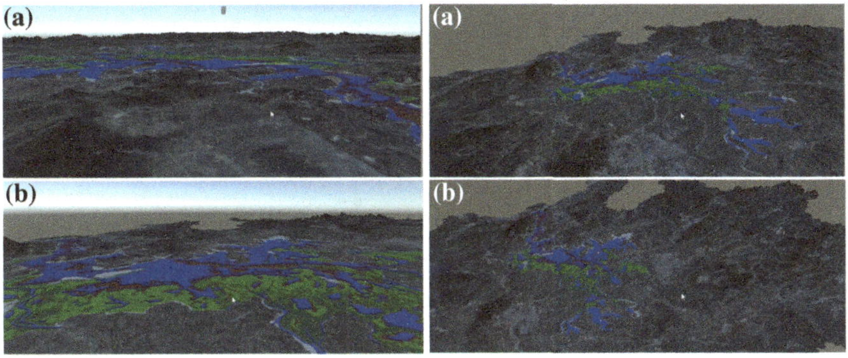

Fig. 18.10 Left: Unlighted view of Poyang Lake, **a** is a scene from the north of the lake and **b** shows the southeastern part. Right: Unlighted views of Poyang Lake, **a** shows the scene from an area with higher elevation in the south-west of the lake. **b** Shows the lake from a central-top view

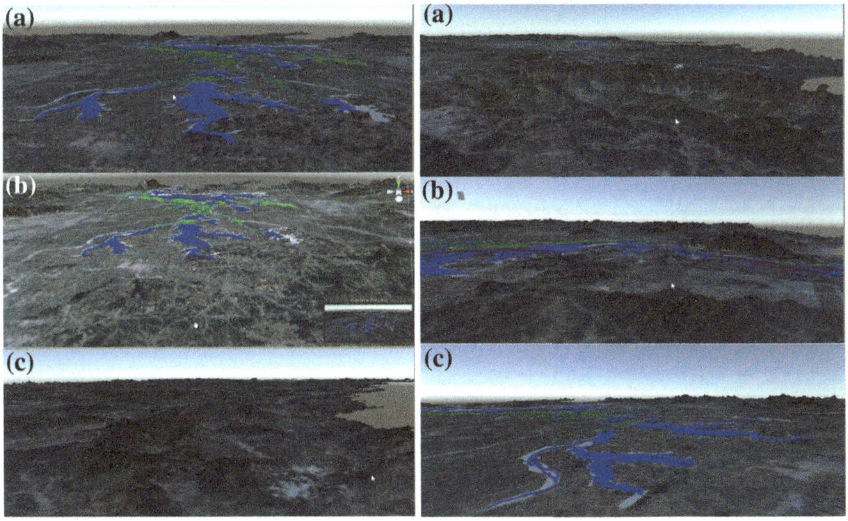

Fig. 18.11 Left: Unlighted view of the basin. **a** and **b** are scenes of the lake in different distances while **c** is a scene from the northern part of the Poyang Lake Basin. Right: South viewpoint of Poyang Lake **a**, **b** and **c** show scenes of the lake from varying distances from the west and north part of the lake

simulation results of every 30th time step of the simulation. The chosen color scheme colors water areas with shallow water depth red and lake areas with deep water depth blue.

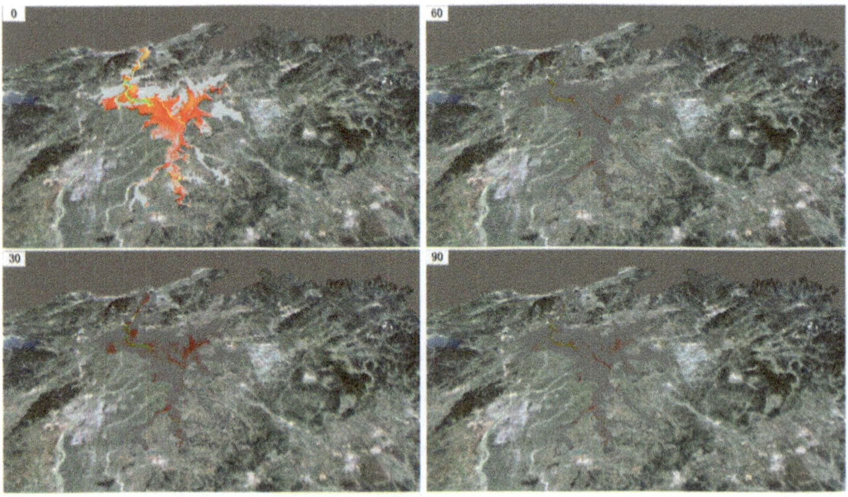

Fig. 18.12 Left: Daily changes of Poyang Lake water level. The scenes show the 0 and 30th time step of the simulation from a top-left view. Right: Daily changes of Poyang Lake water level. The scenes show the 60 and 90th time step of the simulation from a top-left view

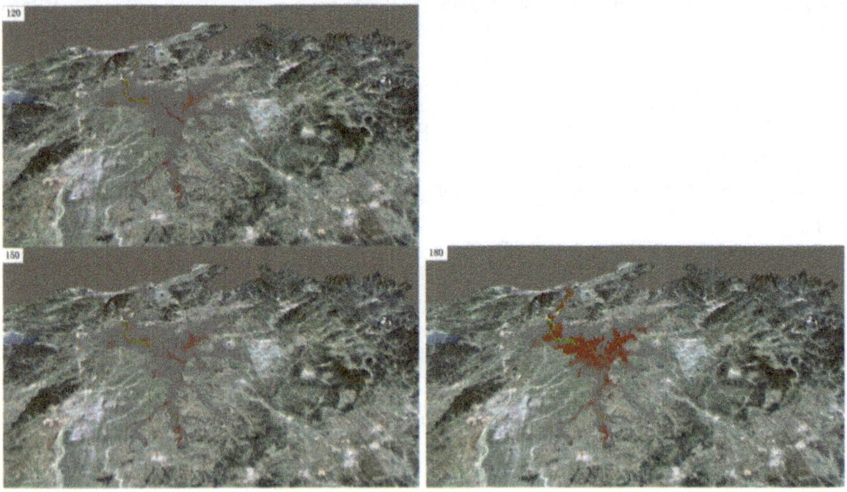

Fig. 18.13 Left: Daily changes of Poyang Lake water level. The scenes show the 120 and 150th time step of the simulation from a top-left view. Right: Daily changes of Poyang Lake water level. The scene shows the 180th time step of the simulation from a top-left view

18.3.3 River Network

The river network was extracted from the HydroSheds database, which provides hydrographic information as raster files based on the SRTM DEM dataset. The river network was overlaid on the 3D visualization scene using different colors to show the distribution of rivers in the Poyang Lake Basin. It should be noticed that replacing the current line-structure of the river network with a more realistic representation of the rivers could be a better solution for visual effect, which will be addressed in future work.

18.3.4 Climate Datasets

Climate data sets, including precipitation, temperature, and ecology system were generated employing a surface modeling method named HASM [8]. High accuracy surface modeling (HASM) method is a novel method for surface modeling from scattered points, which has been proved to achieve higher accuracy than classical methods such as kriging, spline and idw. HASM has been utilized for a variety of surfaces modeling in geosciences due to its high accuracy.

HASM was applied to generate the surfaces of climate data sets from meterological stations, which thereafter were integrated in the EIS. These data sets are all time series data ranging from the year 1961 to 2010. To visualize these data sets, just as aforementioned outline for data integration, all data were georeferenced to the coordinate system used for the EIS, exported as texture images, and then integrated as textures for presentation in the EIS. Each texture represents the data distribution for a single year. It should be noticed that all the texture images should be named properly. The name of each image is divided into two parts, the first part is the same for all the images representing the same parameter and some meaningful labels such as "texture" and "precipitation" are used. The second part of the image name consists of a number representing the chronological order. This labelling allows to show the time series data dynamically in the EIS, by pressing the relevant button item.

18.3.4.1 Ecosystem

The distribution of the ecology system quality is presented in Fig. 18.14(left). A green color represents a high ecological value, while a red color is associated with a low parameter value. The two scenes are taken from a west view and a central lake view. It can be observed that the area with a higher altitude has a lower ecology value than the plain areas of the basin.

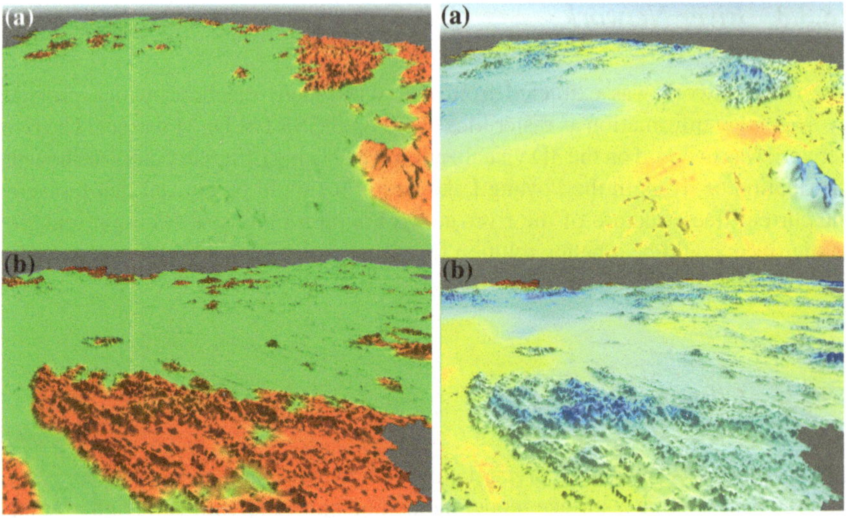

Fig. 18.14 Left: Visualization of the ecology system. **a** Shows a scene from the lake view and **b** is a view from the west of the Basin. Right: Precipitation visualization. **a** Shows a scene from the lake view and **b** is a view from the west of the Basin

18.3.4.2 Precipitation and Temperature

The precipitation distribution is visualized in Fig. 18.14(middle). Blue and yellow color represent high and low precipitation, respectively. It can be observed that the mountainous areas receive more precipitation than the lake itself.

Figure 18.15 shows the variation of mean temperatures across Poyang Lake Basin area. Blue color denotes areas with lower temperature and the red color those with higher temperature. The visualization results indicate that the mountainous area has relatively low temperature, while the lake area and other areas with relatively low elevation have higher average temperature.

18.4 Discussion and Conclusions

We built a virtual geographic environment based environmental information system for Poyang Lake Basin. Plenty of different environmental information including realistic topography, river network, time-dependent (daily and annual) data on water level and water depth changes of Poyang Lake and climate data were integrated into the environmental information system. This system provides required information to researchers and the government in an intuitive way, which supports decision-making based on scientific facts. Additionally, an important feature of our system is its high

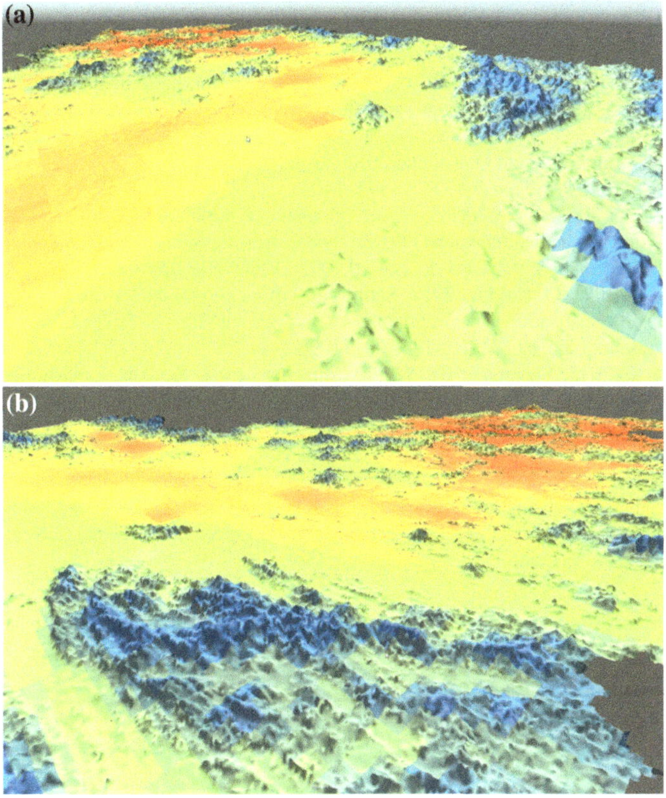

Fig. 18.15 Temperature visualization. **a** Shows a scene from the lake view and **b** is a view from the west of the Basin

scalability which allows the ongoing integration of more environmental data sets and modeling results.

Although the system shows a good performance and provides a rich-information platform, it should be noticed that there remain problems which need a further improvement. Due to the large scale of our research area, the resolution of texture imaginary is not sufficiently high. This could be solved by adopting high resolution imaginary data from satellites such as Sentinel or RapidEye satellite data in the future work. Furthermore, unlimited scale realistic area visualization will also be our research focus in the future.

References

1. Chen, M., H. Lin, O. Kolditz, and C. Chen. 2015. Developing dynamic virtual geographic environments (VGEs) for geographic research. *Environmental Earth Sciences* 10 (74): 6975–6980.
2. Rink, K., T. Kalbacher, and O. Kolditz. 2012. Visual data exploration for hydrological analysis. *Environmental Earth Sciences* 65 (5): 1395–1403.
3. Kolditz, O., et al. 2012. OpenGeoSys: An open-source initiative for numerical simulation of thermo-hydro-mechanical/chemical (THM/C) processes in porous media. *Environmental Earth Sciences* 67 (2): 589–599. https://doi.org/10.1007/s12665-012-1546-x.
4. Rink, K., L. Bilke, O. Kolditz. 2017. Setting up virtual geographic environments in unity. In Workshop on visualisation in environmental sciences (EnvirVis), vol. 5. The Eurographics Association https://doi.org/10.2312/envirvis.20171096.
5. Bilke, Lars. (2015, September 3). VtkFbxConverter: 1.0.2. Zenodo. http://doi.org/10.5281/zenodo.30345.
6. Chen, L. 2014. Dynamic monitoring of Poyang Lake by spatio temporal and spatial-spectral blending models (in Chinese). PhD thesis. Beijing Normal University.
7. Du, Y., et al. 2018. Modeling of water quality evolution and response with the hydrological regime changes in Poyang Lake. *Environmental Earth Sciences* 77: 265.
8. Yue, T. 2011. Surface modeling: High accuracy and high speed methods, vol. 2. New York: CRC Press. ISBN: 1439817588.

Part VI
Sino-German Research Centre

Chapter 19
Research Centre for Environmental Information Science (RCEIS)

Cui Chen, Carsten Montzka, Juliane Huth, Claudia Kuenzer, Harald Kunstmann, TianXiang Yue and Olaf Kolditz

19.1 Concept and Approach

The Research Centre for Environmental Information Science (RCEIS) was established in March 2014. In the long-term vision RCEIS shall become a Sino-German competence centre and research platform for Earth systems observation and modelling by combining expertise in the fields of environmental and information sciences using modern information technology. As a scientific goal, RCEIS will develop novel concepts by using infrastructures for Earth system observation and analysis in order to better understand the evolution and dynamics of environmental systems under global change and globalization (anthropogenic pressure). An important methodology is the further development and combination of different infrastructures for Earth system observation, e.g. combining information from remote sensing (satellites and airborne) and "ground-truth" measurements. More-

C. Chen (✉) · O. Kolditz
Helmholtz Centre for Environmental Research, Leipzig, Germany
e-mail: cui.chen@online.de

C. Montzka
Forschungszentrum Juelich, Jülich, Germany

J. Huth · C. Kuenzer
German Aerospace Center (DLR), German Remote Sensing Data Center (DFD), Oberpfaffenhofen, Germany

H. Kunstmann
Karlsruhe Institute of Technology, Karlsruhe, Germany

T. X. Yue
Institute for Geographical Sciences and Natural Resources Research, Chinese Academy of Sciences, Beijing, China

O. Kolditz
Technical University Dresden, Dresden, Germany

T. Yue et al. (eds.), *Chinese Water Systems*, Terrestrial Environmental Sciences,
https://doi.org/10.1007/978-3-319-97725-6_19

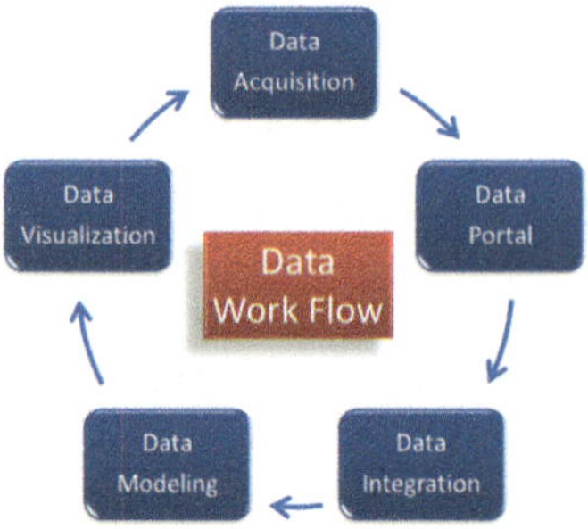

Fig. 19.1 RCEIS workflow concept

over, monitoring (remote sensing and monitoring) and modelling platforms (multi-sphere/compartment, cycle-oriented) need to be combined (Fig. 19.1).

A large variety of tools for data processing and environmental modelling is available from different sources as commercial, scientific or open-source software. For optimal and efficient use of acquired data provided by novel observation and exploration strategies, linking those tools is a challenging task.

For this purpose, the next generation of data and modelling tools—Environmental Information Systems (EIS)—must be embedded into entire continuous workflows, incorporating data acquisition, data management, integrative modelling and visualisation instruments to prevent data stream bottlenecks. In order to achieve this objective, common purpose interface protocols and new methods for heterogeneous data integration must be implemented.

Fig. 19.2 Opening Ceremony of the RCEIS project (*Source* RCEIS)

Fig. 19.3 Top: 1st Workshop linked to the Poyang Lake Symposium. Middle: 2nd Workshop in Leipzig/Dresden in August 2015. Bottom: 3nd Workshop in Tianjin in December 2016 (*Source* RCEIS)

The Opening Ceremony was conducted in November 2014 in Beijing (Fig. 19.2). Three workshops have been held so far in conjunction with the Cooperation Group "A modeling platform prototype for environmental system dynamics" (Fig. 19.3). Two schools are organized in 2018: Winter School (01/2018) in Germany which has already been held from 28.Jan to 07.Feb.2018 and Summer School (10/2018) in China (Nanchang).

19.2 Work Program and Study Areas

Due to the urgent environmental problems and political actions in China, RCEIS is focusing on "Water and Soil" in the first stage as the starting topic for the Helmholtz Network development. RCEIS activities will be related to the "Major Water Program of Science and Technology for Water Pollution and Governance" of China (2006–2020). Huge investments are scheduled by the Chinese Government for improving waste water treatment technology and drinking water supply as well as rehabilitation of rivers and lakes. More specifically, RCEIS will concentrate on the large rivers and lakes in China: Songhua-Liaohe, Yellow, Yangtze and Pearl Rivers (Fig. 19.4), which are not only important transportation ways through the country but also most important sources for drinking water supply in many mega-cities and relevant to energy supply (e.g. Three Gorges) as well as unique ecosystems. RCEIS will also incorporate two smaller catchments (for example Heihe and Haihe River) exemplary for water and soil management problems. We selected river basins also because of the on-going projects and activities by both the Chinese and German partners, guaranteeing the availability of data and infrastructures as well as the involvement of stakeholders and authorities. The project structure is depicted in the graphic below Fig. 19.4. Work packages are organized related to the candidate river basins from north to south (columns). Cross-cutting activities (rows) are organized according to

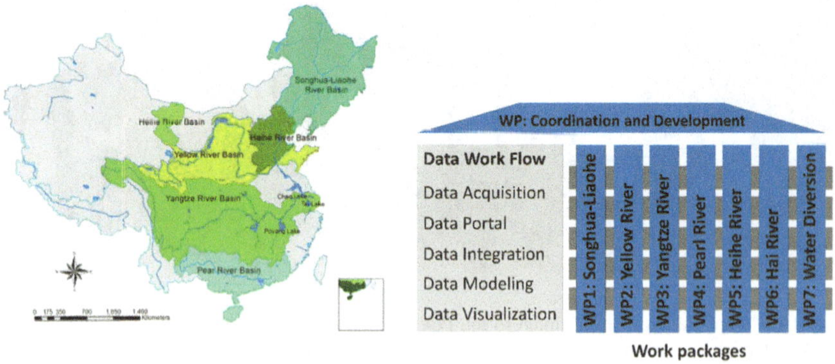

Fig. 19.4 Left: RCEIS study areas in China. Right: RCEIS project metrics (*Source* RCEIS)

the methodology and continuous work flow approach to be implemented for various investigation areas in order to demonstrate the transferability of the general approach. We also include the "South-to-North Water Diversion" project in China for the water supply of Beijing—as this project will impact the Yangtze and Yellow Rivers to a large extent.

A subsequent work step for designing EIS along those river-lake systems (from headwaters to mouth/delta) is the combination of the Delta and Terrestrial Information Systems.

19.3 Networking

The first stage of RCEIS was mainly dedicated to develop the network between the two disciplines: Environmental and Information Sciences towards Environmental Information Science. This have been followed in two ways (1) developing new methodologies (i.e. workflow concepts for Environmental Information Systems) and (2) applied in various investigation areas (Fig. 19.4).

19.3.1 Project Related Publications

UFZ researchers from all research units are involved in project and education activities with China. In order to build a basis for a strategic approach and focus to future research questions, a comprehensive ChinaConcept has been developed and published recently.[1] Focal areas in water, biodiversity, and renewable energies, social as well as data and information sciences have been discussed. Building joint research centres will be appropriate to leverage the Sino-German cooperation in environmental science and technology.

ChinaConcept 2016: Strategic cooperation with China: As a consequence of the enormous economic growth in recent years, China is facing numerous environmental challenges recognized by the countries leadership and counter measured by so-called Major Programs. As an example, the Major Water Program is dedicated to rigorous pollution control and improvement of the water quality with huge investments. In addition, the so-called 100/1000 Talents Programs attracted leading Chinese scientist returning to their home country from abroad and tackling China's grand challenges.

[1] http://www.ufz.de/export/data/454/190226_CC2016-16x24_compressed.pdf.

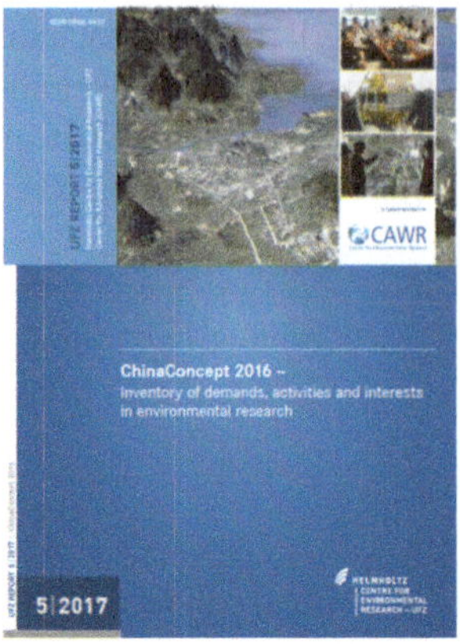

The following selected papers have been published by German and Chinese scientist involved in RCEIS activities in the last four years:

- X. Han, R. Jin, X. Li, and S. Wang (2014): Soil Moisture Estimation Using Cosmic-Ray Soil Moisture Sensing at Heterogeneous Farmland. IEEE Geoscience and Remote Sensing Letters 11, NO. 9, 1659–1663.
- Chen, C., Boernick, H., Cai, Q., Dai, X., Jaehnig, S.C., Kong, Y., Krebs, P., Kuenzer, C., Kunstmann, H., Liu, Y., Nixdorf, E., Pang, Z., Rode, M., Schueth, C., Song, Y., Yue, T., Zhou, K., Zhang, J., Kolditz, O., (2015): Challenges and opportunities of German-Chinese cooperation in water science and technology. Environ. Earth Sci. 73 (8), 4861–4871.
- Chen M., Lin H., Kolditz O and Chen C (2015): Developing dynamic Virtual Geographic Environments (VGEs) for geographic research, Environ Earth Sci., 74:6975–6980. https://doi.org/10.1007/s12665-015-4761-4.
- Chen C, Sun F, Kolditz O (2015): Design and integration of a GIS based data model for the regional hydro-logic simulation in Meijiang watershed, China. Environ Earth Sci., https://doi.org/10.1007/s12665-015-4734-7.
- Lin, H., Batty, M., Joergensen, S.E., Fu, B., Konecny, M., Voinov, A., Torrens, P., Lu, G., Zhu, A.-X., Wilson, J.P., Gong, J., Kolditz, O., Bandrova, T., Chen, M., (2015): Virtual environments begin to embrace process-based geographic analysis. Editorial, Trans. GIS 19 (4), 493–498.
- Nixdorf E, Chen C, Sun Y.Y and Kolditz O (2015): Persistent organic pollutants contaminate Chinese water resources: overview of the current Status, challenges and European strategies. Environ Earth Sci., https://doi.org/10.1007/s12665-015-4448-x.
- Song Y, Liu R, Sun YY, Lei K, Kolditz O (2015): Waste Water Treatment and Pollution Control in the Liao River Basin. Environ. Earth Sci., https://doi.org/10.1007/s12665-015-4333-7.
- Wang CL, Zhao N, Yue TX, Zhao MW and Chen C (2015): Change trend of monthly precipitation in China with an improved surface modeling method. Environ Earth Sci., https://doi.org/10.1007/s12665-014-4012-0.

- Wang S.R., Zheng B.H., Chen C., Dohmann M., Kolditz O (2015): Thematic issue: water of the Erhai and Dianchi Lakes, Environ Earth Sci., 74:3685–3688. https://doi.org/10.1007/s12665-015-4727-6.
- X. Han, H.-J.H. Franssen, R. Rosolem, R. Jin, X. Li, and H. Vereecken (2015): Correction of systematic model forcing bias of CLM using assimilation of cosmic-ray Neutrons and land surface temperature: a study in the Heihe Catchment, China. Hydrol. Earth Syst. Sci. 19, 615–629, https://doi.org/10.5194/hess-19-615-2015.
- Yue, T.-X., Xu, B., Zhao, N., Chen, C., Kolditz, O., (2015): Thematic Issue: Environment and Health in China—I. Environ. Earth Sci. 74 (8), 6361–6365.
- Clauss, K., Yan, H., and Kuenzer, C. (2016): Mapping Paddy Rice in China in 2002, 2005, 2010 and 2014 with MODIS Time Series. Remote Sensing, Vol. 8 (5).
- Dohmann, M., Chen, C., Grambow, M., Kolditz, O., Krebs, P., Schmidt, K.R., Subklew, G., Tiehm, A., Wermter, P., Dai, X.H., Liao, Z.L., Meng, W., Song, Y.H., Yin, C., Zheng, B.H., (2016): German contributions to the Major Water Program in China: "Innovation Cluster-Major Water" Environ. Earth Sci. 75 (8), art. 703.
- Huth, J., Ottinger, M., Eisfelder, C., Liu G., und Kuenzer, C. (2016): Landnutzungswandel im Delta des Gelben Flusses, China - eine Fernerkundungsanalyse. Geographische Rundschau, Themenheft: Deltaregionen Asiens, Vol. 7/8, S. 36–43.
- Kuenzer, C., Moder, F., Jaspersen, V., Dech, Stefan (2016): A Water related Information System for the Sustainable Development of the Mekong Delta: Experiences of the German-Vietnamese WISDOM Project. In: Integrated Water Resources Management: Concept, Research and Implementation Springer International Publishing. 11–22.
- Kuenzer, C., Moder, F., Jaspersen, V., Ahrens, M., Fabritius, M., Funkenberg, T., Huth, J., Vo Khac, T., Trinh Thi, L., Lam Dao, N., and Dech, S. (2016): A Water Related Information System for the Sustainable Development of the Mekong Delta: Experiences of the German-Vietnamese WISDOM Project. In: Integrated Water Resources Management: Concept, Research and Implementation Springer. p. 377–412.
- Ottinger, M., Clauss, K., and Kuenzer, C. (2016): Aquaculture: Relevance, Distribution, Impacts and Spatial Assessments—A Review. Ocean & Coastal Management, 119, p. 244–266.
- Rink K, Chen C, Bilke L, Liao Z.L, Rinke K, Frassl M, Yue T.X and Kolditz O (2016): Virtual geographic environments for water pollution control. International Journal of Digital Earth, https://doi.org/10.1080/17538947.2016.1265016.
- Wohlfart, C., Kuenzer, C., Chen, C., Liu, G., (2016): Social-ecological challenges in the Yellow River basin (China): A review. Environ. Earth Sci. 75 (13), art. 1066.
- Wei, J., H. R. Knoche, and H. Kunstmann (2016), Atmospheric residence times from transpiration and evaporation to precipitation: An age-weighted regional evaporation tagging approach, J. Geophys. Res. Atmos., 121, 6841–6862. https://doi.org/10.1002/2015JD024650.

- Wohlfart, C., Liu, G., Huang, C., and Kuenzer, C. (2016): A River over the Course of Time—Multi-temporal Analyses of Land Surface Dynamics in the Yellow River Basin (China). Remote Sensing, Vol. 8 (186).
- Wolters, M., Sun, Z., Huang, C., and Kuenzer, C. (2016): Environmental awareness and vulnerability in the Yellow River Delta: Results based on a comprehensive household survey. Ocean & Coastal Management, 120, p. 1–10.
- Kolditz et al. (2017): ChinaConcept 2016—Inventory of demands, activities and interests in environmental research. Helmholtz Centre for Environmental Research UFZ, ISBN: 978-3-944280-42-4.

19.3.2 Sino-German Lysimeter Workshop at Beijing Normal University

RCEIS organized and (co)funded a number bilateral Sino-German workshops on specific issues of environmental research. As an example for networking, the RCEIS-funded Sino-German Lysimeter Workshop, took place at Beijing Normal University from 14th to 15th August 2016 (Fig. 19.5) with the aim to share research results and discuss evapotranspiration ground observations technology, such as lysimeters. The workshop was organised by Prof. Harry Vereecken (Forschungszentrum Jülich) and Prof. Shaomin Liu (Beijing Normal University) as chairman and deputy chairman as well as by Dr. Zhongli Zhu (Beijing Normal University) and Dr. Thomas Pütz (Forschungszentrum Jülich). 40 scientists, managers, students, and representatives of instrument companies with background in lysimeter research, hydrology, meteorology, ecology, and remote sensing technology from China, Austria, Germany, and the US took part in the workshop. Eight speakers from China and nine speakers from Austria, Germany, and the US presented their work (Fig. 19.6).

Fig. 19.5 Lysimeter workshop at Beijing Normal University (*Source* RCEIS)

Fig. 19.6 Presentations and discussion about lysimeter research and long-term climate change records (*Source* RCEIS)

Background: Actual evapotranspiration (ET) is after precipitation the second largest flux of the terrestrial water cycle. Recent studies showed that actual (as well as potential) ET has been influenced by global climate change even though the changes differ in sign and size depending on the region. Actual ET can be estimated in situ either by measuring micrometeorological variables (e.g. Eddy covariance method or Bowen ratio method) or by measuring the soil water balance (Lysimeters). Actual ET values obtained using the eddy covariance method are typically around 20% smaller than values obtained by lysimeters because the energy balance cannot be entirely closed.

Presentations: The presented works showed that lysimeter systems are globally operated, multi-purpose instruments. The technical equipment is still evolving in the search of new solutions for new research questions (e.g. Jane Li, Aozuo Ecology Instrumentations Ltd. presented large weighable lysimeters built and produced in China; Pütz, TERENO-SOILCan Lysimeter Network in Germany). The following paragraph will give a short overview about the presented usages, research experiments, and developments.

Lysimeters are used in agriculture in Bushland (semi-arid region in Texas/US, Evett), in different places in China (e.g. on the North China Plain, YCES/CAS, Zhigang Sun; Zhongli Zhu; etc.), and in Brandis (Germany, Haferkorn) to optimise water, fertilizer, pesticide and plant management (crop water consumption, water use efficiency) with regard to high yields, environmental protection and global climate change. Lysimeters have been also used in anti-desertification studies in Northern China with the aim to characterize and quantify the actual ET of shrub communities in an arid desert ecosystem (Xinping Wang/Yafeng Zhang).

Lysimeter data can also be used as input and verification data for models and, thus, as basis for scenario simulations displaying, for example, the impacts of land-use and climate changes (Klöcking, ArcEGMO). Data assimilation provides an opportunity to improve the lysimeter soil hydraulic parameter estimation, which leads to improved flux-estimates at the top and bottom boundaries of the lysimeter (Montzka).

Two studies presented lysimeter-based approaches to quantify the impact of climate change by transporting soil to other climatic regions (Pütz, TERENO-SOILCan) and by, for example, applying CO_2 or altering the surrounding air temperature on the lysimeters (Herndl). Long-term actual ET series (length of minimum 20 years) obtained on lysimeters can provide an insight into the impact and changes in ET caused by global climate change on this flux (Haferkorn, Röseler).

It was also highlighted that lysimeters have limitations (soil disturbance, high staff costs, etc.; Pütz, Evett, Yanjun Shen, Zhongli Zhu), but, if maintained and managed correctly lysimeters systems can be ideal tools to determine actual ET. A topic mentioned by several presenters was the lack of defined quality standards for lysimeter equipment, management, monitoring, and data processing (Vereecken, Evett, Herndl, Klöckner, Röseler, etc.). So far several solutions and approaches exists parallel.

Comparisons between lysimeter and eddy covariance ET measurements typically show a deviation of 20% (as mentioned in the background section; Zhongli Zhu, Sien Li). In a study at a test site in Rollesbroich (Germany) the difference could be reduced to 3.8% (Hendricks-Franssen). A study by the IWHR compared different methods for measuring actual ET on different spatial scales (EC, lysimeter, Bowen-ratio, sap-flowmeter, scintillometer, photosynthesis apparatus; Yu Liu). This point was strengthened by the fact that redundant measurements with different equipment analysing different temporal or spatial scales or relying on different approaches are highly useful to get a better estimate of the "true" value, to understand the underlying processes better and the measuring systems (Evett, Yu Liu).

Future Plans: To establish a database of long-term lysimeter and meteorological data with a minimum length of 20 years from stations all over the world:

- It should include measured actual ET, leachate, and precipitation time series as well as meta-data information about the station, the equipment, and the monitoring.
- It should also include time series of additional weather variables of the same length (Minimum: sunshine duration, solar radiation (if available), precipitation measured with another technique than the lysimeter, wind speed, wind direction, cloud cover (if available), air temperature (minimum, maximum, mean), and relative air humidity).
- To strengthen the exchange in lysimeter research between China and Germany by further collaborations (Ph.D. students, etc.)
- To agree on standards in equipment, station management, meta-data monitoring, as well as processing lysimeter data. A lack in systematic and broadly applied standards is a hindrance for the broader acceptance of lysimeter usage.
- Exchanging experience on station and equipment management, on quality checking/pre-processing routines, as well as on the various possibilities that lysimeter research offers (long-term evaluations of ET data with focus on climate change, water and fertilizer management studies in agriculture, experiments with xerophytic shrub communities for preventing further desertification).

19.4 Contributions to Projects

In the last four years, RCEIS has supported for the following projects:

- BMBF—"Managing Water Resources for Urban Catchments"—Urban Catchments coordinated by UFZ,
- BMBF—DELIGHT (Delta Information System for Geoenvironmental and Human Habitat Transition), coordinated by DLR,
- DFG/NSFC—Cooperation group "A modeling platform prototype for environmental system dynamics" coordinated by UFZ,
- Lysimeter network coordinated by Research Centre Jülich (FZJ).

In the following we briefly present some of the RCEIS contributions to various projects in the corresponding study areas. The mentioned activities were funded by related third-party projects but are closely linked to RCEIS due to its research concepts and methodologies.

19.4.1 Songhua-Liaohe River Basin

RCEIS contributed to knowledge and technology transfer by supporting publication of results within the 12th Five-Years-Plan. Several Thematic Issues have been published in Environmental Earth Sciences (EES) dealing with river basins and lakes in order to improve the water quality. An an example EES Thematic Issue (73/09) "Waste Water Treatment and Pollution Control in the Liao River Basin" was dedicated to introduce new technologies in waste water treatment in industrial zones (e.g. Shenyang). In addition constructed wetlands have been built for cleaning surface water of the Liao River.

Within the EU-China Environmental Sustainability Programme the "SUSTAIN H2O" project dealt with demonstration of pollution discharge management in the

Songhuajiang-Liaohe River Basin. RCEIS supported the development of teaching and training material for groundwater modelling as a management tool. The OpenGeoSys tutorial gives a step-by-step instruction for building comprehensive groundwater models.[2] The complete project results are published in a sub-series "Chinese Water Systems" of Terrestrial Environmental Sciences.[3]

19.4.2 Yellow River Basin

The Yellow River Delta is one of the three major Chinese river deltas and one of the fastest changing river deltas in the world. The exploration of the natural resources oil and gas in the middle of the 20th century led to a continuously increasing anthropogenic influence in the coastal region. Impacts of industrial exploitation, urbanization, agricultural transformation and intensification can be observed in the delta. This delta region needs to find a balance between the growing delta city of Dongying with its oil and gas industries and large protected wetlands as resting place for migratory birds that are increasingly becoming a tourist attraction.

Objective of the Sino-German DELIGHT Project led by PD Dr. Claudia Kuenzer from DLR, funded for 3 years by BMBF Germany and MOST China, was to provide local stakeholders with necessary information on change processes in the delta tailored to their needs for supporting planning processes.

The design, development and implementation of an Environmental Information System (Fig. 19.7) that visualizes and provides all information generated in the project was one of the main tasks that included training of local stakeholders and

[2]http://www.springer.com/us/book/9783319528083.

[3]http://www.springer.com/series/13468.

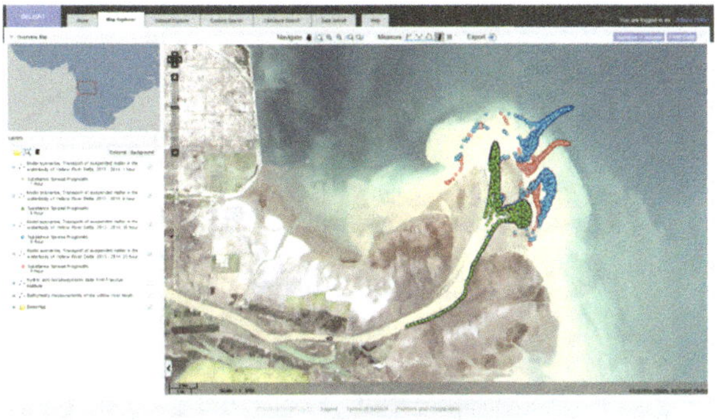

Fig. 19.7 DELIGHT Yellow River Delta Environmental Information System—Map viewer (*Source* DLR)

scientific staff. DLR supported RCEIS with its experience from the implementation of environmental information systems, even without the UFZ's involvement in this project.

19.4.3 Yangtze River Basin

The Yangtze River with a length of 6380 km is the longest river in Asia and forming the natural habitat of the Yangtze-River-Economic-Zone (YREZ) which belongs to one of the special economic zones in China benefiting from more free market-oriented) economic policies and flexible governmental measures.

RCEIS supports a longer term training concept "RIVERCHALLENGE" will develop teaching stations along the Yangtze River Basin (YRB) in order to raise the awareness of water quality changes using modern monitoring concepts. In addition to existing training stations in Chaohu (Fig. 19.8), Nanjing, Shanghai, UC-Wuhan will establish a comprehensive training centre for "Urban Lakes". The Cooperation Group is dealing with Poyang Lake and developing an Environmental Information System for the Lake Basin. The training concept is supported by CRAES, CAS, MEP and related ongoing projects.

Several Thematic Issues related to water management have been published in Environmental Earth Sciences: "Environment and Health in China I/II" (74/08),[4] "Water of the Erhai and Dianchi Lakes/Water of the Taihu Lake" (74/05),[5] "Environmental Research of the Three Gorges Reservoir".[6]

[4]https://link.springer.com/journal/12665/topicalCollection/AC_3272fcfe5090198e9cb55fa7c6de4d82/page/1.

[5]https://link.springer.com/journal/12665/74/5/page/1.

[6]https://link.springer.com/journal/12665/topicalCollection/AC_0ea350c70d54666fa7cc9843bd7d86b5/page/1.

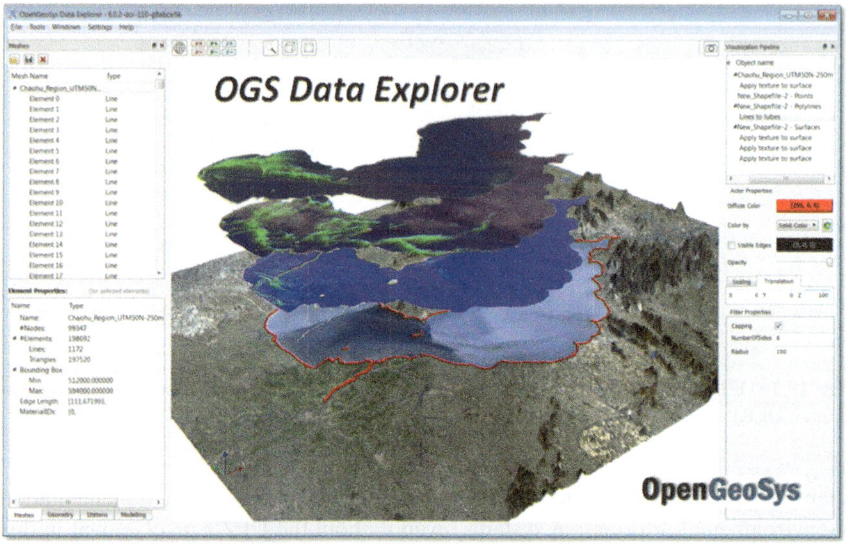

Fig. 19.8 An Environmental Information System has been developed for Chaohu within the "Managing Water Resources for Urban Catchments" project which is supported by the Federal Ministry of Education and Research (BMBF) (*Source* Urban Catchments)

19.4.3.1 Poyang Lake Research at Karlsruhe Institute of Technology (KIT)

Jianhui Wei, a Chinese scholar, did his Ph.D. study at Karsruhe Institute of Technology—Institute of Meteorology and Climate Research, Atmospheric Environmental Research (KIT/IMK-IFU) and has successfully finished the defense of his doctoral thesis in April 2016 at University of Augsburg. Under the supervision of Prof. Dr. Harald Kunstmann and Dr. Hans-Richard Knoche, both from KIT/IMK-IFU, Jianhui Wei developed an age-weighted regional evaporation tagging approach to investigate the information on the timescales for the atmospheric branch of the hydrological cycle, i.e. atmospheric water residence times. This novel approach has been applied to a case study for the Poyang Lake region in Southeast China, the largest freshwater lake in the country. The key result of their recently published study is that for the whole year 2005 the annual-aggregated, area-averaged atmospheric residence time (10 h) of transpired water from the Poyang Lake region until falling out as precipitation in Southeast China is longer than that (6.6 h) of directly evaporated water from the same region (Fig. 19.9). The result reveals different fate of transpired/directly evaporated water from the Poyang Lake region. This newly developed approach has the potential to be used for addressing how atmospheric water residence times change regionally under global warming and land-use/land-cover change, which is particularly relevant for a better understanding of environmental system dynamics in the Poyang Lake basin.

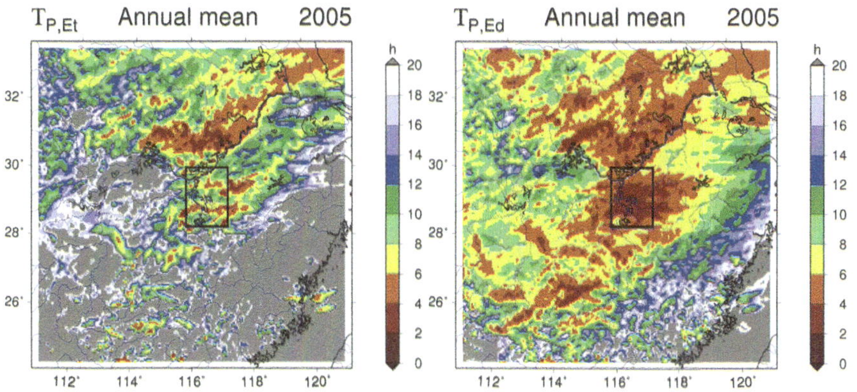

Fig. 19.9 Left: Annual mean atmospheric residence times (in hours) of transpired water from the Poyang Lake region (the area lies within the black rectangle) until falling out as precipitation in Southeast China for the whole year 2005. Right: The same as the left figure but for directly evaporated water (*Source* KIT)

Dr. Dominikus Heinzeller is a senior scientist at KIT/IMK-IFU specialising on atmospheric modelling in both contexts numerical weather prediction and climate research. Besides standard modelling approaches used widely in the communities, he is working intensively with the new Model for Prediction Across Scales (MPAS). This model is mainly developed by the National Centre for Atmospheric Research (NCAR) and the Los Alamos National Laboratory (LANL) in the US. Based on an unstructured Voronoi mesh, it allows global meshes with arbitrary regions if refinement in areas of interest and smooth transitions between them. Our work with MPAS previously concentrated on the African monsoon regions and shows that such an approach leads to an improved model accuracy because of the smaller distortion of large-scale flows and the awareness of teleconnection features. It can therefore be expected that other monsoon regions in the world, as for example the Asian monsoon regions including China, will also benefit from this novel approach. The purpose of Dr. Heinzeller's business trip to the AGU 2016 was to present our work and future research plans for a joint modelling of the monsoon regions and to establish contacts with researchers in and on the Asian monsoon regions. Dr. Heinzeller also presented and discussed work on the influence of anthropogenic aerosols on local and regional weather and climate, a research field he has recently worked on, with Ph.D. students and postdocs from China and surrounding countries. This work raised a high level of interest due to the latent problems with air pollution in Asian megacities.

19.4.3.2 Poyang Lake Research at German Aerospace Center (DLR)

In China the two Yangtze floodplain lakes Poyang and Dongting are complementary in terms of ecosystem composition and functions, etc. Furthermore, both lakes face similar threads from anthropogenic induced development upstream and within the project region itself. DLR will contribute gained experience and information on Dongting Lake related research and support future projects at Poyang Lake such as planned for a Sino-German Cooperation Group project mentioned above.

DLR previously conducted further analyses and satellite data processing to reveal inundation dynamics on Poyang and Dongting Lakes and further worldwide inland waters based on a daily-resolved intra-annual time series of global MODIS satellite data [1].

New analyses were conducted due to availability of new, innovative Sentinel-1A satellite data. The Copernicus Programme of the European Union is providing the Sentinel-1 datasets (also Sentinel-2 and Sentinel-3) via a scientific data hub—https://scihub.copernicus.eu—free of charge. In the picture (Fig. 19.10) first results of data analyses using Sentinel-1A data for the Poyang Lake are displayed—satellite data have been available for selected regions on Earth since October 2014.

A semi-automated approach (WaMaPro—Water Mask Processor; [2, 3]) based on empiric threshold derivation from the 15 available Sentinel-1A SAR images in 2014 was used to derive surface water extent and coverage. Details of generated inundation frequency maps of Poyang Lake—see box A, B, C in Fig. 19.10–reveal a highly dynamic lake environment, as occurs in the Nature Reserve area (detail A). Dark blue color displays areas where water was detected in all 15 analyzed satellite

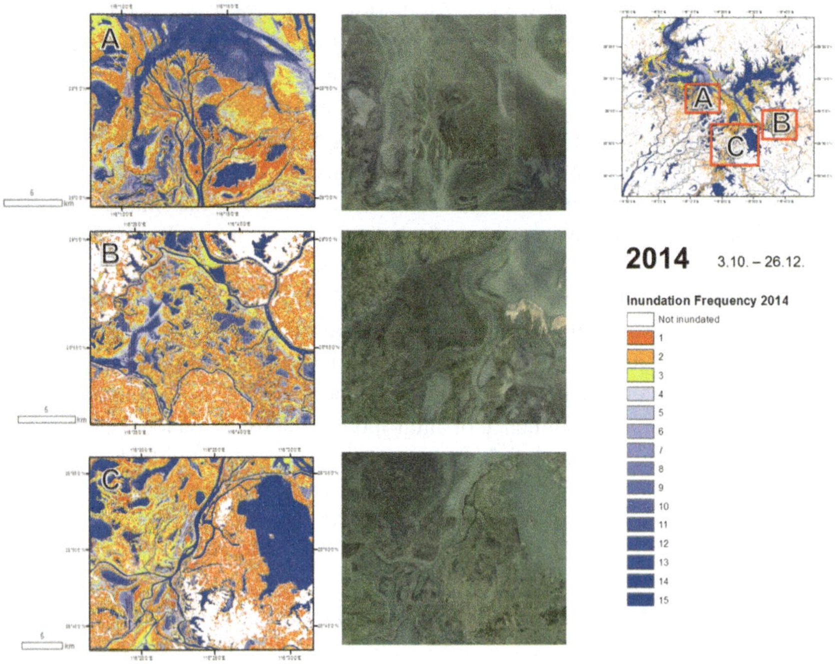

Fig. 19.10 Inundation pattern of Poyang Lake 2014 derived from first available Sentinel-1A data for the region (*Source* DLR)

scenes—such as permanent water bodies and main lake body. Whereas, reddish to yellow colors in the maps reveal locations of temporal water coverage, such as dynamically changing wetlands or flood prone areas. For comparison the boxes on the right (see Fig. 19.10) provide a mono-temporal Google Earth view on the same spots.

19.4.4 Heihe River Basin

RCEIS has developed a research concept for Heihe River Basin in 2015 (Belmont proposal). Due to the natural uniqueness of the Heihe River Basin but also with respect to the limited resources for the project, we developed the following research concept for the Belmont HEIHERIVER proposal. As the HRB is reaching from the Qilian Mountains (M) through the oases (O) of the Silk Road and a transition zone (X) until the Gobi Desert (D) with terminal lakes, we select four specific research areas following the Heihe River downstream. This investigation path from the mountains towards the desert is representing a Gradient Approach with regard to the varying aspects of climate, ecology, biodiversity, hydrology and land use.

Another project "Water resource management of Heihe River: Optimising irrigation in oasis desert ecosystems" was also conducted in the framework of (WP5), which was coordinated by Research Centre Jülich.

The recent development of the non-invasive cosmic-ray soil moisture sensing technique contributes to filling the gap between point-scale and regional scale soil moisture measurements. In this study, it was tested whether neutron counts allow correcting for a systematic error in the model forcings. Lack of data for water management often causes systematic input errors to land surface models. Here, a data assimilation procedure was tested for an irrigated corn field where no irrigation data were available as model input although for the area a significant amount of water was irrigated. The measured cosmic-ray neutron counts and Moderate-Resolution Imaging Spectroradiometer (MODIS) land surface temperature (LST) products were jointly assimilated into the Community Land Model (CLM) with the local ensemble transform Kalman filter. Different data assimilation scenarios were evaluated, with assimilation of LST and/or cosmic-ray neutron counts, and possibly parameter estimation of leaf area index (LAI). The results show that the direct assimilation of cosmic-ray neutron counts can improve the soil moisture and evapotranspiration (ET) estimation significantly, correcting for lack of information on irrigation amounts. The joint assimilation of neutron counts and LST could further improve the ET estimation, but the information content of neutron counts exceeded the one of LST. Additional improvement was achieved by calibrating LAI, which after calibration was also closer to independent field measurements. It was concluded that assimilation of neutron counts was useful for ET and soil moisture estimation even if the model has a systematic bias like neglecting irrigation. However, also the assimilation of LST helped to correct the systematic model bias introduced by neglecting irrigation and LST could be used to update soil moisture with state augmentation.

19.4.5 Haihe River Basin

The RCEIS-funded Sino-German Lysimeter Workshop took place in Beijing at the Beijing Normal University from 14th to 15th August 2016 with the aim to share research results and discuss evapotranspiration ground observations technology, such as lysimeters. The workshop was organised by Prof. Harry Vereecken (Forschungszentrum Jülich) and Prof. Shaomin Liu (Beijing Normal University) as chairman and deputy chairman as well as by Dr. Zhongli Zhu (Beijing Normal University) and Dr. Thomas Pütz (Forschungszentrum Jülich). 40 scientists, managers, students, and representatives of instrument companies with background in lysimeter research, hydrology, meteorology, ecology, and remote sensing technology from China, Austria, Germany, and the US took part in the workshop. Eight speakers from China and nine speakers from Austria, Germany, and the US presented their work.

Several lysimeter systems in the Haihe basin, e.g. that in Huailai driven by Beijing Normal University, and the TERENO SoilCan lysimeters in Germany will serve as the core for intensive research on hydrological fluxes. By incorporating lysimeter facilities in other parts of the world, a global lysimeter data base will be established with the aim to support long-term evapotranspiration analysis for climate change research.[7]

19.4.6 South-to-North-Water Diversion

More recently, RCEIS has started activities related to the South-to-North-Water Diversion (SNWD) project, in particular concerning groundwater management concepts. A new EES Topical Collection describing the status and progress of SNWD has been launched, recently.

Background: The South-North Water Transfer Project (also translated as the South-to-North Water Diversion Project[8]) is a multi-decade infrastructure mega-project in the People's Republic of China. Ultimately it aims to channel 44.8 billion cubic meters of fresh water annually from the Yangtze River in southern China to the more arid and industrialized north through three canal systems[9,10]:

- The Eastern Route through the course of the Grand Canal[11];
- The Central (middle) Route flowing from the upper reaches of the Han River (a tributary of Yangtze River) to Beijing and Tianjin[12];

[7]For further information see "Ecohydrology—Observation and Management" https://link.springer.com/referencework/10.1007/978-3-662-47871-4.

[8]http://www.nsbd.gov.cn/zx/english/.

[9]https://en.wikipedia.org/wiki/South%E2%80%93North_Water_Transfer_Project.

[10]The South-to-North Water Diversion Project. Engineering, 2: 265-267, http://dx.doi.org/10.1016/J.ENG.2016.03.022.

[11]http://www.nsbd.gov.cn/zx/english/erp.htm.

[12]http://www.nsbd.gov.cn/zx/english/mrp.htm.

- The Western Route which goes from three tributaries of Yangtze River near the Bayankala Mountain to provinces like Qinghai, Gansu, Shaanxi, Shanxi, Inner Mongolia and Ningxia.

19.5 Partner Network

19.5.1 Helmholtz RCEIS Partners

Helmholtz Centre for Environmental Research—UFZ
The UFZ is a national centre of excellence for integrative environmental research. It was established in 1991 as the only centre in the Helmholtz Association exclusively devoted to environmental research in a great variety of fields. Founded in response to the severe pollution prevailing in East Germany, the UFZ has become a world-wide acknowledged centre of expertise in the remediation and restoration of contaminated landscapes, as well as the preservation of biodiversity, natural landscapes and water resources.

German Aerospace Centre (DLR)
DLR is the national aeronautics and space research centre of the Federal Republic of Germany. The German Remote Sensing Data Center is an institute of the German Aerospace Center and part of the Earth Observation Center, which is known as a centre of competence for remote sensing from satellites and aircrafts. The research and development work conducted by DLR in the field of Earth observation covers virtually the entire range of satellite-based Earth observation topics, from innovation in sensor systems and evaluation of data to the preparation and development of new missions, their ground operations and data processing for applications.

Forschungszentrum Jülich (FZJ)
FZJ develops technologies that benefit research in Germany and worldwide in the areas of health, energy, and climate, as well as information technology. Land use and climate change bring about long-term changes to terrestrial ecosystems. The increasing demand for food and energy has necessitated the intensification of land use and agriculture, resulting in greater strain on fertile arable land and water resources. The Agrosphere Institute, IBG-3, is part of the Institute of Bio- and Geosciences. It analyses transport and conversion processes in soils and surface near groundwater systems with the aim to contribute to a sustainable use of resources in agro-ecosystems. Agrosphere is coordinating TERENO.

Karlsruhe Institute of Technology (KIT)
KIT was founded 2009 as a merger of Forschungszentrum Karlsruhe (KIT Campus North) and University of Karlsruhe (KIT Campus South). KIT/Campus North is a large-scale research institution of the Helmholtz Association conducting program-oriented provident research on behalf of the Federal Republic of Germany. Its remote lab KIT/IMK-IFU is located in Garmisch-Partenkirchen, and for many years it has

had several successful collaborations with Chinese research institutes (e.g. Institute of Atmospheric Physics, Chinese Academy of Sciences) and universities (e.g. Hohai University in Nanjing—the leading university for hydrology, and the China Agricultural University in Beijing—the leading university for agricultural research). KIT/IMK-IFU has organized two Sino-German symposia funded by Sino-German Center for Research Promotion for the topics "Steppe Ecosystems and Climate and Land Use Changes—Vulnerability, Feedbacks and Possibilities for Adaptation" in 2009 and "Modelling approaches and observational techniques for the quantification of fast environmental changes and its impacts on land use and water resources" in 2011.

19.5.2 Chinese Partners

The Helmholtz International Research Network "Helmholtz-CAS Research Centre for Environmental Information Science" (RCEIS) will involve leading Chinese institutions in environmental and landscape research, from the Chinese Academy of Sciences, Universities and Authorities. The Chinese part of the network is coordinated by the Institute for Geographical Sciences and Natural Resources Research of the Chinese Academy of Sciences (CAS-IGSNRR).

19.5.3 Related Partners

Involving Universities
RCEIS explicitly invites universities for participation. TU Dresden, as a member of the BMBF-CLIENT initiative "Urban Catchments" and several CSC Ph.D. projects with different Chinese research institutions (e.g. TONGJI University in Shanghai) are already involved. This process involving universities will be continued.

Centre for Advanced Water Research—CAWR
CAWR is one of Europe's largest centres for water research with a workforce that includes over 500 researchers in water area. It is a strategic cooperation between UFZ and TU Dresden and its competence covers a wide range of topics: water quality, water resources management, urban water as well as social-scientific aspects of water policy, societal change and environmental change. The partners of CAWR are successfully cooperating with China since many years. In addition, CAWR was actively participating in the recent BMBF-CLIENT project for "Managing Water Resources for Urban Catchments" in Chaohu City, China.

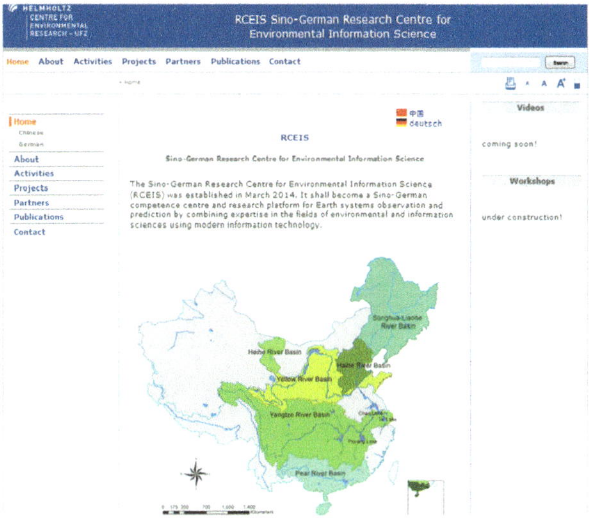

Fig. 19.11 RCEIS webpage

19.6 Webpage

The Internet webpage (Fig. 19.11) for RCEIS has been constructed and will be regularly updated. The content of the webpage will be presented in English and Chinese (under preparation). The RCEIS webpage can introduce the aim, scientific methods, vision of the RCEIS project. It also presents the ongoing projects, workshops, Sino-German events and activities. It lists the cooperation partners and related projects which are strongly supported by RCEIS. The scientific outcomes are also included in the context of RCEIS. The webpage was set-up by Leslie Jakobs and has been continuously updated by the project coordinator Dr. Cui Chen and Chengzi Zhou.

The RCEIS webpage is accessible using the following link: www.ufz.de/rceis.

The RCEIS flyer is presented as following Fig. 19.12.

19.7 Future Directions

With the progress of environmental protection, the environmental quality in China has been overall approved; however China's environmental pollution situation is still severe. According to the newest "Report on the work of the Government"[13] by Chinese Premier Minister Keqiang Li on 05.03.2017, China will further put significant efforts on controlling air, water and soil pollutions, and based on this in-depth imple-

[13] http://news.xinhuanet.com/politics/2017lh/2017-03/16/c_1120638890.htm.

Fig. 19.12 Top: The front page of the RCEIS flyer. Bottom: The back page of the RCEIS flyer (*Source* RCEIS)

mentation of the air,[14] water[15] and soil pollution control action plans.[16] In addition to that, China will also fully implement the 13th five-year ecological environmental protection plan.[17] Forecast and early warning of environmental pollution risks and pollution sources will be more targeted to help the pollution control. In March, 2016, the Chinese Ministry of Environmental Protection already published the "Master plan for constructing eco-environmental big data system"[18]. The implementation of

[14]http://www.gov.cn/jrzg/2013-09/12/content_2486918.htm.

[15]http://www.gov.cn/zhengce/content/2015-04/16/content_9613.htm.

[16]http://www.gov.cn/zhengce/content/2016-05/31/content_5078377.htm.

[17]http://www.gov.cn/zhengce/content/2016-12/05/content_5143290.htm.

[18]http://www.mep.gov.cn/gkml/hbb/bgt/201603/t20160311_332712.htm.

big data technology together with Environmental Information Systems (EIS) will change the traditional way of our thinking about the environmental management.

Within the starting phase of RCEIS (04/2014-06/2017), it mainly focused on network building and bilateral research initiatives in the area of water science and technology. The concept of RCEIS currently is being developed and implemented within several third-party funded projects (BMBF-CLIENT, DFG#NSFC), e.g. "Managing Water Resources for Urban Catchments": Visual Environmental Information System for Chaohu (Early warning system for drinking water supply)[19], DELIGHT: Yellow River Delta Environmental Information System for the support of local stakeholders planning processes,[20] and a Modeling Platform for Poyang Lake. Along with the recent development in the area of information and data sciences, Environmental Information Systems integrating big data (different types of remote sensing data, data from ground-truth UAVs,[21] and satellite missions (such as SENTINEL-2[22] from ESA[23]; SuperView-1 and global CO2-observatory satellite from China) related system processes will be further developed and applied to build the Environmental Information Systems in China within future activities of RCEIS.

An essential prerequisite for mitigating and solving environmental problems is the availability of information about the current environmental situation and prediction of dynamic environmental processes at different scales of interest. Therefore, the establishment of long-term terrestrial Earth observatories in China[24] similar to Helmholtz Infrastructures such as TERENO,[25] ACROSS[26] and MOSES[27] (international perspective) is important to create required monitoring platforms for collecting data. The concept of RCEIS by combining monitoring and modelling platforms through Environmental Information Systems (EIS) could provide information in a comprehensive way and develop management tools which can be used for planning and decision-making processes by the responsible authorities. This EIS is challenging to be built in the complex environmental situation and therefore the major purpose of RCEIS is to continuously develop the system. By means of improving the online monitoring data transmission network technology, real-time monitoring information system construction and making full use of big data, cloud computing and other modern information technology, it helps the decision making for the comprehensively improving the ecological environment protection.

The RCEIS Future Concept is "Combining Environmental Research with Information Technology" in various areas, water resources, soil protection, geo-energy

[19]www.ufz.de/urbancatchments.

[20]www.delight.eoc.dlr.de/en.

[21]UAVunmannedaerialvehicle.

[22]http://www.esa.int/Our_Activities/Observing_the_Earth/Copernicus/Sentinel-2.

[23]ESA:EuropeanSpaceAgency.

[24]HiWATER Project—Heihe Watershed Allied Telemetry Experimental Research http://card.westgis.ac.cn/hiwater.

[25]TERENO—TERestrial ENvironmental Observatories www.tereno.net.

[26]ACROSS—Advanced Remote Sensing Validation and Test Facilities.

[27]MOSES—Modular Observation Solutions for Earth Systems.

resources and therefore mitigating air pollution as well as environmental friendly waste management in subsurface repositories. A “Sino-German Institute for Environmental Science and Technology” would a suited platform being accompanied by structured education and training programs.

References

1. Klein, I., U. Gessner, A.J. Dietz, and C. Kuenzer. 2017. Global WaterPack—A 250 m resolution dataset revealing the daily dynamics of global inland water bodies. In *Remote sensing of environment*, vol. 198, 345–362. ISSN: 0034-4257. https://doi.org/10.1016/j.rse.2017.06.045. http://www.sciencedirect.com/science/article/pii/S003442571730305X.
2. Kuenzer, C., et al. 2013. Flood mapping and flood dynamics of the Mekong Delta: ENVISAT-ASAR-WSM based time series analyses. *Remote Sensing* 5(2): 687–715.
3. Gstaiger, V., J. Huth, S. Gebhardt, T. Wehrmann, and C. Kuenzer. 2012. Multisensoral and automated derivation of inundated areas using TerraSAR-X and ENVISAT ASAR data. *International Journal of Remote Sensing* 33(22): 7291–7304.

The manufacturer's authorised representative in the EU is Springer Nature Customer Service Centre GmbH, Europaplatz 3, 69115 Heidelberg, Germany. If you have any concerns regarding our products, please contact ProductSafety@springernature.com

Printed and bound by CPI Group (UK) Ltd, Croydon, CR0 4YY
30/09/2026
02225230-0009